Environmental Hazards and Disasters

Environmental Hazards and Disasters

Contexts, Perspectives and Management

Bimal Kanti Paul

Kansas State University

WILEY-BLACKWELL
A John Wiley & Sons, Ltd., Publication

This edition first published 2011
© 2011 by John Wiley & Sons, Ltd.

Wiley-Blackwell is an imprint of John Wiley & Sons, formed by the merger of Wiley's global Scientific, Technical and Medical business with Blackwell Publishing.

Registered Office
John Wiley & Sons Ltd, The Atrium, Southern Gate, Chichester, West Sussex, PO19 8SQ, UK

Other Editorial Offices
9600 Garsington Road, Oxford, OX4 2DQ, UK
111 River Street, Hoboken, NJ 07030-5774, USA

For details of our global editorial offices, for customer services and for information about how to apply for permission to reuse the copyright material in this book please see our website at www.wiley.com/wiley-blackwell

The right of the author to be identified as the author of this work has been asserted in accordance with the Copyright, Designs and Patents Act 1988.

All rights reserved. No part of this publication may be reproduced, stored in a retrieval system, or transmitted, in any form or by any means, electronic, mechanical, photocopying, recording or otherwise, except as permitted by the UK Copyright, Designs and Patents Act 1988, without the prior permission of the publisher.

Designations used by companies to distinguish their products are often claimed as trademarks. All brand names and product names used in this book are trade names, service marks, trademarks or registered trademarks of their respective owners. The publisher is not associated with any product or vendor mentioned in this book. This publication is designed to provide accurate and authoritative information in regard to the subject matter covered. It is sold on the understanding that the publisher is not engaged in rendering professional services. If professional advice or other expert assistance is required, the services of a competent professional should be sought.

Library of Congress Cataloging-in-Publication Data

Paul, Bimal Kanti.
  Environmental hazards and disasters : contexts, perspectives and management / Bimal Kanti Paul.
      p. cm.
  Includes bibliographical references and index.
  ISBN 978-0-470-66002-7 (cloth) – ISBN 978-0-470-66001-0 (pbk.)
  1. Hazardous geographic environments. 2. Disasters. 3. Natural disasters. 4. Human ecology.
 I. Title.
  GF85.P38 2011
  363.34′06–dc23
                                                                                          2011023520

This book is published in the following electronic formats: ISBN: ePDF 9781119979623; Wiley Online Library 9781119979616; ePub 9781119951025; Mobi 9781119951032

A catalogue record for this book is available from the British Library.

Set in 10.5/13pt Sabon by Aptara Inc., New Delhi, India.
Printed and bound in Malaysia by Vivar Printing Sdn Bhd

First Impression   2011

In memory of my mother, Parbati Sundari Paul, and father, Basanta Kumar Paul.

Contents

Preface ix

1 **Introduction** 1
   1.1 Hazards and disasters: definitions and distinctions 2
   1.2 Types of hazards 15
   1.3 Physical dimensions of natural hazards 21
   References 33

2 **History and Development of Hazard Studies in Geography** 37
   2.1 Historical perspective 37
   2.2 The focus of hazard research 44
   2.3 Conclusion 60
   References 61

3 **Vulnerability, Resiliency, and Risk** 67
   3.1 Vulnerability 67
   3.2 Resilience 85
   3.3 Risk 93
   References 110

4 **Disaster Effects and Impacts** 119
   4.1 Disaster effects 119
   4.2 Disaster impacts 138
   References 151

5 **Disaster Cycles: Mitigation and Preparedness** 157
   5.1 Mitigation 158
   5.2 Preparedness 169
   References 192

6 Disaster Cycles: Response and Recovery — 197
  6.1 Response — 197
  6.2 Recovery — 217
  References — 233

7 Disaster Relief — 237
  7.1 Providers and distributors of disaster aid — 237
  7.2 Conclusion — 263
  References — 264

8 Hazards/Disasters: Special Topics — 269
  8.1 Gender and natural disasters — 269
  8.2 Global warming and climate change — 277
  References — 302

Author Index — 307

Subject Index — 315

Preface

Having been born and brought up in Bangladesh, I have experienced many natural disasters and have also participated in a dozen post-disaster relief operations. Despite this first-hand experience and active participation, I did not become interested in hazard research for many years. My interest in disasters began to grow after Bangladesh experienced two consecutive devastating floods in 1987 and 1988. These two extreme natural events were soon followed by a category VI tropical cyclone in 1991. This cyclone was called Cyclone Gorky and it killed nearly 140 000 coastal residents in Bangladesh. Receipt of a Quick Response Research Grant (QRRG) from the Natural Hazards Center, University of Colorado at Boulder in 1994 acted as a catalyst in directing my research focus toward natural hazards and disasters. Receipt of seven subsequent QRRGs made me a hazard researcher.

Soon I realized that teaching and research are complementary to each other, and my research has tremendously benefited from teaching hazard courses. I started my hazard teaching with a course (GEOG 765: Geography of Natural Hazards) on natural hazards in 2000. I now offer this course once a year, and it is designed for upper-level undergraduate and graduate level students in social, physical or behavioral sciences, as well as interested students in business and engineering disciplines. In teaching this course, I confronted a serious problem; the absence of any suitable text that can be used in this course. Although several texts are available for an introductory collegiate-level hazard course, these texts emphasize individual hazards, such as earthquakes, hurricanes, floods and tornadoes, and their focus is primarily on the physical dimensions of these hazards. These texts cannot fully cater to the needs of upper-level undergraduate and/or graduate students.

The lack of an appropriate text for a post-introductory level hazard course, and the 2004 Indian Ocean tsunami, Hurricane Katrina in 2005 and the Pakistan earthquake, also known as the Kashmir earthquake, in the same year provided the necessary impetus for writing this book. Additional impetus came from Ms. Peggy McGhehey – the Wiley Publisher Representative in our area – who has always encouraged me to author a text for my hazard course. My exposure to a vast literature related to natural hazards and disasters as a result of teaching this hazard course for more than 10 years has both inspired and helped me to write this book. A one-semester sabbatical leave in the spring of 2006, and release from teaching one course in the fall of 2010, also facilitated writing this book. For these reasons, I am sincerely grateful to the Dean of the College of Arts and Sciences here at Kansas State university (KSU) and to Dr. Richard Marston, University Distinguished Professor and Head of the Department of Geography at KSU.

The focus of this book is primarily on manifested threats to humans and their welfare due to natural hazards and disasters. It uses an integrative approach to provide a comprehensive context for natural hazards and disasters along with important emergency management issues related to extreme events. It also addresses socio-cultural, political, and physical components of the disaster process, and assesses social vulnerability as well as the risk that natural hazards pose. This book offers an overview of the key issues related to natural hazards and disasters, and provides operational definitions, and methodologies that are useful for addressing hazard-related issues.

In writing this book, attempts have been made to provide examples of extreme events from all over the world. However, because of my direct experience of disasters and involvement in hazard research and teaching for more than a decade, more examples in this text are drawn from Bangladesh and the United States. In terms of individual extreme events, the same is true for the 2004 Indian Ocean tsunami, Hurricane Katrina, the Pakistan earthquake in 2005, and 2007 Cyclone Sidr in Bangladesh.

As indicated, this book is designed for classroom instruction of upper-level undergraduate and graduate students. Although not targeted for an introductory hazards course, students in such a course should find it very useful as well. In addition, emergency managers, planners, and both public and private organizations involved in disaster response, and mitigation could benefit from this book – along with hazard researchers and/or instructors.

Although the major part of this book was written during the last year, my teaching and research experience in this field has helped immensely in completing this task in a timely manner. However, its successful completion was contingent on the support received from many colleagues and students here at KSU and beyond. I am particularly grateful to my colleagues in the Department of Geography at Kansas State University for providing a rich institutional environment for me to explore and write about hazards and disasters. Some of the issues and topics I have included in this book were first explored in the classroom with my students, and/or funded research projects where the preliminary findings were presented at professional meetings.

I would also like to thank Fiona Woods, Project Editor, Life and Health Sciences for Wiley-Blackwell and John Wiley & Sons, Ltd., as well as Miss Izzy Canning, Assistant Editor, Life and Earth Sciences at John Wiley & Sons Ltd., for their knowledge, guidance, and patience during the writing of this book. I am also indebted to several anonymous reviewers for providing detailed and insightful comments and suggestions which helped to strengthen the quality of the material presented. For preparing several illustrations for this book and helping me in all stages of publication of this book, I gratefully acknowledge the assistance of Michel Stimers, a former graduate student of Geography at Kansas State University. Finally, my deepest gratitude goes to my wife, Anjali Paul, daughters – Anjana and Archana, and son Rahul – for their patience, love, and support over the years, but particularly during the time I was busy writing this book.

1
Introduction

To clearly understand the essential dimensions of hazards and disasters, it is necessary to introduce basic concepts regarding these extreme events. This chapter is, therefore, devoted to providing both traditional and contemporary definitions of hazards as well as disasters. It will be evident from this and subsequent chapters that human actions play a major role in causing and/or exacerbating the effects of extreme events. The outcomes of such events arise in large part from human conditions and processes that differentially exposed or protect people and either limit or enhance their ability to appropriately respond to and recover from hazardous events or conditions.

Since both hazards and disasters are related to many disciplines and professional communities, there are many definitions of these two terms. An attempt will be made to provide definitions from an interdisciplinary perspective, which will help in arriving at a suitable definition of a term accepted by all dealing with extreme events. In most instances, the terms hazards and disasters, as used in this book, refer to natural hazards or disasters. In defining these two terms, a distinction will be made between them because people often use these terms interchangeably. In the hazard literature, however, each term has a precise and distinct meaning.

Subsequent to this discussion of definitions, an overview of hazard typologies will then be presented, followed by a discussion of important physical dimensions of natural hazards. Although three important components (definitions and distinctions of hazards and disasters, types of hazards, and physical dimensions of natural hazards) of extreme events are considered in this chapter, emphasis has been given to definitions and distinctions between hazards and disasters. Topics included in this introductory chapter will aid in understanding material presented in subsequent chapters.

Environmental Hazards and Disasters: Contexts, Perspectives and Management, First Edition. Bimal Kanti Paul.
© 2011 John Wiley & Sons, Ltd. Published 2011 by John Wiley & Sons, Ltd.

1.1 Hazards and disasters: definitions and distinctions

1.1.1 Hazards

There are many definitions of hazards, but only one definition makes a distinction between hazards and disasters. According to Alexander (2000, p. 7), "a hazard is an extreme geophysical event that is capable of causing a disaster." The word "extreme" is used here to signify a substantial departure (either in the positive or the negative direction) from a mean or a trend. Although Alexander did not specify the distinction between these two terms in his definition, it does suggest that hazards may transform into disasters and thus become sequential events. That is, every disaster starts with a hazard (Thywissen, 2006). Alexander's definition, however, ignores the fact that human actions often play a major role in causing and/or exacerbating the effects of extreme events. Hazards represent the potential occurrence of extreme natural events, or likelihood to cause the severe adverse effects, while disasters result from actual hazard events (Tobin and Montz, 1997). Only after such an extreme event occurs may we term it a "natural disaster." A hazard is a threat not the actual event.

One of the earliest definitions of hazards is provided by Burton and Kates (1964) who maintain that natural hazards are those elements of the physical environment that are harmful to humans and caused by forces extraneous to them. That is, these events originated in the physical environment and are not caused by humans, but have consequences harmful to them. Burton and Kates' definition is correct in the sense that hazards are harmful to people, but it fails to recognize people's role in causing or amplifying the impacts associated with hazards. For example, floods can originate either from a natural variability in meteorological conditions or from human actions, such as deforestation, intensive use of land, or failure of dams constructed to control flooding (Haque, 1997). Similarly, landslides are commonly triggered by heavy rainfall (sometimes on hills and mountainsides previously denuded by wildfires), earthquakes, and volcanic eruptions, but also result from human activities such as logging, road building, and home construction, all of which can expose bare soil (Wisner et al., 2004).

There is no doubt that humans affect natural processes in many ways and thus often contribute to hazards. Many physical aspects of natural hazards, however, are out of their control. This does not mean that people are just passive in facing hazards; they can and do construct defenses against and implement measures to mitigate the impacts of hazards. Still, many people, even in the United States, consider natural hazards to be "acts of God" or the result of some external force (Mitchell, 2000, 2003). A careful reading of digital and print media reports covering more recent events, including Cyclone Nargis in Myanmar (May 2, 2008), the outbreaks of tornadoes in the Midwest and Southeast United States (April 29 and early May, 2008, respectively), and the earthquake in China (May

12, 2008) all occurring within a time span of one month, found that these extreme events are typically characterized as acts of God or of Mother Nature that leave human victims in their wake. This perspective suggests – quite wrongly – that people have no role to play in creating/exacerbating disasters, nor in mitigating their impacts (Smith, 2001). This interpretation of natural hazards as acts of divine will or forces of Mother Nature exonerates people from sharing in the responsibility for the creation of such extreme events.

According to Cutter (1993), hazards are threats to people and the things they value (such as their homes and belongings and environment). Oliver (2001, p. 2) has provided a similar definition; to him "A hazard is a threat posed to people by the natural environment." In its broadest sense, a hazard reflects a potential threat to humans and their welfare. A hazard is defined by the United Nations as an agent or threat that is "A potentially damaging physical event, phenomenon or human activity that may cause the loss of life or injury, property damage, social and economic disruption or environmental degradation" (UNISDR, 2004). This definition directly acknowledges the role of humans in causing/exacerbating hazards. It also makes the distinction between a hazard and a disaster by including the word "potentially," and it accounts for all possible hazard manifestations.

Hazards are also often defined in the context of vulnerability, which is a complex outcome of many factors, such as affluence, education, gender, demography, technology, and, above all, preparedness. Location can also act as a vulnerability factor. African Americans living in New Orleans were the most vulnerable segment of the population impacted by Hurricane Katrina in 2005, not only because they were poor but also because they inhabited (almost exclusively) areas below sea level. Low-income residents simply could not afford housing in areas above sea level. Another example of location as an important factor of vulnerability can be cited from Rio de Janeiro, Brazil. Many of the poor in this city occupy steep hillside locations above the reach of sewer, water, and power lines. Frequent heavy rains saturate the hillside and cause mudslides that destroy their rudimentary shelters. A combination of different factors, such as socio-economic conditions, geographic locations, political influences, and demographic characteristics, shapes the differing levels of vulnerability of different groups of people when an extreme natural event occurs (Wisner et al., 2004).

Chapman (1999, p. 3) argues,

> A natural hazard should be defined as the interaction between a human community with a certain level of vulnerability and an extreme natural phenomenon, which may be geophysical, atmospheric, or biological in origin, resulting in major human hardship with significant material damage to infrastructure and/or loss of life or disease.

He further claims that a particular level or severity of a natural event becomes a hazard only in relation to the capacity of society or individuals to cope with the impacts of such events. In addition to incorporating an ability to cope with the

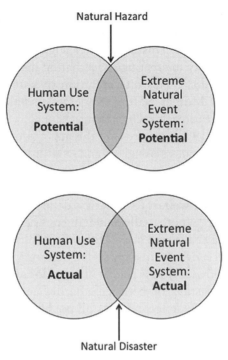

Figure 1.1 Distinction between natural hazards and natural disasters. *Source:* Modified after Erickson (n.d.).

impacts of hazards, this definition also suggests further classification of hazards based on their source or origin.

Strategies adopted to cope with hazards can be classified in two different ways: by making a distinction between indigenous and modern coping mechanisms, as well as between those strategies adopted purposefully and those which are incidental. Planting floating rice, which grows with the rise of flood water in lowland areas of rural Bangladesh, is an example of an indigenous coping strategy, while the flood warning system introduced in the early 1990s by the government in this South Asian nation can be considered a modern coping strategy. Further, coping adaptations can be separated into those that are purposefully undertaken and those resulting from incidental actions. Cultivating different rice crops in different elevation zones within the floodplain of Bangladesh is an example of a purposefully undertaken coping strategy, while constructing houses in the floodplain using materials such as thatch and bamboo, which can be readily moved by boats to safer places during abnormal flooding, is an example of incidental action.

Since the 1970s, hazards have been increasingly viewed as acts of human agency as well as the potential interaction or conflict between humans and one or more extreme natural events. Natural hazards lie at the interface between the (potential) natural event system and the (potential) human use system (Figure 1.1). Hazards thus exist within social, political, historical, and environmental contexts (Cutter, 2001). Hewitt (1983) and other researchers believe that hazards result more from social than geophysical processes. They eloquently argue that a "natural" hazard is a misleading term, as very little is natural about phenomena in which the danger results largely from human decision-making, land use, and/or socio-economic activities. Traditional explanations tend to emphasize hazards as processes of nature, their wrath inflicted randomly on unfortunate people. Contemporary hazard researchers blame many of the impacts associated with hazards on society and its varied institutions.

Hazards exist because humans and/or their activities are usually exposed to natural forces. As noted, we often create/exacerbate hazards, or modify hazard effects. For example, many wildfires are started by humans, either deliberately or unintentionally (Chapman, 1999). Hazards are in part socially constructed by individual perceptions and experiences. Thus, hazards can vary by culture, gender, age, race, socio-economic status, and political structure as well. For example, although the West Nile virus is usually fatal to the elderly, it is not to young adults, who often survive it. Normal flooding, which may inundate up to one-third of the total land area of Bangladesh, is seen by people of this country as a necessary and expected part of life; however, in the United States or most other Western countries, floods are generally seen as hazards.

The above example makes an important distinction between hazards and resources and supports the human ecology perspective of natural hazards proposed by Burton *et al.* (1978). They noted that when a natural event exceeds a threshold point or upper limit, a resource turns into a hazard. In the context of flooding in Bangladesh, 33% of the total land area of the country under water would be considered a threshold point. The specification of a threshold level for a given place or society, however, poses significant problems for hazard researchers. A threshold level for crippling snowfall in Toronto, for example, is much higher than the threshold for Norfolk, Virginia. Similarly, a temperature of 20°C is considered "freezing" in Tonga, but "scorching" in Baffin Island, Canada.

Any discussion of hazards must cover several essential attributes of such events. First, extreme natural events that do not affect (harm) people do not constitute a hazard. An earthquake in an uninhabited desert area, for example, cannot be considered a hazard, no matter how strong its intensity or how long its duration. Second, what is considered a hazard in one society may not be considered as such in another. It will be evident from the excerpt (below) from the poem "Floods in Bulozi," by O.K. Sibetta [translated by Charles and Slater (1995)], that floods are not considered hazards in Zambia, but they are in Western countries. Third, a hazard to some is to others a business opportunity or even a joyful experience. Drought is a hazard to farmers, but not to those engaged in construction – they can build houses and repair roads without interruption. Similarly, global climate change is likely to benefit some regions of the world while detrimentally impacting others.

Floods in Bulozi (Western Zambia)

It is floodtime in Bulozi
There is the floodplain clothed in
the water garment
Everywhere there is water!
there is brightness!
there are some sparkles!
Waves marry with the sun's glory
Birds fly over the floods slowly,
they are drunken with cold air
they watch a scene which comes
but once a year
floods are tasty (nice, beautiful)
Bulozi is the floods' place of abode
every year the floods pay us a visit.
we do not fear floods . . .
A Lozi does not beg for floods
We do not turn the herbs to have floods
We do not practice witchcraft whatsoever
They are floodwaters indeed!
The floods know their home area.
Floods are ours
the floods themselves
they (floods) know their own route
they know their home area
they know where they're needed
they know where they are cared for
And when we ourselves see them
we are inflated with happiness
our hearts become lighter

To say that a hazard to some may be a completely different experience for another is a subjective position. Whether or not one perceives driving without a seatbelt as hazardous, it can result in fatalities; whether or not one perceives smoking as a hazard, countless studies show that health is negatively affected by smoking (Mitchell and Cutter, 1997). However, a final attribute is that natural hazards constitute a threat to all societies, and these societies must cope with the hazards they encounter in one way or another. The type of hazards may vary from place to place, but no area is free from extreme natural events. Californians, for example, may not experience hurricanes, but they are vulnerable to earthquakes, mudslides, and wildfires.

In defining hazards, some researchers (e.g., Cutter, 2001; Cutter *et al.*, 2000) advocate inclusion of a technological dimension. They consider hazards on a continuum of potential interactions among physical, social, and technological systems (Montz *et al.*, 2003). According to them, this dimension is important because technology may affect the extent of damage – particularly the number of deaths. A tropical cyclone in Bangladesh killed more than 500 000 people in November 1970, while Hurricane Agnes along the east coast of the United States caused the deaths of only 12 people in June 1972. Bangladesh introduced a cyclone warning system after the 1970 cyclone. For this reason, not a single person was evacuated from coastal areas prior to the landfall of the 1970 cyclone. In contrast, prior to the landfall of Hurricane Agnes, 250 000 people were evacuated and this evacuation undoubtedly saved many lives.

Another example in support of the effect of technology on hazard impacts is the comparison between the 1998 Armenian earthquake and the 1989 Loma Prieta earthquake in California. Despite being of similar magnitude – the Armenian earthquake measured 6.9 on the Richter scale, while the California quake exceeded 6.9 – these two events caused very different numbers of deaths. At least 24 000 people died in Armenia, while the Californian quake killed only 63 people. The difference is often explained in terms of the different levels of technology in Armenia and the United States.

It is important to note that hazards are also a consequence of development and industrialization, which are directly associated with technology. In Europe, experts believe that countries such as France and Germany are more adversely affected by floods today because major rivers, such as the Rhine, have been straightened to ease commercial traffic. Interference with the natural flow of river water is thought to increase flood frequency in these countries. In addition, some hazard researchers stress that culture should be identified as an independent component, not a part of the human use system as shown in Figure 1.1.

1.1.2 Disasters

The term "disaster" is multifaceted, and many divergent definitions of the term appear in hazards literature (Mesjasz, 2011; Thywissen, 2006). There is extensive debate on the definition of disaster. At least two edited volumes are dedicated solely to this task (Quarantelli, 1998; Perry and Quarantelli, 2004). Fundamentally, disasters are actual threats to humans and their welfare. Disasters are generally conceived as adverse events, the negative impacts of which cannot be overcome without outside assistance, or support from many outside sources, including state and national governments, and even governments from other countries. One of the earliest definitions of disaster is provided by Charles Fritz. He defined a disaster as:

> an event, concentrated in time and space, in which a society, or a relatively self-sufficient subdivision of a society, undergoes severe danger and incurs such losses to its members and physical appurtenances that the social structure is disrupted and

the fulfillment of all or some of the essential functions of the society is prevented. (Fritz, 1961, p. 655)

According to Degg (1992), natural disasters result from spatial interaction between a hazardous environmental process (i.e., an extreme physical phenomenon such as an earthquake) and a population that is sensitive to that process and likely to experience human and/or economic loss from it. Disasters disrupt "normal" life, affect livelihood systems, and halt individual and/or community functions at least temporarily. Disasters often erase decades of development in a matter of minutes and push an affected community and/or nation years back in its quest for development. For example, the 2004 Indian Ocean tsunami caused a 20-year setback in the development of Maldives, an island nation off the coast of India (Coppola, 2007). Burton and Kates (1964) consider disasters as abrupt shocks to the socio-economic and environmental system (see also Turner, 1976).

To Cohen and Werker (2004, p. 3), "shock" refers to the natural phenomenon itself – the volcanic eruption, earthquake, drought, etc. – and "disaster" refers to the net impact of the shock on a population, when there is sufficient suffering. Thus, whether or not a shock translates into a disaster depends on the nature of emergency response, and on the mitigation measures and the level of preparedness (public and private) prior to an extreme natural event. For example, the extent to which a drought actually causes famine depends, to a great degree, on the ability of the government to bring adequate food from other areas. As a result, "natural" disasters are never 100% natural. In conflict zones, extreme events are even less "natural" because governments in such areas allow or encourage disasters in order to inflict suffering on their enemies (Cohen and Werker, 2004). Though earthquakes, typhoons, and tsunamis may be "acts of God," their impacts on a population are certainly nonrandom. For other disasters, such as floods and droughts, humans often contribute to their occurrence by draining wetlands, clearing forest, and using intensive farm practices, especially on marginally arable land. By choosing an optimal level of disaster prevention and mitigation, including preparedness and emergency response, governments are able to lessen the human impact of natural disasters.

Disasters are often defined as large-scale, stressful and traumatic events (Briere and Elliott, 2000; Dominici et al., 2005). Disasters cause widespread destruction of property, infrastructure, and natural resources and are responsible for a substantial number of deaths and injuries. As a result, a disaster is usually defined as an event that leaves a large and tragic impact on society, which disrupts established ways of life, interrupts socio-economic, cultural, and sometimes political conditions, and slows the pace of development in the affected communities. Unfortunately, it is not easy to define a *large* impact.

There are no agreed-upon boundaries to determine exactly when a threshold has been reached such that one can categorically say, "This constitutes a disaster." However, hazard researchers have made attempts to quantify disaster impacts in terms of the number of deaths, injuries, or the extent of damage experienced. For example, Sheehan and Hewitt (1969) defined disasters as events leading to at least 100 deaths, 100 injuries, or US$1 million in damages. Glickman *et al.* (1992) used 25 deaths as their threshold to consider an event a disaster.

Researchers at Resources For the Future (RFF) in Washington, D.C. have compiled a database of events that occurred between 1945 and 1986, including both natural and industrial disasters. This database excludes drought, and industrial accidents are limited to those in which a hazardous material caught fire, exploded, or was released in a toxic cloud. The recording threshold was set at a minimum of 25 fatalities for natural disasters, compared with 5 for industrial disasters (Smith, 2001). Use of differential weights was justified on the grounds that the public perceives industrial hazards/accidents more severely than the death toll alone would suggest.

The Center for Research on the Epidemiology of Disasters (CRED) at the Catholic University of Louvain in Brussels, Belgium, has developed the Emergency Events Database (EM-DAT), starting in 1988. EM-DAT organizes and counts disaster events by country. Thus disasters affecting many nations generate multiple registrations. It is the goal of EM-DAT to systematically define and routinely report disasters in a timely manner with high fidelity and consistency worldwide. However, CRED defines a natural disaster as "a situation or event which overwhelms local capacity, necessitating a request to the national and international level for external assistance, or is recognized as such by a multilateral agency or by at least two sources, such as national, regional, or international assistance groups and the media" (Guha-Sapir *et al.*, 2004, p. 16). By placing the event within the context of "local capacity," this definition indirectly encompasses the notion of scale, since local capacity can refer to places and populations of varying size. This is a very useful definition since some disasters occur in small towns and impact only a small number of people.

In order for a disaster to be entered into CRED's database, at least one of the following criteria has to be fulfilled (Smith, 1996): 10 or more reported killed by the extreme event; at least 100 people reportedly affected; a call for international assistance; or the declaration of a state of emergency. Among these four criteria, any appeal for international assistance from the country concerned usually takes precedence over all other criteria (Smith, 1996). These criteria do not include, however, the number of people injured or the monetary value of any property damage.

It is important to note that people are considered to be affected by an extreme natural event if they require immediate assistance (during the period of emergency) to fill basic survival needs, such as food, water, shelter, sanitation, and immediate medical attention (Picture This, 2003). Donor countries and aid agencies use the number of people affected by a disaster to decide whether they will provide emergency assistance to victims of these extreme events. For this reason, governments of disaster-impacted countries, hoping to receive external aid, often deliberately exaggerate this figure.

There are, however, several problems with the criteria used by CRED to enter a disaster in its database. First, 10 fatalities is a fairly low figure for some societies. The deaths of 10 people would constitute a relatively significant figure for more developed nations such as the United States, but in lesser developed societies many more people tend to be killed by disasters. Another problem is with the declaration of a state of emergency. In the United States, people will not receive any funding from public sources until the state and federal governments declare a state of emergency. This declaration is often politically motivated and not necessarily based on need (Schmidtlein et al., 2008). For example, if it is close to election time, it may be relatively easier for an affected area to be declared a disaster than would otherwise be the case. As mentioned, another problem is that CRED criteria do not take into consideration monetary losses or the number of people injured.

CRED data can, however, be used to identify what they term "significant" natural disasters, for which the more stringent criteria are: (1) number of deaths per event – 100 or more; (2) significant damage – 1% or more of the total annual national gross domestic product (GDP); and (3) affected people – 1% or more of the total national population (Hewitt, 1997; Smith, 2001). While the first criterion uses an absolute measure, the last two criteria represent relative measures. For this reason, relative criteria adopted for "significant" natural disasters can be used easily for comparative purposes across countries with disasters so labeled. The CRED database is now regarded as the primary reference among hazard researchers because it is more comprehensive than the others and also because it is used by disaster agencies themselves.

The United States Office of Foreign Disaster Assistance (OFDA) has maintained a record of disasters affecting all overseas countries since 1900. This organization relies heavily on officially declared disasters – those in which the Chief of the US Diplomatic Mission in the affected country decides that a US government response is required (Smith, 1996). Even within the United States, disaster victims are not eligible for federal emergency assistance if the president does not declare an area disaster-affected. According to the Disaster and Emergency Assistance Act, governors of affected states may request that the president issue a declaration indicating that an emergency or a major disaster exists.

A request for presidential disaster declaration is based on the finding that the disaster is of such severity and magnitude that effective response is beyond the capabilities of the affected state and local governments, and that federal assistance is necessary (GAO, 1995). Once a declaration is issued by the president, different types of assistance are available to disaster victims, including temporary housing, public assistance program funds for the repair of public facilities, public assistance grants for such activities as debris removal, and loans and grants to restore homes and businesses.

As indicated, while hazards have traditionally been represented as the work of natural forces, recent research indicates that society plays an important part in determining which natural events become disastrous. Thus, natural disasters

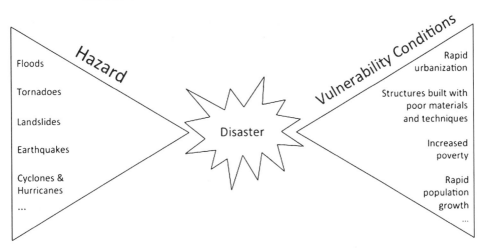

Figure 1.2 The disaster as the interface between hazards and the vulnerable conditions. *Source:* Modified after Mauro (2004), p. 243.

are widely considered the outcome of an interaction between the natural events system (geophysical processes) and the human use system (socio-economic, cultural, and physical conditions). This, as mentioned previously, means that not all natural disasters are exclusively natural (Tobin and Montz, 1997). For this reason, many hazards researchers prefer using the term "environmental" disaster rather than "natural" disaster (Oliver, 2001). Others stress that while hazards are natural, disasters are not. Natural events may become disasters if they occur in vulnerable human environments (Wisner *et al.*, 2004). Mauro (2004) considers disasters as the interface between hazard and the vulnerable conditions at community, household, and individual levels (Figure 1.2).

Similar to Mauro (2004) and Wisner *et al.* (2004), El-Masri and Tipple (1997) consider that natural disasters occur because of the interaction between natural hazards and humans living in vulnerable conditions that can arise from socio-economic, cultural, and/or political conditions and are usually created by human action. This implies that impacts of disasters are not uniform across society. As previously indicated, natural disasters are complex events with multidimensional outcomes: socio-economic, cultural, demographic, political, physical, and humanitarian. The interrelated and complicated nature of natural disasters renders any manageable definition of the term contentious and difficult. However, the United Nations Center for Human Settlements (UNCHS) provides a holistic framework for natural disasters, in terms of creation, effects, outcomes, and responses:

> [A] natural disaster could be defined as the interaction between a natural hazard, generated in most cases from a sudden and unexpected natural event, and vulnerable conditions which cause severe losses to man and his environment (built and natural). These losses create suffering and chaos in the normal patterns of life, which lead

to socio-economic, cultural, and sometimes, political disruption. Such a situation requires outside intervention at international and national levels in addition to individual and communal responses. (UNCHS, 1994)

Outside intervention is also evident in many other definitions of disaster. Porfiriev (1996) defines a disaster as a breaking of the routines of social life in such a way that extraordinary measures are needed for survival. Spiegel (2005, p. 1915) defines a disaster as "a serious event that causes an ecological breakdown in the relationship between humans and their environment on a scale that requires extraordinary efforts to allow the community to cope, and often requires outside help or international aid."

Wisner *et al.* (2004, p. 50) claim that a "disaster occurs when a significant number of vulnerable people experience a hazard and suffer damage and/or disruption of their livelihood system in such a way that recovery is unlikely without external aid." They also note the need for outside intervention in the form of emergency relief to cope with the impacts of disaster at household levels in their access model. This model "deals with the amount of 'access' that people have to the capabilities, assets and livelihood opportunities that will enable them (or not) to reduce their vulnerability and avoid disaster" (Wisner *et al.*, 2004, p. 88). As access to resources is essential to maintain livelihoods, less access means an increased level of vulnerability. Thus, a disaster results from the combination of hazards, conditions of vulnerability and insufficient capacity to reduce the potential negative consequences of risk.

Quarantelli (1985), on the other hand, provides a definition of disaster in the context of demand and capabilities of a community to handle an extreme event. He defines a disaster as a crisis occasion where demands exceed capabilities, and he offers the following continuum of labels for disasters of varying magnitude relative to community resources (Quarantelli, 2006):

- Crisis: Capacity exceeds demand – with capacity to spare
- Emergency: Capacity meets or somewhat exceeds demands
- Disaster: Demand exceeds capacity
- Catastrophe: Demands overwhelmed and may destroy capacity.

Various impacts of natural extreme events of different magnitude are listed in Table 1.1. Similar to Quarantelli (1985), Gad-el-Hak (2010) recently classified disasters into five groups. His classification is based on one of the two criteria: either number of person deaths/injuries/displaced/affected or the size of the affected area of the event. He classifies disaster types as Scope I–V according to the scale illustrated in Table 1.2.

Although there is a lack of consensus on a definition for a natural disaster, two important points emerge from the previous discussion. First, extreme

Table 1.1 Comparison of event magnitude

	Crisis	Emergency/disaster	Calamity/catastrophe
Injuries	Many	Scores	Hundreds/thousands
Deaths	Many	Scores	Hundreds/thousands
Damage	Moderate	Major	Severe
Disruption	Moderate	Major	Severe
Geographic impact	Dispersed	Dispersed/diffuse	Diffuse
Availability of resources	Sufficient	Limited	Scarce
Number of responders	Many	Hundreds	Hundreds/thousands
Time to recover	Days/week	Months/years	Years/decades

Source: McEntire (2007), p. 3.

natural events invariably cause extensive damage, which is usually expressed in monetary terms. In calculating damage, only direct material damage is included and valuation of costs of human life is rarely considered. Extreme events may or may not lead to deaths, but they usually impose a severe economic burden on the victims and communities of both developed and developing countries. This is particularly true for developing countries where disasters create demands that typically exceed resources available at the household and/or community levels (Wenger, 1978).

Inherently, disasters necessitate more than just a localized response in order for a community to survive and recover. This means that not all adverse events are disasters but rather only those that overwhelm the response capability of

Table 1.2 Classification of disaster by Gad-el-Hak

Class	Number of persons killed/injuries/displaced/affected	Area of impact (in square km)
Scope I (small disaster)	<10	<1
Scope II (medium disaster)	10–100	1–10
Scope III (large disaster)	100–1000	10–100
Scope IV (enormous disaster)	1000–10^4	100–1000
Scope VI (gargantuan disaster)	>10^4	>1000

Source: Based on information provided in Figure 1 by Gad-el-Hak (2010), p. 2.

the affected community. For example, a simple house fire requires response by a jurisdictional fire department. This fire may cause property damage, death, and injury. However, as house fires are routine occurrences that are easily managed locally, they are not considered disasters, but wild forest fires are (Coppola, 2007).

It follows from the above that disaster-impacted communities require the assistance of outside donors of all types, including governments, non-government organizations (NGOs), and private donors. Not only do disaster victims need outside support to help cope with losses, hardships, and distress, but many of the impacts of natural disasters are also influenced by the effectiveness of relief and emergency operations.

As indicated, the extent of damage and disruption caused by extreme natural events can be minimized by adopting hazard mitigation and preparedness measures at various levels before events occur. People do not have to be merely reactive when coping with disasters; they can adopt measures to reduce disaster risk and/or losses. From subsequent events, such protective measures seek to improve the household, community, and government's abilities to absorb the impact of a disaster with a minimum amount of damage. Mitigation measures include land-use regulations, building codes, and construction practices designed to ensure that structures are able to resist or withstand the physical forces generated by hazards, such as wind or seismic forces (Tierney *et al.*, 2001).

Preparedness relates to individual and/or community readiness to take precautionary measures in response to threats posed by an impending disaster. Preparedness activities may include devising disaster plans, gathering emergency supplies, training response teams, and educating residents and others about disaster vulnerability and what to do to reduce it (Mileti, 1999). For example, in the case of an impending hurricane, preparedness activities include, among other things, buying bottled water and/or placing shutters on windows.

In the post-disaster period, victims generally adjust to losses by bearing and spreading losses among various organizations and individuals (White, 1974; White and Haas, 1975; Burton *et al.*, 1978; Hewitt, 1997; Chapman, 1999; Smith, 2001). Although disaster victims all over the world absorb the major share of losses, they expect, and usually receive, outside support in mitigating disaster losses. This support usually comes as emergency or disaster aid. Thus, the outcome is determined not only by the severity of the disaster but also by the overall response, timeliness, and adequacy of any emergency relief received. However, hazard researchers differ in their opinions over whether disasters should even be addressed by relief measures (Susman *et al.*, 1983; Hewitt, 1997; Bolin and Stanford, 1999; Wisner *et al.*, 2004).

There are several problems with the definitions of disasters provided here. As noted, in the context of fatalities caused by disasters, several definitions use a threshold number, which may be interpreted differently in different societies, cultures, or countries. Secondly, most of the definitions presented ignore the concept of a scale factor, or an "impact ratio," which simply means that a disaster that causes harm to 100 persons signifies a very different level of impact for a community of 500 versus a community of 50 000 (Norris et al., 2002a, 2002b). Similarly, these definitions also ignore the fact that US$1 million in disaster damage will affect developed and developing countries differently. In addition, events of identical type and equal magnitude may pose very different challenges for populations of different sizes or for communities with ample versus limited response capacity (UNDP, 2004).

Even when disasters are defined in the context of property damage, in many instances such definitions ignore inflation. Moreover, disaster definitions do not consider that damage and death/injury data may be unreliable, depending on the country/entity reporting them; such data are not reliable for most developing countries – often, as mentioned, these figures are inflated to receive international aid. Further complicating the issue is the fact that many natural disasters arise from a combination of individual geophysical events. For example, on April 2, 2007 at 7:39 AM local time, a magnitude 8.1 earthquake hit the Solomon Islands' Western Province. The earthquake triggered over 1000 landslides, and generated a sizable tsunami. The tsunami was responsible for 50 reported deaths (McAdoo et al., 2009). McAdoo et al. (2009) reported that of all fatal earthquakes occurring during the past 40 years in this island, 21.5% had deaths due to secondary (non-shaking) causes. Thus, it may be difficult to allocate all fatalities or losses to a single event. Finally, little or no attempt has been made in these definitions to assess or even acknowledge more indirect losses, such as those resulting from injury, loss of employment, or forced relocation after a disaster.

Wisner and his colleagues (2004) classified available definitions of extreme events into three broad groups: those that emphasize the trigger role of geotectonics, climate or biological factors arising in nature; definitions that focus on the human response, psychological and physical trauma, economic, legal and political consequences; and a final group that does not deny the importance of extreme events as triggers, but puts the main emphasis on the various ways in which social systems operate to generate/exacerbate disasters by making people vulnerable. The first two groups assume that disasters are departures from "normal" social functioning and that recovery means a return to normal. Definitions from all three of these classes have been included in this section.

1.2 Types of hazards

Hazards arise from numerous sources. In order to establish some order in an ever-increasing list of hazards, many efforts have been made to develop typologies of hazards. Classification of hazards provides a useful framework for identifying similarities among and making generalizations about hazardous events. Typologies also promote sound management practices. Most of the existing hazard typologies use the causes or origins of hazard events as the classifying principle. Most hazards, however, arise from interrelated causes and the allocation of a hazard to one class is often difficult. For this reason most researchers now refrain from proposing a single cause-based typology of hazards. Still, it is possible to classify hazards according to the area in which they usually originate (Mitchell and Cutter, 1997).

1.2.1 Natural hazards

Natural events, those originating from extreme and/or common physical processes, are referred to as natural hazards. Earthquakes, volcanic eruptions, floods, hurricanes, tsunamis, blizzards, and tornadoes that originate in the lithosphere, hydrosphere, or atmosphere are some examples of natural hazards. Natural hazards can be further categorized into hydro-meteorological or atmospheric hazards (typically weather-related) such as floods, droughts, forest fires, and tornadoes, and geophysical or geologic hazards such as earthquakes, tsunamis, and volcanic eruptions.

Based on the nature of the physical processes involved, Tobin and Montz (1997) classified natural hazards into four categories: (1) meteorological (tropical cyclones/hurricanes, thunderstorms, tornadoes, lightning, hailstorms, windstorms, ice storms, snowstorms, blizzards, cold waves, heat waves, avalanches, fog, and frost), (2) geological (earthquakes, volcanoes, tsunamis, landslides, subsidence, mudflows, and sinkholes), (3) hydrological (floods, droughts, and wildfires), and (4) extraterrestrial (meteorites). Natural hazards originating from different (and even similar) physical processes may have few or no similarities. For example, avalanches typically have little in common with earthquakes or tornadoes, and volcanic eruptions and tropical cyclones would seem to be completely dissimilar.

Natural hazards often vary regionally, even seasonally, and they also frequently trigger secondary hazards. For example, landslides, tsunamis, and fire may follow earthquakes. Thunderstorms may be accompanied by heavy rains that can cause mudflows, flash floods, and conventional flooding. The volcanic eruption of Mt. St. Helens in 1980 led to an earthquake, landslides, floods, and wildfires. Table 1.3 lists selected secondary hazards and their causing extreme natural events. Note that depending on time of initiation, a hazard can be a primary or a secondary hazard.

Closely related to secondary hazards are compound disasters, which can occur either sequentially or simultaneously with one or more disasters. Compound disasters have a tendency to exacerbate consequences and increase victims' sufferings. However, natural hazards also arise from rather common natural events, such as hail, coastal erosion, and heat waves – all of which can cause considerable damage to the natural environment and society. As indicated, the focus of this text is on natural hazards and disasters.

Table 1.3 Secondary hazards

Primary hazard	Secondary hazard
Earthquakes	Landslides, tsunamis, fires, floods
Storm surges	Coastal floods
Volcanic eruptions	Earthquakes, wildfires, floods
Wildfires	Landslides
Severe storms	Tornadoes, flash floods
Landslides	Tsunami
Extreme summer weather	Wildfires
Floods	Fires
Hurricanes/cyclones	Storm surges

1.2.2 Social hazards

Some hazards, such as famine, warfare, acts of terrorism, and civil disorders originate in social systems. These types of hazards are also termed "intentional hazards" as these are caused or exacerbated by human action. Contrary to popular belief, famine is a social hazard because it is not necessarily caused by drought. In analyzing the Great Bengal famine of 1943 on the basis of his "entitlement approach," Sen (1981) concludes that the famine was not caused by drought, but rather by the government's inability to take proper action against the massive hoarding of food grains by corrupt politicians and their associates, whose primary aim was to gain wealth through inflated prices. Sen (1981) concluded that famine is dependent on the pattern of individual legal entitlement to acquire enough food to avoid starvation. He further claims that famine is not entirely caused by drought; a weak and insufficient political system may also create such an outcome.

Some social hazards, such as acts of terrorism, are created with the sole intention of provoking widespread fear that often extends beyond those directly targeted (Butler et al., 2003). Civilians, rather than military forces, are increasingly targeted and represent a growing proportion of the casualties, from all types of mass violence, including acts of terrorism. Government policy can also cause social hazards, such as religious and ethnic strife.

Certain social hazards, such as civil or armed conflicts, are often termed "complex emergencies," also known as complex humanitarian emergencies (CHEs) (Spiegel, 2005). The IASC (Inter-Agency Standing Committee) defines a CHE as a "humanitarian crisis in a country or region where there is a total or considerable breakdown of authority resulting from the internal and/or external conflict and which requires an international response that goes beyond the mandate or capacity of any single agency" (IASC, 1994). These create large number of refugees and internally displaced persons (IDPs). Of all the types of disaster response, none is more difficult and involved than that directed at treating and ending a CHE.

There are more similarities than differences between complex emergencies and natural hazards. Both often have increased mortality, morbidity, and displacement. Victims of both events need external aid as well as immediate assistance, such as water, food, shelter, health care, and protection. Sexual exploitation and violence against women and children may occur in varying degrees following both types of events. Furthermore, complex emergencies and natural hazards can occur concurrently. As noted, the 2004 Indian Ocean Tsunami (IOC) affected areas in Sri Lanka and Aceh Province, Indonesia that had existing rebel insurgencies. Aceh is subject to a 30-year-long violent conflict between the Free Aceh Movement (GAM) and the national government.

Between complex emergencies and natural hazards (or disasters), both public and private response is quicker and overwhelming in the case of hazards that become disasters. These events also typically receive more media attention than complex emergencies, because natural disasters can occur suddenly anywhere in the world, and their causes cannot easily be attributed to people. Complex emergencies are more political and complicated; depending on one's perspective, there is always someone or some group to hold accountable. Thus, it is often difficult for foreign and multinational agencies to remain neutral. Furthermore, providing aid in a complex emergency can sometimes exacerbate the situation or tensions among polarized groups (Spiegel, 2005).

1.2.3 Biological hazards

Hazards that originate for biological reasons (e.g., epidemics) are called biological hazards or biohazards. Sources of biological hazards include bacteria, viruses, medical wastes, insects, plants, birds, animals, and humans. These sources can cause a variety of health effects ranging from skin irritation, allergies, and infections (e.g., AIDS and tuberculosis) to deaths. The United States Centers for Disease Control and Prevention (CDC) categorizes various diseases in levels of biohazard – Level 1 being minimum risk and Level 4 being extreme risk.

Biological hazards are often divided into two categories: pathogens and toxins (McEntire, 2007). Pathogens are organisms that spread disease and may include anthrax, smallpox, influenza, plague, hemorrhagic fever, and rickettsiae, while toxins are poisons created by plants and animals. While pathogens could kill many people, toxins are not likely to do so. For example, the 1918 Spanish influenza pandemic killed more people in the United States than had died in combat in World War I (McEntire, 2007). In recent years, public health officials have been very much concerned with several diseases (e.g., foot-and-mouth disease, HIV/AIDS, the hantaviruses, severe acute respiratory syndrome (SARS), the West Nile virus, and the Avian "bird" flu) associated with biological hazards.

1.2.4 Technological hazards

Technological hazards originate from the interaction of society, technology, and natural systems (e.g., explosions, releases of toxic materials, and oil spills) and constitute a relatively new kind of threat. Events of this group are also termed (anthropogenic) non-intentional hazards, based on the absence of purposeful human causation. Terrorism, including bioterrorism, is an example of anthropogenic hazard of intentional origin. This topic will be covered in Chapter 2. However, industrial and transportation-related accidents and progressive or precipitous destruction of ecosystems reflect failures or side-effects of human-devised technologies, failures of judgment, or even flagrant human neglect. Serious consequences associated with technological hazards are largely the result of site-specific accidents, rather than a low-dose exposure of the general public to certain toxic substances.

Early concern about technological hazards was confined to developed countries, but the risks have now extended to developing countries through industrialization (Smith, 2001). Several particularly devastating technological disasters include the accidental toxic gas release from a Union Carbide pesticides plan in Bhopal, India in 1984, and the nuclear power plant accident in Chernobyl, Russia in 1986. The Bhopal disaster is frequently cited as one of the world's worst industrial disasters; it killed almost 3000 people. Another 300 000 were affected by exposure to the deadly gas, and perhaps 150 000 of those suffered long-term permanent disabilities, including blindness, sterility, kidney and liver infections, and brain damage (Fellmann *et al.*, 2008).

McEntire (2007) divided technological hazards into five categories: industrial hazards, structural collapse hazards, nuclear hazards, computer hazards, and transportation hazards. Industrial hazards are hazard agents produced by the extraction, creation, distribution, storage, use, and dispersal of chemicals, such as chlorine, benzene, insecticides, and fuel. Structural hazards occur when buildings, roads, and other construction projects collapse due to poor engineering. A nuclear hazard results from the presence of radioactive material, while computer hazards are disruptive hazards associated with computer hardware and software. Transportation hazards occur on roads, railways, air, or water bodies. Such hazards may result from adverse weather conditions, human error, and/or mechanical failure.

1.2.5 Chronic hazards

These hazards arise from long-term events such as continuous discharges and occupational exposure. Chronic hazards do not stem from one event, but arise from continual or long-term conditions or problems (e.g., famine, resource degradation, and pollution). These hazards, also termed 'elusive' hazards, can lead to health or environmental effects that occur gradually over a long period of time following continuous and repeated exposure to relatively low levels or concentrations of a hazardous agent. Chronic hazards are the type that will be most affected by changes in the global environment. Unlike natural hazards, their effects are generally less concentrated in time and space. For example, industrial pollution rarely poses an immediate threat to human life on a community scale. Technological and chronic hazards may be called human-induced hazards.

There are many other hazard classifications based on origin. Chapman (1999), for example, classified hazards into three broad categories: hazards originating primarily from the atmosphere and hydrosphere (e.g., tropical cyclones, tornadoes, blizzards, and droughts), hazards originating primarily from the lithosphere (e.g., earthquakes, volcanic eruptions, and tsunamis) and hazards originating primarily from the biosphere (e.g., wildfire, bacterial, and protozoan hazards). Another origin-based classification is as follows: endogenous origin (e.g., earthquakes and volcanic eruptions), exogenous origin (e.g., floods and droughts) and anthropogenous origin (e.g., flood caused by dam failure). In a similar vein, the World Health Organization (1999) offered the following five-fold typology:

- Natural-physical: hydro-metrological and geophysical
- Natural biotic, biological: pest infestations, epidemics, pandemics
- Socio or pseudo natural: human transformation of the natural environment
- Man-made technological: contamination, explosions, conflagrations
- Social: conflict including war, civil strife, violence.

In addition to origin-based classification, extreme events can also be grouped using other criteria. Some attempt has been made to scale the range of hazards and disasters according to whether the impacts are intense and local (e.g., earthquake and tornado) or diffuse and widespread (e.g., flood and drought) within society (Smith, 2001). Classification is also based on the extent to which extreme events are voluntary (e.g., civil strife and mountaineering) or involuntary (e.g., earthquake and tsunami). Disasters can also be divided into two major categories according to speed of occurrence: abrupt and slow onset. Earthquakes and tornadoes, for example, happen quickly, whereas drought and desertification occur over an extended period of time.

Noji (1997) uses a classification that distinguishes between natural disasters and human-generated disasters. His classification facilitates an understanding of the loose boundary that exists between these areas, as well as the potential for synergy. For example, an earthquake (natural disaster) may cause electrical fires and destruction of buildings (components generated by humans). A disaster that results from interaction between natural and human phenomena is referred to as NA-TECH, a portmanteau of "natural" and "technological" disaster types.

The EM-DAT of CRED distinguishes two generic categories for disasters: natural and technological. The former is divided into two groups:

- Hydro-meteorological disasters: avalanches/landslides, droughts/famines, extreme temperatures, floods, forest/scrub fires, windstorms and other disasters, such as insect infestations and wave surges.
- Geophysical disasters: earthquakes, tsunamis and volcanic eruptions.

The technological disasters comprise three groups:

- Industrial accidents: chemical spills, collapses of industrial infrastructure, explosions, fires, gas leaks, poisoning, and radiation.
- Transport accidents: by air, rail, road or water means of transport.
- Miscellaneous accidents: collapses of domestic/nonindustrial structures, explosions, and fires (IFRC and RCS, 2007).

A major shift in the classification of hazards and disasters has recently been noted by Montz and her colleagues (2003). They claim that instead of neatly dividing these events into natural and technological as was done a decade earlier, hazards and disasters are now viewed as a continuum of interactions among physical/environmental, social, and technological systems, ranging from extreme natural events to technological failure to social disruption to terrorism. This shift is also reflected in the hazards/disasters typologies presented above.

1.3 Physical dimensions of natural hazards

Natural hazard researchers have learned much from examining the physical forces that compose disasters (Tobin and Montz, 2007). There are many common features among extreme events, even those as apparently divergent as floods and droughts, or tornadoes and earthquakes. For example, it is possible to classify these events according to measures of hazard frequency, duration, and speed of onset. There is no doubt that the study of physical characteristics of natural hazards is required in order to gain a complete understanding of these events, particularly their ability to cause damage and destruction.

Understanding the physical characteristics of hazards and disasters is needed because the mitigation measures prepared and implemented for different disasters vary according to these characteristics. Before the work of Gilbert White (1964) on adjustments, the physical aspects of extreme events were the main focus in preparing for and mitigating against these events (Tobin and Montz, 1997). While the focus has since changed, or at the very least expanded, knowledge of the physical parameters of hazards and disasters remains an important tool in understanding and reducing their impacts. Such an understanding also improves disaster response, helps detect similarities among and make generalizations about hazardous events, and promotes the adoption of sound management practices.

Important physical characteristics of hazards will now be discussed. Following Tobin and Montz (1997), these characteristics can be classified into four groups: (1) physical mechanism (magnitude, duration, and spatial extent), (2) temporal distribution (frequency, seasonality, and diurnal patterns), (3) spatial distribution (geographic location), and (4) countdown interval.

Physical mechanism refers to how destructive an extreme event is in the context of its force and areal extent. Such parameters are not always strongly associated with the vulnerability of an affected community or population. Temporal distribution includes the time dimension of occurrence of an event, while spatial distribution refers to the spatial dimension of such an event. Examination of both of these broad categories helps researchers make probability statements about the likelihood of occurrence of events at specific time and locations. Countdown interval is a measure of how long it takes for a given extreme event to occur, and thus this parameter has bearing on the success or failure of actions undertaken to mitigate the effects of the event.

1.3.1 Magnitude

The most important physical property of extreme events is magnitude, which describes the strength or force of such events. In general, the greater the magnitude of the hazard, the greater the potential for fatalities, injuries, and damage to property. High-magnitude hazards usually generate media attention. It is possible to measure the magnitude of each type of hazard. Such a measure of magnitude may permit comparisons of extreme events over time and space, but they do not necessarily allow direct comparison of impacts. Measures of magnitude are neither simple nor separate from their context. For example, a high-magnitude flood confined to a small geographical area may cause limited damage, while a low-magnitude flood that inundates a large area may cause widespread damage to property.

Often the extent of damage caused by a particular disaster depends on the size of the vulnerable population, as well as the nature of preparedness and mitigation measures adopted prior to event occurrence. Cyclone Gorky, a high-magnitude storm that struck coastal Bangladesh in 1991, killed 138 000 people, while a lower magnitude cyclone that occurred in 1970 in the country caused an estimated 500 000 deaths. After this 1970 cyclone, the Bangladesh government introduced a number of public mitigation measures, such as the establishment of cyclone warning systems and the construction of public cyclone shelters and coastal embankments to reduce the impacts of tropical cyclones (Paul, 2009). The much lower death toll after the 1991 cyclone was largely attributed to the implementation of these mitigation measures.

In order to measure the strength or magnitude of a hazard, one must first have a baseline for comparison. For example, in the case of floods, magnitude is often expressed as the maximum height reached by flood waters above average sea level, or simply above ground. Alternatively, the maximum discharge of water at a given point is also used to measure flood magnitude. In the United States, flood stage is widely used to measure flood magnitude. It is an arbitrarily fixed and generally accepted gauge height or elevation at which the water level of a body of water overtops its banks and begins to cause damage to any portion of the defined reach. Flood stage is usually higher than or equal to bank-full stage. Flood stages for each United States Geological Survey (USGS) station are usually provided by the National Weather Service (NWS). River flood warnings are issued by the NWS when water levels are forecast to reach flood stage.

For seismic events, magnitude is measured on the Richter scale, which is an estimate of the amount of energy released by an earthquake. Developed in 1935 by Charles Richter in partnership with Beno Gutenberg, it is a base-10 logarithmic scale obtained by calculating the logarithm of the maximum amplitudes of waves recorded by seismographs. Because of the logarithmic basis of this scale, each whole number increase in magnitude represents a tenfold increase in measured amplitude. For this reason, an earthquake that measures 5.0 on the Richter scale releases 10 times the energy of one that measures 4.0. Typically, people can feel a magnitude 4.5 earthquake and up. A magnitude of 6 or 7 earthquake can completely destroy a building if it is not constructed following appropriate building codes (Phillips, 2009).

The strength of an earthquake can also be measured by estimating the effect of an earthquake, called its intensity. Intensity provides a useful relative indication of the severity of an earthquake based on subjective human experience; it should never be confused with magnitude. While each earthquake has a single magnitude value, its effects will vary from place to place, and there will be many different intensity estimates. One popular way to characterize earthquake effects is through the use of the Mercalli Intensity scale. It measures the intensity of earthquakes based on damage to structures (e.g., broken windows and cracked walls) and human experiences of the event. Table 1.4 shows typical effects of earthquakes of various magnitudes near the epicenter.

For hurricanes, the Saffir–Simpson Hurricane Scale is a measure of both intensity and magnitude. It evaluates hurricane strength and impact based on a five-point scale with Category 5 hurricanes listed as the most severe and destructive, and Category 1 the weakest (Table 1.5). The Saffir–Simpson Hurricane Scale is used to give an estimate of the potential property damage and flooding expected along the coast from a hurricane landfall. Wind speed is the determining factor in the scale, as storm surge values are highly dependent on the slope of the continental shelf and the shape of the coastline in the landfall region.

The Palmer Drought Severity Index (PDSI) is used for measuring the conditions in a region stricken with drought. The PDSI was developed by Wayne Palmer in 1965 and uses temperature and rainfall information in a formula to determine

Table 1.4 Earthquake magnitude and effect

Richter number (description)	Earthquake effects	Mercalli intensity value
1–2 (micro)	Usually not felt, but can be recorded by seismograph	I
3 (minor)	Felt quite noticeably indoors, especially on upper floors of buildings, but rarely causes damage	II–III
4 (light)	Noticeable shaking of indoor items, rattling noises. Significant damage unlikely	IV–V
5 (moderate)	Felt by nearly everyone. Can cause major damage to poorly constructed buildings	VI–VII
6 (strong)	Felt by all, damage slight	VIII
7 (major)	Can cause serious damage over larger areas	IX–X
8 (great)	Can cause serious damage in areas several hundred miles across	XI
9 (great)	Devastating in areas several thousand miles across	XII
10 (great)	Never recorded	

Source: Compiled from various sources.

dryness. It is the first comprehensive drought index developed in the United States. Palmer based his index on the supply-and-demand concept of water balance equation, taking into account more than just the precipitation deficit at specific locations. The Palmer index is most effective in determining long-term drought (a matter of several months) and is not as good with short-term forecasts (a matter of weeks); it is also not well suited for mountainous land or areas of frequent climatic extremes. It uses a 0 as normal, and drought is shown in terms of minus numbers. For example, minus 2 is moderate drought, minus 3 is severe drought, and minus 4 is extreme drought.

From 1971 through 2007, tornado intensity was measured on the Fujita scale or F-Scale. This scale was introduced by Theodore Fujita in 1971 and it ranged from F-0 through F-5. Since February 1, 2007, the Enhanced Fujita Scale or EF Scale has replaced the Fujita scale. The EF Scale has the same basic design as the original Fujita scale, with six categories from 0 to 5 representing increasing degrees of damage. It was revised to better reflect examinations of tornado damage surveys, specifically to align wind speeds more closely with associated storm damage. It adds more types of structures as well as vegetation, expands degree of damage, and better accounts for variables such as differences in construction quality than does the original scale (Table 1.6). However, magnitude alone is

Table 1.5 Saffir–Simpson Hurricane Scale

Category One Hurricane	Winds 74–95 mph (119–153 km/h) or storm surge generally 4–5 ft (1.2–1.5 m) above normal. No real damage to building structures. Damage primarily to unanchored mobile homes, shrubbery, and trees. Some damage to poorly constructed signs. Also, some coastal road flooding and minor pier damage
Category Two Hurricane	Winds 96–110 mph (154–177 km/h) or storm surge generally 6–8 ft (1.8–2.4 m) above normal. Some roofing material, door, and window damage of buildings. Considerable damage to shrubbery and trees with some trees blown down. Considerable damage to mobile homes, poorly constructed signs, and piers. Coastal and low-lying escape routes flood 2–4 h before arrival of the hurricane center. Small craft in unprotected anchorages break moorings
Category Three Hurricane	Winds 111–130 mph (178–209 km/h) or storm surge generally 9–12 ft (2.7–3.7 m) above normal. Some structural damage to small residences and utility buildings with a minor amount of curtainwall failures. Damage to shrubbery and trees with foliage blown off trees and large trees blown down. Mobile homes and poorly constructed signs are destroyed. Low-lying escape routes are cut by rising water 3–5 h before arrival of the center of the hurricane. Flooding near the coast destroys smaller structures with larger structures damaged by battering from floating debris. Terrain continuously lower than 5 ft (1.5 m) above mean sea level may be flooded inland 8 miles (13 km) or more. Evacuation of low-lying residences with several blocks of the shoreline may be required
Category Four Hurricane	Winds 131–155 mph (210–249 km/h) or storm surge generally 13–18 ft (4.0–5.5 m) above normal. More extensive curtainwall failures with some complete roof structure failures on small residences. Shrubs, trees, and all signs are blown down. Complete destruction of mobile homes. Extensive damage to doors and windows. Low-lying escape routes may be cut by rising water 3–5 h before arrival of the center of the hurricane. Major damage to lower floors of structures near the shore. Terrain lower than 10 ft (3 m) above sea level may be flooded requiring massive evacuation of residential areas as far inland as 6 miles (10 km).
Category Five Hurricane	Winds greater than 155 mph (249 km/h) or storm surge generally greater than 18 ft (5.5 m) above normal. Complete roof failure on many residences and industrial buildings. Some complete building failures with small utility buildings blown over or away. All shrubs, trees, and signs blown down. Complete destruction of mobile homes. Severe and extensive window and door damage. Low-lying escape routes are cut by rising water 3–5 h before arrival of the center of the hurricane. Major damage to lower floors of all structures located less than 15 ft (4.5 m) above sea level and within 500 yards (460 m) of the shoreline. Massive evacuation of residential areas on low ground within 5–10 miles (8–16 km) of the shoreline may be required

Source: Compiled from NOAA/NWS (http://www.nhc.noaa.gov/aboutsshs.shtml – accessed June 1, 2008).

Table 1.6 Enhanced Fujita (EF) Scale for tornado damage

Category	Wind speed, mph (km/h)	Relative Frequency (%)	Expected damage
EF0	65–85 (105–137)	53.5	Light damage: peels surface off some roofs; some damage to gutters or siding; branches broken off trees; shallow-rooted trees pushed over
EF1	86–110 (138–178)	31.6	Moderate damage: roofs severely stripped; mobile homes overturned or badly damaged; loss of exterior doors; windows and other glass broken
EF2	111–135 (179–218)	10.7	Considerable damage: roofs torn off well-constructed houses; foundations of frame homes shifted; mobile homes completely destroyed; large trees snapped or uprooted; light-object missiles generated; car lifted off ground
EF3	136–165 (219–266)	3.4	Severe damage: entire stories of well-constructed houses destroyed; severe damage to large buildings; trains overturned; trees debarked; heavy cars lifted off the ground and thrown; structures with weak foundation blown away some distance
EF4	166–200 (267–322)	0.7	Devastating damage: well-constructed houses and whole frame houses completely leveled; cars thrown and small missiles generated
EF5	>200 (>322)	0.1	Incredible damage: strong frame houses leveled off foundations and swept away; automobile-sized missiles fly through the air; steel reinforced concrete structure badly damaged; high-rise buildings have significant structural deformation

Source: Compiled from various sources.

often a poor guide for estimating disaster impacts. This is largely due to the great differences in the degree of physical exposure and human vulnerability of the communities at risk.

It is worth mentioning that magnitude alone does not govern the hazardousness of an extreme event. Without impact on humans, there is no difference between a high and a low magnitude event. The Sherman landslides that occurred in central Alaska in 1964 can be used an example to support the above contention. This landslide displaced 29 million cubic meters of rock debris which traveled more than 3 miles (5 km) at an estimated maximum speed of 120 miles (180 km) per hour. However, as the event took place in an uninhabited area, it had no real human consequences. In contrast, the landslide of October 21, 1966 at Aberfan, South Wales, involved one-hundredth as much debris traveling one-twelfth of the distance at one-twentieth of the speed, but it killed 144 people and destroyed many structures. This illustrates that it is not just magnitude that is important, human vulnerability is a fundamental determinant of disaster potential (Alexander, n.d.).

1.3.2 Duration

Duration of an extreme event represents a temporal dimension that describes how long an extreme event persists or length of time over which a hazard event persists. This can range from a few seconds (earthquakes and landslides), minutes (as is common in the case of tornadoes and tsunamis), hours and days (as in the cases of flash floods and hurricanes), or weeks and months (in the case of river floods) to years (as in the case of drought). It is important to mention that although an extreme event such as a tornado may last for an hour or so, its duration is shorter at a given location or site.

Preparation and planning for different hazards may show many similarities based on duration. Short duration events often require immediate emergency action, cause severe damage to life and property, and are difficult to prepare for. On the other hand, long duration events (e.g., flood and drought) offer opportunities for adjustments, even as these events are occurring. Most events of this type do not cause loss of life but often result in significant property damage. Some hazards, such as volcanic activity, can vary in duration. Although landslides and debris flows represent events that usually have a very short duration, they normally need a long period of onset. In general, events of longer duration tend to generate greater losses because they cover a larger area than those of shorter duration. When the duration of an extreme event is prolonged and of a high magnitude, losses usually increase significantly.

1.3.3 Frequency

Frequency represents one of the three temporal dimensions of hazards; it describes how often an event of a given magnitude or intensity occurs over a given period of time. This can be expressed in qualitative terms such as "frequent" or "rare," or in more quantitative terms such as recurrence interval. The recurrence interval (or return period) is the time which, on average, elapses between two events that equal, or exceed, a particular magnitude. For floods, a recurrence interval of 20 years suggests that in any given year, a flood of that magnitude has a 1 in 5 (20%) chance of occurring. In general, the extent of damage experienced increases with increasing event frequency.

Hazard event frequency is estimated through careful examination and analysis of historical records and incorporation of that information into predictive models. Using models, these records can be reported or displayed as averages over time, as the probability of an event occurring at a given time or within a given timeframe, or as the interval between events of a given magnitude. Technological advances and expansion of international communications have increased the completeness (and standardization) of disaster reporting in recent decades. As a result, on an international scale, more complete data are now available on disaster frequency than in the past. Gaining an understanding of how often a particular type of hazard will occur assists in planning and mitigation processes (Tobin and Montz, 1997).

1.3.4 Seasonality or temporal spacing

Hazard seasonality encompasses the notion that certain hazards are more prominent and likely during a given season. For example, a heat wave is not expected to occur during December in Wisconsin, and just as improbable is a blizzard striking Florida during summer or even in winter. Many types of natural disasters, particularly hydro-meteorological events, tend to be seasonal. Some hazards (e.g., industrial accidents and volcanic eruptions) are quite random in their timing, while other hazards, such as tornadoes, hurricanes, and floods, have a seasonality or regular periodicity. Winter storms are seasonal by definition, and flooding usually occurs with spring thaws or in the rainy season.

The implications of temporal spacing for hazard management are quite clear. Randomly occurring hazards are much more challenging to emergency response agencies because they require a low level of preparedness at all times, which involves significant costs. These hazards also require rapid, efficient, and effective responses. Regularly or seasonally occurring hazards, in contrast, are generally less challenging to emergency response agencies because these types of hazards are predictable in terms of timing as well as location, hence emergency management personnel are typically better prepared for them. Further, the allocation of resources can be better managed and distributed season by season, depending on the predominant hazard at that time of year.

Meteorological and hydrological events may exhibit a distinct seasonal pattern: snow storms and blizzards are clearly winter phenomena and hurricanes are typically summer events. Some events may occur in two distinct seasons. For example, coastal Bangladesh is prone to violent tropical cyclones and associated storm surges during pre-monsoon (April–May) and post-monsoon (October–November) seasons (Chowdhury, 2002). In addition, the frequency of such cyclones may differ in these two seasons. Between 1891 and 2007, 860 cyclones occurred in the northern Indian Ocean; about 33% occurred in the pre-monsoon period and 67% in the post-monsoon period (Shamsuddoha and Chowdhury, 2007).

Seasonality of events can be obscured by inadequate consideration of spatial location. For instance, global flooding incidence will show little seasonal variation because summer does not occur during the same time period in the northern and southern hemispheres. Tornadoes have a distinct seasonal component in the United States, but consideration must be given to spatial location. For example, Grazulis (1991) showed that North Dakota and Mississippi have distinctly different temporal distributions of monthly incidences of tornadoes. Mississippi has a bimodal distribution with peaks in April and November, while in North Dakota tornadoes peak in June.

It is important to note that the severity of several extreme events varies over multiyear periods. One notable example is the 20- to 40-year cycle of hurricane frequency and intensity in the United States. Increases in the number of named tropical storms and hurricanes occurred in the Atlantic Hurricane Basin during the 1940s, 1970s, and again during the early 2000s.

1.3.5 Areal or spatial extent

Areal extent refers to the area over which a hazard event occurs. It is usually associated with the amount of damage incurred, and frequently with the number of deaths. Some hazards, like a tornado, may have a small areal extent; others, such as a drought or a major nuclear accident (like the one at Chernobyl in 1986), affect large geographic regions. When assessing the spatial extent of an extreme event, consideration should also be given to the location of the area under scrutiny. As noted, a flood in Bangladesh is generally considered hazardous only if it inundates more than a third of the total land area of the country. However, if only 10% of all land area was inundated in the United States, it would be considered a major catastrophe.

1.3.6 Spatial distribution or dispersion

Spatial distribution or dispersion refers to the distribution of hazards over the space in which they can occur. This is a very important spatial component of extreme events because all places are not subject to the same types of hazards. For example, hurricanes occur between 5° and 25° latitude north and south of the Equator. Similarly, snowstorms and ice storms occur only in temperate latitudes, floods are generally confined to floodplains and coastal areas, and some hazardous geophysical processes are restricted to certain geographical regions of the world, such as earthquakes and volcanic eruptions which occur most often along tectonic plate boundaries.

Different criteria are used to show spatial distribution of each hazard. For example, yearly occurrence of tornadoes in a particular country or state can

Table 1.7 The top ten states in five categories, based on 1953–1989 official tornado statistics

Rank	Total number of tornadoes	Incidence per 25 000 km²	Total deaths	Deaths per 25 000 km²	Deaths per million population
1	Texas	Florida	Texas	Massachusetts	Mississippi
2	Oklahoma	Oklahoma	Mississippi	Mississippi	Arkansas
3	Florida	Indiana	Alabama	Indiana	Oklahoma
4	Kansas	Iowa	Michigan	Alabama	Kansas
5	Nebraska	Kansas	Indiana	Michigan	Alabama
6	Iowa	Louisiana	Oklahoma	Ohio	Indiana
7	Missouri	Mississippi	Ohio	Arkansas	Texas
8	Illinois	Delaware	Kansas	Oklahoma	North Dakota
9	South Dakota	Texas	Arkansas	Illinois	Nebraska
10	Mississippi	Illinois	Illinois	Kentucky	Kentucky

Source: Grazulis (1991).

be expressed in the following ways: frequency of tornadoes, annual incidence of tornadoes per 10 000 square miles, absolute number of deaths, deaths per 10 000 square miles, or deaths per million people. If annual tornado incidence is compared across countries or states, each one of these criteria will produce a different result. This complexity is shown in Table 1.7, where the ranking of any particular state differs by the criterion used. Although the table is based on dated information, use of more recent data will not significantly change the ranking.

Spatial distribution or dispersion is an important component of hazards since it differentiates between those that occur within a particular region and those that are more widespread. Although tornadoes can occur just about anywhere in the United States, they primarily occur in an area called the "tornado alley" of the Central Plains from Texas to Nebraska (Mitchell and Cutter, 1997). Similarly, Central Texas is called "flash flood alley" because this part of the state experiences frequent flash floods from heavy rain that occurs within a short time (Phillips, 2009). Landslides and mudslides require steep terrain and thus they occur in mountainous and other high-relief areas. Because of the differential distribution of extreme events, it is possible to make probability statements about the likelihood of their occurrence at specific locations.

1.3.7 Speed, rate of onset

Rate of onset, also called countdown interval, is the speed at which an extreme event transforms from its first appearance to peak strength, that is, the length of time between the first appearance of the event and its peak. This can be very rapid, as with earthquakes, landslides, tornadoes, and flash floods, or fairly slow, as with most hurricanes and river flooding, to extremely slow, as is the case with droughts. The former type of disaster is also called sudden-onset disasters and the latter type "creeping" disasters.

Sudden-onset or quick-onset disasters generally occur with little or no warning, and most of their damaging effects are sustained within hours or days. Creeping disasters occur when the ability of responders to support survivors' needs degrades over weeks or months, and they can persist for months or years once discovered (Coppola, 2007). Thus, this temporal parameter of hazards has a direct bearing on the success or failure of remedial action. In general, the more rapidly occurring events allow less time for remedial action. The opposite is true for the slow-onset hazards.

1.3.8 Diurnal factor

The diurnal (time of day) factor, like rate of onset, is temporal in nature. It is an important temporal factor, which refers to occurrence of an event at a specific time of day or night. For example, thunderstorms usually occur during the late afternoon in response to heat buildup during the day. Using all tornado touchdowns between 1950 and 1991 in Kansas, Frank (1998) reported that the incidence of weak (F-0 and F-1) and violent (F-4 and F-5) tornadoes peaks between 5 and 6 o'clock in the evening, while the incidence of strong (F-2 and F-4) tornadoes peak nearly an hour later. Tornadoes, however, occur at night too. Response to a tornado can be more immediate if it occurs during the day, since most people are active during this time. However, if it occurs during the night when most people are sleeping, more injuries and/or deaths could result.

Diurnal distribution, however, is not a factor for other geophysical events such as floods, droughts, or earthquakes. This does not mean that timing is not crucial for other geophysical events. The potential degree of damage and deaths from a severe earthquake is likely to be very high in larger cities like San Francisco if such an event occurs during rush hour when people are commuting rather than at night.

It is important to note that physical characteristics of individual hazards may differ from place to place because of spatial differences in occurrence. Blizzards, for example, would be more frequent in the Great Plains region of the United States than in southern Australia. Thus, while the characteristics of different hazards can be used to compare each other, they cannot be generalized over space. In addition, it is more meaningful to study a number of physical parameters of a given hazard together rather than focus on only one parameter. In addition to magnitude, how damaging a tornado will be to a large extent depends (as mentioned) on the time of day it occurs. Tobin and Montz (1997) argue that magnitude, intensity, frequency, and areal extent are interrelated and only when

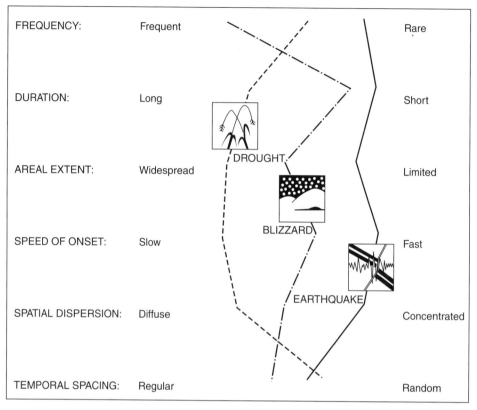

Figure 1.3 Hazard events can be classified based on selected physical characteristics. In this figure, drought, blizzard, and earthquake are classified on the basis of their selected physical characteristics. *Source:* Burton *et al.* (1978), p. 29.

taken together provide a comprehensive view of the physical dimensions of extreme natural events.

Comparison and classification of different hazards can be possible by organizing these events according to their magnitudes, frequencies, duration, areal extent, speed of onset, and seasonality. Figure 1.3 shows how such events might be integrated and evaluated. According to Tobin and Montz (1997), such a system allows analysis of extreme events from a broader conceptual perspective, avoiding the bias toward geophysical processes and hazard-specific concepts. There are, however, several problems with this system of analysis. The distribution of a particular event, such as tornadoes, is not identical at global, continental, and regional scales. In addition, some characteristics, such as speed of onset and duration, are virtually universal, while others are not.

As indicated, destruction wrought by an extreme event does not depend on a single physical characteristic of such an event. A combination of several characteristics determines the impact of a particular hazard or disaster. For example, the risk of death from a flood is influenced by several physical parameters of the event, such as its magnitude (depth), scale (spatial extent), duration, and frequency (Tobin and Montz, 1997). The number of deaths is thought to increase with increasing flood frequency, magnitude, and duration. However, these characteristics are neither simple nor isolated. For instance, a flood that inundates a large area may not cause any fatalities because of its short duration and/or shallowness of the flood water. In contrast, when flood duration and magnitude are substantial, deaths can increase significantly (Hofer and Messerli, 2006). While all physical parameters of hazards are important, magnitude delimits the essential characteristic of the extreme event.

References

Alexander, D. (2000) *Confronting Catastrophe: New Perspectives on Natural Disasters*. New York: Oxford University Press.

Alexander, D. (n.d.) Understanding Hazards and Disasters. Pollution Issues. http://www.pollutionissues.com/Co-Ea/Disasters-Natural.html (accessed January 10, 2011).

Bolin, R. and Stanford, L. (1999) Constructing vulnerability in the first world: the Northridge Earthquake in Southern California. In *The Angry Earth: Disaster in Anthropological Perspective* (eds A. Oliver-Smith and S.M. Hoffman). London: Routledge, pp. 89–112.

Briere, J. and Elliott, D. (2000) Prevalence, characteristics, and long-term sequelae of natural disaster exposure in the general population. *Journal of Traumatic Stress* **13** (4): 661–679.

Burton, I. and Kates, R.W. (1964) The perception of natural hazards in resource management. *Natural Resources Journal* **3**: 412–441.

Burton, I. et al. (1978) *The Environment as Hazard*. New York: Oxford University Press.

Butler, A.S. et al. (2003) *Preparing for the Psychological Consequences of Terrorism: A Public Health Approach*. Washington, D.C.: National Academic Press.

Chapman, D. (1999) *Natural Hazards*. Oxford: Oxford University Press.

Charles, N. and Slater, F. (1995) Floods: friends or enemies? *Geographical Education* **8** (3): 57–62.

Chowdhury, K.M.M.H. (2002) Cyclone preparedness and management in Bangladesh. In *Improvement of Early Warning System and Response in Bangladesh towards Total Disaster Risk Management Approach* (ed. BPATC). Dhaka, Savar: BPATC, pp. 115–119.

Cohen, C. and Werker, E. (2004) Towards an understanding of the root causes of forced migration: the political economy of "natural" disasters. Working Paper no. 25. Harvard University, Cambridge, MA.

Coppola, D.P. (2007) *Introduction to International Disaster Management*. Amsterdam: Elsevier.

Cutter, S.L. (1993) *Living with Risk: The Geography of Technological Hazards*. London: Edward Arnold.

Cutter, S.L. (2001) *American Hazardscapes: The Regionalization of Hazards and Disasters*. Washington, D.C.: Joseph Henry Press.

Cutter, S.L. *et al.* (2000) Revealing the vulnerability of people and places: case study of Georgetown County, South Carolina. *Annals of the Association of American Geographers* **90** (4): 713–737.

Degg, M. (1992) Natural disasters: recent trends and future prospects. *Geography* **77** (3): 198–209.

Dominici, F., Levy, J.I., and Louis, T.A. (2005) Methodological challenges and contributions in disaster epidemiology. *Epidemiologic Reviews* **27** (1): 9–12.

El-Masri, S. and Tipple, G. (1997) Urbanisation, poverty and natural disasters: vulnerability of settlements in developing countries. In *Reconstruction after Disasters, Issues and Practices* (ed. A. Awotona). Burlington, VT: Ashgate, pp. 1–12.

Erickson, N.J. (n.d.) Natural Hazards: Basic Concepts. Hamilton, New Zealand: Department of Geography, University of Waikato.

Fellmann, J.D. *et al.* (2008) *Human Geography: Landscapes of Human Activities.* Boston: McGraw-Hill.

Frank, K.L. (1998) A geographical examination of multiple tornado touchdowns in Kansas. *Papers and Proceedings of the Applied Geography Conferences* **21**: 323–329.

Fritz, C. (1961) Disaster. In *Contemporary Social Problems* (ed. R.K. Merton and R.A. Nisbet). New York: Harcourt Press, pp. 651–694.

Gad-el-Hak, M. (2010) Facets and scope of large-scale disasters. *Natural Hazards Review* **11** (1): 1–6.

Glickman, T.S. *et al.* (1992) Acts of God and acts of man: recent trends in natural disasters and major industrial accidents. Center for Risk Management, Discussion Paper 92-02. Washington, D.C.: Resources for the Future.

GAO (Government Accounting Office) (1995) *Disaster Assistance Information on Declaration for Urban and Rural Areas.* Washington, D.C.: GAO.

Grazulis, T.P. (1991) *Significant Tornadoes, 1880–1989: Volume I: Discussion and Analysis.* St. Johnsbury, VT: The Tornado Project of Environmental Films.

Guha-Sapir, D. *et al.* (2004) *Thirty Years of Natural Disasters: 1947–2003: The Numbers.* Center for Research on the Epidemiology of Disasters. Louvain-la- Neuve, Belgium: Presses Universitaries de Louvain.

Haque, C.E. (1997) *Hazards in a Fickle Environment: Bangladesh.* Boston: Kluwer Academic Publishers.

Hewitt, K. (1983) The idea of calamity in a technocratic age. In *Interpretations of Calamity* (ed. K. Hewitt). Boston: Allen and Unwin, pp. 3–32.

Hewitt, K. (1997) *Regions of Risk: A Geographical Introduction to Disasters.* London: Longman.

Hofer, T. and Messerli, B. (2006) *Floods in Bangladesh: History, Dynamics and Rethinking of the Role of the Himalayas.* Tokyo: University Nations University Press.

IASC (Inter-Agency Standing Committee) (1994) 10th Meeting FAO Field Programme Circular. December.

IFRC (International Federation of Red Cross) and RCS (Red Crescent Societies) (2007) *World Disaster Report 2007: Focus on Discrimination.* Geneva: IFRC and RCS.

Mauro, A. (2004) Disaster, communication and public information. In *Natural Disasters and Sustainable Development* (ed. R. Casale and C. Margottini). Berlin: Springer, pp. 239–246.

McAdoo, B.G. *et al.* (2009) Indigenous knowledge and the near field population response during the 2007 Solomon Islands Tsunami. *Natural Hazards* **48** (1): 73–82.

McEntire, D.A. (2007) *Disaster Response and Recovery: Strategies and Tactics for Resilience.* Hoboken, NJ: John Wiley & Sons, Inc.

Mesjasz, C. (2011) Economic vulnerability and economic security. In *Coping with Global Environmental Change, Disasters and Security: Threats, Challenges, Vulnerabilities and Risks* (eds H.G. Brauch *et al.*). New York: Springer, pp. 123–156.

Mileti, D.S. (1999) *Disasters by Design: A Reassessment of Natural Hazards in the United States*. Washington, D.C.: Joseph Henry Press.

Mitchell, J.T. (2000) The hazards of one's faith: hazard perceptions of South Carolina Christian Clergy. *Environmental Hazards* **2** (1): 25–41.

Mitchell, J.T. (2003) Prayer in disaster: case study of christian clergy. *Natural Hazards Review* **4** (1): 20–26.

Mitchell, J.T. and Cutter, S.L. (1997) *Global Change and Environmental Hazards: Is the World Becoming More Disastrous?* Washington, D.C.: AAG.

Montz, B.E. *et al.* (2003) Hazards. In *Geography in America: at the Dawn of the 21st Century* (ed. G.L. Gaile and C.J. Willmott). Oxford: Oxford University Press, pp. 479–491.

Noji, E.K. (ed.) (1997) *The Public Health Consequences of Disasters*. New York: Oxford University Press.

Norris, F. *et al.* (2002a) 60,000 Disaster victims speak, part 1: a review of the empirical literature, 1981–2001. *Psychiatry* **65**: 207–239.

Norris, F. *et al.* (2002b) 60,000 Disaster victims speak, part 2: summary and implications of the disaster mental health research. *Psychiatry* **65**: 204–260.

Oliver, S. (2001) Natural hazards: some new thinking. *Geography Review* **14** (3): 2–4.

Paul, B.K. (2009) Why relatively fewer people died? the case of Bangladesh's Cyclone Sidr. *Natural Hazards* **50**: 289–304.

Perry, R.W. and Quarantelli, E.L. (2004) *What is a Disaster? More Perspectives*. Philadelphia: Xlibris.

Phillips, B.D. (2009) *Disaster Recovery*. New York: CRC Press.

Picture This (2003) Hit by disasters. *Finance and Development* **40** (3): 40–41.

Porfiriev, B.N. (1996) Social aftermath and organizational response to a major disaster: the case of the 1995 Sakhalin Earthquake in Russia. *Journal of Contingencies and Crisis Management* **4** (4): 218–227.

Quarantelli, E.L. (1985) Social support system: some behavioral patterns in the context of mass evacuation activities. In *Disasters and Mental Health: Selected Contemporary Perspectives* (ed. B. Sowder). Rockville, MD: National Institute of Mental Health, pp. 122–136.

Quarantelli, E.L. (1998) *What is a Disaster?* London: Routledge.

Quarantelli, E.L. (2006) Emergencies, Disasters, and Catastrophes are Different Phenomena. http://www.udel.edu/DRC/preliminary/pp304.pdf (accessed January 31, 2006).

Schmidtlein, M.C. *et al.* (2008) Disaster declarations and major hazard occurrences in the United States. *The Professional Geographer* **60** (1): 1–14.

Sen, A.K. (1981) *Poverty and Famines: An Essay of Entitlement and Deprivation*. Oxford: Clarendon Press.

Shamsuddoha, M. and Chowdhury, R.K. (2007) *Climate Change Impact and Disaster Vulnerabilities in the Coastal Areas of Bangladesh*. Dhaka: COAST Trust.

Sheehan, L. and Hewitt, K. (1969) A pilot survey of global national disasters of the past twenty years. Natural Hazard Research Working Paper 11. Boulder, CO: Institute of Behavioral Science, University of Colorado.

Smith, K. (1996) Natural disasters: definitions, databases and dilemmas. *Geography Review* **10** (1): 9–12.

Smith, K. (2001) *Environmental Hazards: Assessing Risk and Reducing Disaster*. London: Routledge.

Spiegel, P.B. (2005) Differences in world responses to natural disasters and complex emergencies. *JUMA* **293**: 1915–1918.

Susman, P. *et al.* (1983) Global disasters: a radical interpretation. In *Interpretations of Calamities* (ed. K. Hewitt). London: Allen & Unwin, pp. 274–276.

Thywissen, K. (2006) *Components of Risk: A Comparative Glossary.* Bonn: United Nations University.

Tierney, K. *et al.* (2001) *Facing the Unexpected: Disaster Preparedness and Response in the United States.* Washington, DC: Joseph Henry Press/National Academy Press.

Tobin, G.A. and Montz, B.E. (1997) *Natural Hazards: Explanation and Integration.* New York: The Guilford Press.

Tobin, G.A. and Montz, B.E. (2007) Natural hazards and technology: vulnerability, risk, and community response in hazardous environments. In *Geography and Technology* (ed. S.D. Brunn *et al.*). Dordrecht: Kluwer, pp. 547–570.

Turner, B.A. (1976) The development of disasters: a sequence model for the analysis of the origin of disasters. *Sociological Review* **24** (4): 753–774.

UNCHS (United Nations Center for Human Settlements) (1994) Sustainable human settlements in an urbanizing world, including issues related to land policies and mitigation of natural disasters. 15th Session of the Commission on Human Settlement. Unpublished Draft Theme Paper.

UNDP (United Nations Development Programme) (2004) *Reducing Disaster Risk: A Challenge for Development.* New York: UNDP Bureau for Crisis Prevention and Recovery.

UNISDR (United Nations International Strategy for Disaster Reduction) (2004) *Living with Risk: A Global Review of Disaster Reduction Initiatives.* Geneva: UN.

Wenger, D.E. (1978) Community response to disaster: functional and structural alterations. In *Disasters, Theory and Research* (ed. Quarantelli, E.L.). Beverly Hills, CA: Sage, pp. 17–47.

White, G.F. (1964) Choice of adjustments to floods. Department of Geography Research Paper, 93. University of Chicago Press, Chicago.

White, G.F. (ed.) (1974) *Natural Hazards: Local, National, Global.* New York: Oxford University Press.

White, G.F. and Haas, J.E. (1975) *Assessment of Research on Natural Hazards.* Cambridge. MA: MIT Press.

Wisner, B. *et al.* (2004) *At Risk: Natural Hazards, People's Vulnerability and Disasters.* London: Routledge.

WHO (World Health Organization) (1999) *Emergency Health Training Programme for Africa: Training Modules.* Geneva: WHO.

2
History and Development of Hazard Studies in Geography

Research pertaining to the understanding of the human–environment synergism is a major emphasis of contemporary social science and other disciplines, including geography. In particular, global inquiry into human behavior in stressful situations such as natural hazards, has constituted a considerable portion of this research by scholars across disciplines. Starting more than 50 years ago, research efforts are still continuing but both research strategies and topics have undergone tremendous changes. Using a global perspective, this chapter will present the development of hazard research in geography. Specifically, an historic linage will be outlined. In addition, different approaches and perspectives to hazard and disaster research will be reviewed, and important contributions by geographers and others will be noted. Gaps in hazard and disaster research will also be identified and new arenas for future research will be suggested.

2.1 Historical perspective

Natural hazard research forms part of one of the oldest traditions in geography: the study of relationships between society and the environment. Pattison (1964) called this a culture–environment (man–land) tradition, also known as ecological perspective in geography. Although this perspective is evident in the writings of Alexander von Humboldt (1769–1859), Karl Ritter (1988–1859), and John Perkins Marsh (1801–1882), Cutter *et al.* (2000) claim that the initial birth of hazard research in geography is attributable to Harlan Barrows. In his Association of American Geographers' (AAG) presidential address in 1923, Professor Barrows focused on the human ecological viewpoint, which he borrowed from Marsh who emphasized the importance of the relationship between humanity and nature. In this address, Barrows offered an alternative to the outdated environmental determinism paradigm and defined geography simply as human ecology, arguing that geographers should focus on the mutual relationship between human and physical environment (Barrows, 1923).

Despite Marsh's early contributions and rediscovery of his ecological perspective by Barrows in the early 1920s, natural hazards research by geographers did not gain scholastic prominence until the early 1950s. The early ecological approach failed to achieve widespread acceptance among geographers and other social scientists, primarily because of its narrow definition and a bias towards synthesis of phenomena (Ackerman, 1963). However, in the 1940s and 1950s this approach has enjoyed a revived interest among geographers with the emerging concerns over environmental problems (B. Mitchell, 1979).

Geographical hazard research was first started with the pioneering work on flood hazards by Gilbert White (1945). Joined by his students Ian Burton and Robert Kates in the early 1960s and later by their students, the White–Burton–Kates school of natural hazards emerged as a research perspective opposed to the environmental determinism promulgated by Huntington and Semple in the late nineteenth and early twentieth centuries (Emel and Peet, 1989). Over the course of time, their study methods have grown in sophistication, leading to the development of the "human ecological" model or a behavioral approach to hazards, directly descending from Barrow's notion of geography as human ecology (Mitchell, 1989). Later, other approaches, traditions, or paradigms emerged and these are discussed below along with the traditional behavioral paradigm. This discussion also presents the strengths and shortcomings of each of these approaches.

2.1.1 Human ecological model or traditional behavioral approach

A large body of literature on natural hazards is rooted in the human ecology tradition within geography. Early major works in the human ecology tradition include White's (1964) choice of adjustments to floods; farmers' perceptions of flood risks in the United States and their impact on land use decisions (Burton, 1962); perception of drought on the Great Plains of the United States (Saarinen, 1966); and the perception of hazards and choice in floodplain management (Kates, 1962).

White and his students, in a series of monographs published by the Department of Geography at the University of Chicago, examined diverse aspects of human adjustment and perceptions concerning (flood) hazards. Using the ecological approach and its theoretical premises, many international comparative case studies of hazard perception and behavior have been conducted by geographers and others (Whyte, 1986). Almost all of the above studies give prominence to individual perceptions and responses – attitudes, beliefs, values, and personalities – to expand the explanatory framework in the traditional behavioral and decision-making models.

At least at the beginning, the behavioral approach was based on the principle that the main culprits of disasters are geophysical extremes and phenomena. This logically led to the assumption that an appropriate "cure" to these extreme events is to develop systems of control and prediction of these events, aptly called "technological fixes." Key to this approach is human perception of extreme events – the "range of judgments, beliefs and attitude" an individual holds toward such events (Tobin and Montz, 1997).

There are four basic elements inherent in behavioral approaches to hazard studies: the environment which people perceive may vary markedly from the true nature of the real world; individuals interact with their environment, responding to it and reshaping it; the focus is the individual rather than the group; and this approach requires multidisciplinary interactions. Cutter *et al.* (2000) classified works based on the human ecological tradition of White, Burton, and Kates into three groups: (1) the identification and distribution of hazards, (2) the range of adjustments that are available to individuals and the society at large, and (3) how people perceive and make choices regarding hazard events.

One of the major criticisms of studies following the human ecological paradigm is that these studies overexaggerate the role of the individual in hazards and disasters, either as a decision-maker or as a hazard victim. Exclusive attention to individual responses and strategies for coping with hazards seems to force these studies to pay inadequate attention to other possible social, economic, and cultural factors underlying human adjustments to natural hazards. In addition to this narrow perspective, other criticisms include the lack of a sound theoretical base for such studies and that such approaches view society as largely static. Furthermore, the behavioral approach to environmental hazards does not consider historical and structural features of society in its analysis of human responses to extreme natural events (e.g., Alexander, 1991, 1997; Hewitt, 1983, 1997; Oliver-Smith, 1986; Watts, 1983).

While many of behavioral studies are rigorous in data collection and analysis, such investigations often seem to accept at face value replies to structured questionnaires. Respondent backgrounds are rarely proved nor are they compared among different situations with a different respondent or with the same respondent in different situation. Rahman (1991) maintains that these so-called "superficial" replies probably reflect the frustration of real-world experience. It is worth mentioning that the behavioral approach to natural hazards emphasizes examining human behavior from a positive experience of real-world situations. Similar to Rahman, Leach (1978) raises a methodological problem. He rejects behavioral concepts by maintaining that any attempt to explain human behavior solely in terms of ideas, intentions, or preference is very limited at best. He argues that it is underlying economic forces that mold the cultural and intentional appearance of people, and reflect the complex, intertwined relationship between the historical development of society and individual behavior.

Mustafa *et al.* (2011) point out that despite their shortcomings the perception and human adjustment studies of White and others stressed the need for scientific research and debate to inform public policy, so as to expand the perceived range of practical choices. They further maintain that because of their specific policy focus, behavioral studies influence hazards policy more effectively than most other approaches (also see Platt, 1986). Such studies help in determining weaknesses in structural approaches to flood control, demonstrating that dams, levees, sea-walls, dikes, and similar devices convey a false sense of security, leading to intensification of land use in supposedly protected floodplains. Since the 1930s, a "technological fix" approach has often been advocated, especially in developed countries, to control natural events.

Studies by White and others proved instrumental in the formation of the US National Flood Insurance Program in 1968, marking a shift in national policy away from structural controls towards more nonstructural controls. In the context of floods, nonstructural measures include flood warning and floodplain management. Their studies laid the groundwork for expanding the range of choices for flood control beyond simple dam and levee construction (Mustafa *et al.*, 2011). Thus, the human ecological model has proved successful in providing significant insights into human response toward hazard issues, creating a wide range of public policy choices, and identifying more alternatives for hazard loss reduction.

2.1.2 Political ecology or structuralist approach

From the 1970s the direction of hazard research began to shift toward a more socially oriented context. This shift was largely initiated by Blaikie and Brookfield (1987), Hewitt (1983), Watts (1983), and other hazard researchers who emphasized the influence of social structural factors on differential access to resources, hence creating different susceptibilities to extreme events (Mustafa *et al.*, 2011). Proponents of the political ecological perspective (also called the structural approach to natural hazards) maintain that an individual's exposure to disaster "differs according to their class (which affects their income, how they live and where), whether they are male or female, what their ethnicity is, what age group they belong to, whether they are disabled or not, their immigration status, and so forth" (Wisner *et al.*, 2004, p. 6). This approach provides an alternative explanation of human behavior in relation to natural change and social flux (Haque, 1997).

Followers of this perspective contend that people affected by disasters respond in different ways, and their ability to cope with losses caused by extreme natural events also differs, depending on their economic position, as well as on any social and political linkages involved (Alexander, 2000; Blaikie *et al.*, 1994; Bolin and Stanford, 1999; Hewitt, 1983; Susman *et al.*, 1983). This perspective also claims that natural hazards cannot be fully understood without paying close attention to the political, economic, and social structures that constitute a given society. Stated succinctly, adherents to the political ecological perspective believe hazard responses are influenced by conditions existing beyond the local (individual/micro) level (Palm, 1990). Thus, hazard responses are influenced not only by micro level factors, but also by meso and macro levels factors.

Unlike behaviorists, political ecologists are more concerned with issues of class, type of economic development, international dependency, gender, and deeper social structures in explaining the root causes of extreme natural events. This approach is much broader in the sense that it links environmental disasters with underdevelopment and the economic dependency of developing countries on developed nations and blames the global economy, particularly capitalism, for the widespread suffering caused by geophysical events. Said another way, disasters in developing countries arise more from workings of the global economy, from the spread of capitalism, and the marginalization of the poor than from the direct effects of extreme events. According to the political ecological paradigm, redistribution of wealth is the solution to the mitigation of impacts associated with natural hazards rather than the application of science and technology.

Poor and marginalized groups have little choice but to interact with their environment in ways that tend to increase their vulnerability to hazards and disasters. Because they are more vulnerable, they generally suffer more and disaster impacts compound their difficulties, exacerbating the cycle of poverty in which they exist. As indicated, vulnerability is closely associated with the political ecological perspective of natural hazards and disasters. It refers to the social and economic characteristics of an individual, a household, or group with respect to its capacity to anticipate, cope with, resist, and recover from the impact of a disaster (Blaikie *et al.*, 1994). According to Bolin and Stanford (1999), vulnerability concerns the complex of social, economic, and political circumstances in which people's lives are embedded and these factors structure the choices and opinions they have in coping with environmental hazards. They claim that: "The most vulnerable are typically those with the fewest choices, those whose lives are constrained, for example, by discrimination, political powerlessness, physical disability, lack of education and employment, illness, the absence of legal rights, and other historically grounded practices of domination and marginalization" (Bolin and Stanford, 1999, pp. 9–10). Vulnerability is discussed in greater detail in the next chapter.

Because of differential vulnerability, disasters are no longer viewed as purely natural constructs with geophysical phenomena serving as sole agents of destruction and death. Some authors (e.g., Kapur, 2010) maintain that considering disasters as natural events is a categorical mistake and they emphasize the need to initiate a "vocabulary change" for drawing increased attention to the social dimensions of vulnerability. Both the vulnerability and the political ecological perspectives are concerned with how social structures and discourses make various groups differentially vulnerable to the ill-effects of natural hazards. Both emphasize the constraints which are placed on individual action by broader and more powerful institutional forces. Tragically, this situation also tends to promote dominant class interests. Political ecologists have preferred to examine issues of vulnerability and equity (Liverman, 1990; Mitchell, 1999). They suggest targeting short-term relief assistance to the most vulnerable populations, and also linking that assistance with long-term development; they also champion empowerment by gender and class.

Although the structural paradigm acknowledges the importance of meso and macro levels factors, its emphasis still lies on household level factors – and in that respect it is not much different from the behavioral approach. Regional and global factors are not generally taken into due consideration, leaving the structural paradigm little more than a theoretical perspective with limited real-world application. However, one of the insightful aspects of this approach is its conceptualization of the interaction between the physical environment and socio-economic conditions as well as a holistic understanding and treatment of this interaction. Furthermore, this approach stresses a dynamic view of the social and political matrix, forming the context within which individual behavior may take place. Its concern with rational and contextual understandings of local level vulnerability helps in building explanatory linkages between local, national, regional, and global scales. An additional fundamental contribution of this approach has been the now common realization that although perception is important in determining behavior, it is not determined in a socio-spatial vacuum.

Another advantage of the structural approach is that it is useful for performing analyses at different study units – at the individual, household, or community levels. In many developing countries, rural residents usually make decisions at the household level (Zaman, 1989). Although it is possible to conceptualize hazard adjustments or responses as individual "choices" at the household level, such choices are in fact determined by a number of social and cultural factors, such as composition of households (whether nuclear, joint, or extended), kin-based interdependencies and by other social linkages. This social structural perspective may be extended even further to include variables such as the structure of local landholdings and the nature of the local land tenure system.

The political ecological perspective has provided an opportunity to broaden the research on natural hazards and disasters. In the words of Cutter *et al.* (2000, p. 715), this perspective "has broadened both the definition of hazard and geographers' approaches toward understanding and ameliorating them." Within this perspective, geographers and other hazard researchers are also able to broaden their view of both the causes of hazards and the consequences, and to more fully understand the often-limited options available to people exposed to dangerous conditions. The vulnerability approach helps in expanding the range of theoretical tools available both for studying hazard and for providing policy guidance.

2.1.3 Other approaches

The political economy perspective gave birth to the vulnerability approach, which has dominated hazard research for the last two decades. During this period, the focus has shifted from the hazard itself to the combination and interaction of people's vulnerability to extreme events. The political economy perspective also contributed to the development of other perspectives used in hazard studies. Unlike researchers from the behavioral and political economy perspectives, several geographers (e.g., Cutter *et al.*, 2000; Mitchell, 1989; Palm, 1990) consider both the physical properties of the hazards as well as aspects of the social, political, spatial, temporal, organizational, and economic milieu within which the extreme natural events take place. This approach is known as "hazards in context" and it uses both empirical and social analyses, and acknowledges that hazards are inherently complex physical and social phenomena. Geographic scale is a key component in this approach.

Risk perspective is another approach to hazard research; developed in the 1980s, it is closely related to the vulnerability approach. In their pioneering work, Kasperson *et al.* (1988) maintain that risk intertwines with individual psychological, cultural, social, and institutional factors in ways that may either amplify or attenuate public perceptions of risk. The Social Amplification of Risk Framework (SARF) combines research in many disciplines, such as psychology, sociology, anthropology, and communication theory. According to this framework, amplification occurs in two stages: in the transfer of information about the risk, and in the response mechanisms of society. SARF examines how communications of risk events pass from the sender through intermediate stations to a receiver and in the process serve to amplify or attenuate perceptions of risk. All links in the communication chain – individuals, groups, and media – contain filters through which information is sorted and understood. SARF helps to interpret public perceptions and, ultimately, policy responses to risk and hazards in contemporary society (Cutter *et al.*, 2000). SARF has generated considerable interest outside the geographical community and has helped to stimulate interdisciplinary research into social theories of risks.

Rooted in Hewitt and Burton's (1971) regional ecology of hazards and the concept of vulnerability, Susan Cutter's work in the 1990s contributed to the development of the "hazards of place" or "vulnerability of place" approach (Cutter, 1996; Cutter and Solecki, 1989; Cutter *et al.*, 2000). Hewitt and Burton (1971) maintained that considering the threat from all hazards provides an opportunity to mitigate several hazards simultaneously. Yet, until middle of the 1990s, no one attempted to characterize risk from all hazards or the intersection they share with vulnerable populations (Cutter *et al.*, 2000). In an attempt to bridge gaps between the physical and social scientific perspectives, Cutter and colleagues (Cutter, 1996; Cutter *et al.*, 2000) proposed the concept of "vulnerability of place," where biophysical exposure intersects with political, economic and social factors to generate specific configurations of vulnerability. This concept is discussed in more detail in Chapter 3.

The term "hazardscape" or "riskscape" is closely associated with the "hazards of place" approach, which examines the distributive patterns of hazards and underlying processes that gave rise to them. With a focus on technological hazards, Corson (1999, p. 57) defines hazardscape as "the spatial distribution and attributes of human engineered facilities . . . that contain or emit substances harmful to humans and environment." Even though this definition is viable for most technological hazards, it does not fully apply to hazards in general or especially to natural hazards. In the following year Cutter et al. (2000) used the term "hazardscape," in their study of place vulnerability, to refer to the landscape of all hazards in a particular place or the net result of natural and artificial hazards and the risks they pose cumulatively across a given area. The concept of hazardscape includes the interaction among nature, society, and technology at a variety of spatial scales and creates a mosaic of risks that affect places and the people who live there. This term is normally used in reference to a specific place or region; however, the term can be expanded to apply to a larger area (Cutter, 2001; Khan and Mustafa, 2007; Mustafa, 2005). In examining the gender dynamics of the flood hazards and disaster events in Bangladesh, Sultana (2010) recently used the term "waterscapes" as an alternative to hazardscape or riskscape.

A hazardscape or disasterscape depicts the current situation of hazards or disasters at a place. However, Khan and Crozier (2009) consider it to be a dynamic construct reflecting the physical susceptibility of a place and vulnerability of human life and assets to various hazards in a given human ecological system. They maintain that both susceptibility and vulnerability change through time, as does nature, and the types of hazards a place may be exposed to. Process, people, and place are three essential elements of the hazardscape, which interact and give shape to its three resultant characteristics (i.e., hazards, human vulnerability, and physical susceptibility) (Khan and Crozier, 2009). The different approaches discussed above have broadened our understanding of the nature of hazards and the range of causal or contributing factors. The next section is devoted to discussing the changing focus of hazard research in geography.

2.2 The focus of hazard research

The General Assembly of the United Nations designated the 1990s as the International Decade for Natural Disaster Reduction (IDNDR). The main purpose for this proclamation was to drastically reduce loss of life, property damage, and socio-economic disruption caused by natural disasters by the year 2000. This is an important event because it helped in broadening the research arena for hazard researchers, particularly for hazard geographers. Their emphasis shifted from a response-based analytical perspective, which focuses on the perception and range of adjustments people adopt to compensate for disaster losses, to new topics such as hazard preparedness, mitigation, and recovery from the effects of disasters along with hazard identification, risk assessment, vulnerability, and disaster resiliency at individual, household, and community levels. The post-IDNDR period has seen a shift in focus of most hazard research in geography. The IDNDR also ushered in profound changes in public disaster management efforts.

In their 2003 review for *Geography in America*, Montz et al. (2003) noted an expansion of hazard research agenda within geography. In the context of dissertation research topics, they claim that the number of studies addressing issues associated with multi-hazards and largely ignored hazards (such as global warming) increased in the 1990s. They further claim that:

> New topics are being explored and previous topics are being reworked from different perspectives. Although still dominant, the proportion of dissertations emphasizing human use systems continued to decline, as has the share coming from a physical geographic perspective. Conversely, studies that develop predictive models of hazards, evaluate the utility of Geographic Information Systems in hazards analysis, or explore methods of mapping hazards became more commonplace during the 1990s. Another major trend in hazards dissertation research was the movement away from studying specific types of hazards toward the study of total risk, environmental risk and equity, generic hazard risk perceptions, or disaster preparedness – all topics that incorporate a variety of hazard threats, natural and technological, physical and social. (Montz et al., 2003, p. 482)

Several new areas of research emerged during the 1990s and early 2000s, two of which are discussed below: disaster and development, and terrorism. These two decades have also seen an increased application of geographic information systems (GIS) in hazards research, which will also be covered in this section. Another increasingly important area of research, global climate change, will be discussed in the last chapter.

2.2.1 Disaster and development

In recent years, natural disasters have been integrated into the processes of development. For this reason, disasters are often considered a function of development (Collins, 2009). The disaster and development approach acknowledges that a disaster is the consequence of insufficient development of appropriate means to avoid the impacts of an extreme natural event, or an aspect of development and/or processes associated with it are the cause of the disaster event. The purpose of this subsection is to examine the relationship between disasters and development by exploring how this relationship functions. Examples will be drawn from different countries of the world.

By causing physical damage to property and infrastructure, removing livelihoods, and increasing health risks, disasters also destroy development initiatives, put development gains at risk, and often reverse development for years or even decades. Such damage may set back social investment targeted to ameliorating poverty and hunger, providing access to education, health services, safe housing, drinking water and sanitation, or protecting the environment as well as economic investments that provide employment and income. All may have severe impacts on development initiatives, particularly in developing countries, where natural disasters not only put tremendous pressure on available resources, but also usually reverse rates of development subsequent to a disaster. For example, Hurricane Mitch, which destroyed as much as 70% of the infrastructure in Honduras and Nicaragua in 1998, reversed the rates of development in these two countries for at least a decade and as much as 20 and 30 years in other Central American countries (Coppola, 2007). Developing countries need to divert allocated funds from development projects in order to manage disaster consequences and begin recovery efforts.

For countries with developing economies, the financial setbacks inflicted by extreme natural events can be ruinous compared to those in countries with developed economies. For example, both the United States and El Salvador experienced earthquakes in 2001 resulting in losses of about US$2 billion each. While this amount had little or no noticeable impact on the US economy, this figure represented 15% of El Salvador's GDP for that year (UNDP, 2004). A disaster exacerbates the debilitating root causes of poverty in developing countries, leaving the poor even more vulnerable for the next disaster in what may seem a never ending cycle. Thus, a strong correlation exists among disaster, poverty, vulnerability, and development. The integration of disaster management and holistic development can contribute to a serious reduction in disaster risk, which in turn paves the way for achieving intended development goals, and reducing poverty and vulnerability of prospective victims to natural extreme events.

Although the integration of disaster and development was first proposed by Cuny (1983), it was not pursued until later when the political ecological approach to natural hazards and the concept of vulnerability began to be widely recognized by hazard researchers. As noted, these researchers argued that the impact of natural disasters depends not only on the physical characteristics of such events, but on the capacity of people to absorb the impact and recover from loss or damage. Logically, the focus of hazard research shifted to social and economic vulnerability, with mounting evidence that natural hazards and disasters had widely varying impacts on different social groups and among different countries. The causal factors of disaster thus shifted away from those concerned mainly with the natural event, towards those that favored development processes that generated different levels of vulnerability (UNDP, 2004). By this time, it was clearly evident that poverty is the principal reason why people are most vulnerable to natural hazards (Blaikie *et al.*, 1994). Middleton and O'Keefe (1998, p. 12) succinctly and rightly state that: "people are vulnerable because they are poor and lack resources, and because they are poor and lack resources they are vulnerable." Not surprisingly, alleviating poverty and helping people towards economic self-reliance have become the major goals of development initiatives in

Table 2.1 Millennium Development Goals

Goals
1. Eradicate extreme poverty and hunger
2. Achieve universal primary education
3. Promote gender equality and empower women
4. Reduce child mortality
5. Improve mental health
6. Combat HIV/AIDS, malaria and other diseases
7. Ensure environmental sustainability
8. Develop a global partnership for development

Source: Compiled from UN (2008).

many countries of the world. Thus, vulnerability reduction began to be advanced as a key strategy for reducing disaster impacts.

Declaration of the IDNDR also helped to provide considerable incentives for rethinking disaster risk as an integral part of the development process and for dialogue centered on the social and economic causes of disaster risk. Through these discussions, hazard researchers explored relationships between disasters and development, and realized that reduction of disaster risk required long-term integration with processes of development. Some authors, such as Collins (2009), consider participation in development means participation in disaster reduction activities. When individuals and communities are put at the center of development or disaster reduction strategies there are substantial synergies between the two modes of thinking. To Collins, development is synonymous with the aims of disaster reduction.

Major disasters (e.g., floods in East Africa, Latin America, the Caribbean, and South and Southeast Asia, Hurricanes Georges and Mitch in Central America and the Caribbean, and a cyclone in Orissa, India) occurring at the end of the 1990s together with the Millennium Declaration, which sets forth a road map for development supported by 191 nations, helped to galvanize support for the view that disaster reduction and development are inextricably linked. Eight Millennium Development Goals (MDGs) were agreed upon in 2000, which in turn have been broken down into 18 targets with 48 indicators for progress. The MDGs are listed in Table 2.1. Most goals are set for achievement by 2015 (UNDP, 2004). Table 2.1 clearly indicates that poverty, health, education, human rights, environment, and good governance lie at the heart of the MDGs. Achieving these MDGs will considerably reduce the risk of disaster.

Exploring relationships between disasters and development also benefits from the recurrent debate among hazard researchers over whether natural disasters should be addressed by relief measures (e.g., Blaikie et al., 1994; Hewitt, 1997; Bolin and Stanford, 1999; Susman et al., 1983). Opponents of the provision of emergency aid argue that disaster relief accelerates continued underdevelopment and the marginalization of victims of extreme natural events. This is because disaster relief in the 1990s was not closely linked with development interventions either during or in the pre-, and/or post-disaster period (e.g., Susman et al., 1983; Sollis, 1994; Wisner et al., 2004). This is discussed in detail in Chapter 7. However, the involvement of both domestic and international nongovernmental organizations (NGOs) in the development activities of developing countries in the 1980s and 1990s led to the integration of disasters with development efforts, particularly during the post-disaster periods (Paul, 2006). More specifically, all programs of disaster recovery phases (relief, rehabilitation, and reconstruction) have been aligned with development programs (Frerks et al., 1995).

Since the 1990s, NGOs have actively participated in development activities in many developing countries of the world. After being involved in providing emergency aid to disaster victims for nearly two decades, NGOs realized that relief work causes disruption to normal development activities and often causes beneficiary groups to revert to relief dependence. They also realize that "development" is about reducing vulnerability of people and communities to both anthropogenic and natural hazards (Paul, 2006). In many developing countries, NGOs have programs to alleviate poverty, raise democratic consciousness, and empower the most deprived sections of society. The contributions of NGOs in bringing even the most deprived classes of society under the umbrella of development are widely acknowledged globally (Streeten, 1999). In sum, alleviating poverty and helping people become self-reliant have become important components of development programs, particularly in developing countries (Middleton and O'Keefe, 1998).

As indicated, vulnerability reduction and poverty reduction are largely synonymous, and the reduction of vulnerability or poverty is a crucial path to disaster reduction. Conventional linkages between disasters and development operate on the premise that future disasters will be less severe. Strategic development for vulnerability reduction considers the socio-economic, as well as the physical and infrastructural dimensions (Lewis, 1997). Physical and infrastructural development seeks to save lives by constructing embankments, bridges, public cyclone shelters, rural electrification and energy generation system and other structures, while socio-economic development aims to provide livelihoods and improve the living conditions among people living in hazard-prone areas by community development, disaster preparedness training, empowering women and the poor, preventive medicine and health services, maternity and child care, and microcredit lending mechanisms (e.g., Grameen Bank in Bangladesh).

To examine the relationship between disaster and development, the UNDP (2004) recommends differentiating between economic and social elements of

Table 2.2 Disaster–development

	Economic development	Social development
Disaster limits development	Destruction of fixed assets. Loss of production capacity, market access or material inputs. Damage to transport, communications or energy infrastructure. Erosion of livelihoods, savings and physical capital	Destruction of health or education infrastructure and personnel. Death, disablement or migration of key social actors leading to an erosion of social capital
Development causes disaster risk	Unsustainable development practices that create wealth for some at the expense of unsafe working or living conditions for others or degrade the environment	Development paths generating cultural norms that promote social isolation or political exclusion
Development reduces disaster risk	Access to adequate drinking water, food, waste management and a secure dwelling increases people's resiliency. Trade and technology can reduce poverty. Investing in financial mechanisms and social security can cushion against vulnerability	Building community cohesion, recognizing excluded individuals or social groups (such as women), and providing opportunities for greater involvement in decision-making, enhanced educational and health capacity increases resiliency

Source: UNDP (2004), p. 20.

development. Although these two components are interdependent and overlapping, it is useful to think of the ways that these elements, and their constituent institutional and political components, are shaped, restrained, and sometimes accelerated by disaster. Similarly, one can analyze the ways in which economic and social development (and their constituent processes) work directly and/or indirectly to increase or decrease disaster risk (UNDP, 2004). Table 2.2 illustrates these complex interactions. As is evident from the table, economic development includes economic production and its supporting infrastructure (e.g., transport networks, and access to sources of drinking water), while social development concerns social assets such as social capital, and health and educational infrastructure.

A close examination of Table 2.2 also reveals that disasters limit both economic and social development. A single disaster can wipe out the gains of many years of development. Since this is discussed in the early part of this subsection, discussion here will be limited to how disasters limit social development. Disasters not only impact economic development, these events often destroy gains made in health care provision, sanitation, drinking water, housing, and education sectors that underpin social development. Improper and inequal distribution of emergency aid after a disaster often creates social inequality. Disaster literature clearly suggest that women suffer more stresses in disaster situations than men, and they also bear a disproportionate burden of the additional domestic and income-generating work necessary for survival following an extreme event. This, in turn, often leads to a reduction in the social development of women (UNDP, 2004).

Table 2.2 further shows that economic and social development increases disaster risk. There are many examples of the drive for economic and social growth generating disaster risk at different levels. For example, the massive forest fires in Indonesia in 1997 that caused severe air pollution in neighboring Malaysia were partly caused by the uncontrolled use of fire by farmers wishing to expand production of a major export crop, palm oil (UNDP, 2004). Although it is hard to imagine that social development increases disaster risk, this is often the case where people are forced to expose themselves or others to risk in order to fulfill their needs. Rapid urbanization in developing countries is a case in point. The growth of slums and squatter settlements are one of the outcomes of this urbanization, which initiates large-scale rural-to-urban migration. These migrants are forced to live in makeshift housing built in highly disaster-prone areas, such as on steep slopes, along floodplains, and adjacent to noxious or dangerous industrial or transport infrastructure sites.

Poorly planned development programs can also increase vulnerability. This may happen if development efforts are inappropriate for existing social and environmental conditions, and/or if their impacts are not assessed properly (Ozerdem, 2003). There is also need for careful consideration and/or avoidance of large development projects such as dams, which often increase risk and vulnerability, especially for marginalized groups (Paul, 2006).

As noted, the relationship between disaster and development has both positive and negative aspects (Figure 2.1). There is no doubt that development can reduce both physical and social vulnerability of potential disaster victims by addressing the root causes of disaster risk and lack of access to economic and political tools, as well as by alleviating poverty. As shown in Table 2.2, one way to reduce poverty is through trade and technology. In this context, microfinance has played an important role in alleviating poverty in many developing countries. The Grameen Bank in Bangladesh has a long-standing commitment to supporting small-scale enterprise as well as empowering poor women. The UNDP (2004) maintains that governance is a critical area for innovation and reform in achieving disaster risk reduction within human development. To reduce disaster risk, governance must be sensitive to the needs of those at risk from disaster and able to facilitate timely, strategically coherent, and equitable decisions in resource mobilization and aid distributions. It is also important to identify governance

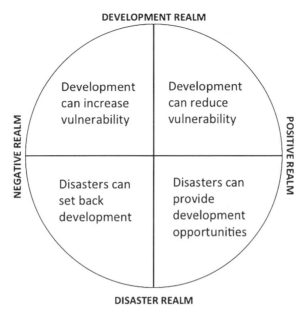

Figure 2.1 The relationship between development and vulnerability to disasters. *Source:* UNDP-DHA (1994), p. 10.

reform that might inadvertently contribute to causing and/or increasing human vulnerability.

Improved health and education status help reduce vulnerability and can limit human losses in a natural disaster. A literate and better-educated population – including women and girls – responds better to warnings and generally complies with evacuation orders. The importance of extending opportunities to women and girls is noted in the MDGs and has been shown to improve the delivery of disaster risk reduction. Social developments usually increase social cohesion and open participation in decision-making across gender, which tends reduce vulnerability and disaster risk (UNDP, 2004).

Disaster can also provide new opportunities for fostering sustainable development, which "means not merely protecting the environment but also mitigating hazards by reducing the degree of potential interaction between hazards agents and large outlays of technology, such as major dams" (Alexander, 2000, p. 89). The occurrence of a disaster provides an opportunity for national governments, communities, and individuals to reassess development strategies in order to build mitigation into those strategies (Cuny, 1983; Hagman, 1985; Varley, 1993). The post-disaster period "presents a host of new opportunities for government, communities and individuals to avoid re-creating the conditions that led to disaster in the first place. Once settlements are established in hazard-prone areas, this 'window-of-opportunity' rarely presents itself, until a major disaster occurs permitting a fresh start" (Parker *et al.*, 1997, p. 25). The United Nations Disaster Relief Organization (UNDRO, 1992, p. 19) add that: " [d]isasters often create a political and economic atmosphere wherein extensive changes can be made more rapidly than under normal circumstances." This is discussed in some detail in Chapter 6.

Figure 2.1 summarizes both positive and negative relationships between disaster and development. The right half of the figure reflects the positive side and the left half deals with the negative aspects of the relationship. Similarly, the upper half of the figure represents the development realm and the lower half represents the disaster realm. The short statement provided in each quadrant sums up the basic concept derived from the overall of the two realms. As noted, disasters can delay development initiatives by many years through the loss of resources and other negative impacts such as increased unemployment and political destabilization. However, disasters can also provide development opportunities. Similarly, development can both reduce and increase human vulnerability to disasters (Figure 2.1). Yet the right type of appropriate development reduces disasters, mitigates their impacts, and aids in sustainable recovery once one has occurred (Collins, 2009). It is important to mention that physical and mental health issues are often considered related to disaster and development. Collins (2009) highlights disease as a disaster in its own right, identifying it as a significant consequence of development as well as an outcome of other disasters. Direct and indirect impacts of disasters on health and diseases are discussed in Chapter 4.

2.2.2 Terrorism

Terrorism, which is a threat or deliberate use of violence to intimidate someone, a group, or even a government, has become one of the most deadly social hazards in recent years. Terrorism is hardly a topic new to hazard geographers and they had been studying it long before the events of September 11, 2001 in the United States (Hewitt, 1987, 1997; J.K. Mitchell, 1979). However, the scope and nature of terrorism have changed after the September 11, 2001 attacks. These deadly events attracted the attention of many hazard researchers and prompted a plethora of research on this important topic. As noted, terrorism existed before the bombing attacks on the World Trade Center in New York, and on the Pentagon in Washington, D.C. in 2001. Terrorist bombing attacks were also undertaken at the World Trade Center in 1993, the Federal Building in Oklahoma City in 1995, and on the US embassies in Kenya and Tanzania in 1998.

Over the past decades, acts of terrorism have included dissemination of aerosolized anthrax spores, intentional food product contamination, release of chemical weapons in major metropolitan systems, and suicide attacks using explosive devices. For example, shortly after 2001 World Trade Center attack, envelopes containing anthrax were mailed to the headquarters of a newspaper in Florida and to congressional leaders in Washington, D.C.; both incidents resulted in the deaths of several people (McEntire, 2007). In addition, the sarin gas attack by the Aum Shinrykio terrorist group in a Tokyo subway in 1995 killed several people, and necessitated medical care for thousands more.

Terrorists usually have ideological and political motives, such as seeking independence, promoting their religion, and/or a having specific political objective. Perpetrators usually use chemical, biological, radioactive/nuclear, or explosive materials (CBRNE), and other means, such as cyber-terrorism, to abduct and/or kill people who do not subscribe to their political or religious ideology. Terrorists disrupt the activities of others. Terrorist group activities have occurred all over the world, and the last decade has seen the United States become one of the main targets of terrorists. It has been found that more than 70% of terrorist attacks involve the use of conventional explosives, which are easier to obtain and difficult to detect because most easily attainable explosive materials are untraceable. Fewer than 5% of actual and attempted bombings are preceded by any kind of threat or warning (Coppola, 2007).

The threat of terrorism has prompted the establishment in the United States of the Department of Homeland Security, which did not exist prior to the September 11, 2001 attacks. This department introduced a color-coded threat level, which was based on risk information (the probability of an attack and its potential consequences) and the credibility of that information. Billions of dollars have been spent on training first-responders and public health professionals, and for airport security (McEntire, 2007). The Madrid train bombings in 2004 and the London subway bombings in 2005 suggest that perhaps the transportation system is becoming a target of choice for terrorists. This leads to an increased need for training of transportation workers on what suspicious activity to look for and report and what actions to take in an emergency.

Terrorists and/or their groups receive funding from various sources to conduct their operations. They gather funds through illegal activities such as drug trafficking, burglary, and arms trading. Their activities are also financed by wealthy individuals, corporations, and/or governments. Hastings (2008) claims that the financial resources of terrorist groups are now fully international. Terrorists recruit people who adhere to their ideology, and they use the Internet to spread their message to others. Sometimes they set up camps and organize conferences where recruits are taught to raise funds, stake out targets, and carry out attacks. Terrorists also open training camp to teach new recruits how to use weapons such as knives, guns, computer viruses, and explosive devices to carry out attacks. Terrorists are using the tools of globalization such as the Internet to conduct attacks in different parts of the world. Information technologies have extended terrorists' communications abilities, and terrorist groups are able to organize across international borders. Terrorist violence has become diffused and internationalized.

However, while modern technology, capital flows, and quick and efficient means of transportation allow terrorist groups to spread out their command and control and to organize multiple simultaneous attacks in different parts of a country (or even in different countries), they do so subject to restrictions that may not be ameliorated by the technologies of globalization. Terrorist groups are not able to take advantage of globalization if they face a hostile political environment. Physical movements of terrorists both within and across national boundaries are not as easy as before 9/11, particularly in Western countries. The same is also true for flow of capital used to fund terrorist groups (Hastings, 2008).

Shortly after the September 11, 2001 attacks, the Association of American Geographers (AAG) began to develop a national research agenda on the geographical dimensions of terrorism to enhance the nation's research infrastructure in addressing this important public policy issue. The Geographical Dimensions of Terrorism project (GDOT), funded by the Geography and Regional Science program at the National Science Foundation, held its first workshop in January 2002 where priority research themes were initially formulated. Further conversations with a broader constituency of geographic researchers and practitioners refined these themes and a consensus agenda emerged centered on three broad areas: geospatial data and technologies infrastructure research; regional and international research related to the root causes of terrorism; and vulnerability science and hazards research. Action items and research questions were developed in each area.

The outcome of the GDOT was publication of an edited volume titled *The Geographical Dimensions of Terrorism* (Cutter *et al.*, 2003). This volume provides a detailed synthesis on the role of geographical inquiry and technologies in combating terrorism. The papers included cover a breadth of topics such as the geographical nexus of public health, law enforcement and hazards, the geographies of inclusion and exclusion and insurrection, the use of geospatial data and technologies in times of crises, defining and delineating critical infrastructure, bioterrorism and agro-terrorism, and how one builds a safer, yet still open society. Several quick response reports were also published by geographers (e.g., Mitchell *et al.*, 2001; Rodrigue, 2002; Thomas *et al.*, 2002) on the 2001 September bombings of the World Trade Center immediately after this tragic event. These studies were supported by the Hazards Center at the University Colorado, Boulder, USA. Using geographers' technological expertise in GIS and remote sensing, these studies assisted rescue and relief efforts undertaken in the aftermath of the tragic events of September 11, 2001.

Unfortunately, interests of hazard geographers on this important topic did not last very long. With few exceptions, geographers seem to show less interest on the topic of terrorism in general. Flint (2003) maintains that geographers are well equipped to provide insightful analyses into terrorism by using three comparative disciplinary advantages: the consideration of geohistorical context, geographic scale, and territoriality. Undoubtedly, geographers need more responsibility in this new arena of research which will allow them to find opportunities to combine risk and spatial analysis in an integrative way.

2.2.3 Geographic information systems (GIS) and other technologies

Understanding the various aspects of hazards and disasters has been facilitated by geo-information technologies (GIT) such as geographic information systems (GIS), remote sensing, and satellite imagery. By providing these relatively new tools, these technologies have facilitated and influenced hazard research, and helped emergency practitioners in efficient management of natural disasters. According to Tobin and Montz (2004, p. 548): "Geographic information system (GIS), remote sensing, and satellite imagery are now fundamental components of hazard research that have created new opportunities and enhanced our scientific approaches." At the same time, many other technological advances have greatly improved the range of hazard forecast, warning, and loss reduction techniques available to emergency managers and personnel.

The application of GIS in hazard studies began in the early 1990s with both natural and technological hazard identification and risk assessment (Hodgson and Palm, 1992; Palm and Hodgson, 1992; Sorensen et al., 1992). Dash (1997) claims that GIS was used after Hurricane Andrew in 1992. He maintains that the initial use of GIS was in mapping damage and analyzing community demographics. Later, as the potential of GIS was better understood, its use expanded in many areas such as public assistance, and tracking debris and debris removal, and mapping damaged homes and the location of trailers used for temporary housing. Other applications of GIS after a disaster include: showing tornado paths, plotting the location of fire stations, assessing the extent of areas impacted by disasters, and many others. GIS has also been used to combine geographically coded data with survey questionnaires to look at locations, attitudes toward earthquake hazards in California, and the purchase of earthquake insurance (Hodgson and Palm, 1992). This study provided a new insight into hazard-zone behavior that may eventually lead to improved mitigation. Potential use of GIS and other geo-technologies in disaster mitigation and management activities are listed in Table 2.3.

In recent years GIS has gained popularity among hazard scholars as a research tool for studying different types of natural hazards, planning and managing emergency activities, assessing potential and actual hazard risk, monitoring emergency responses, estimating disaster losses, and recovery mapping (Curtis and Mills, 2009). GIS technology has been used in all phases of the emergency response cycle, although in some phases more than others (Cutter, 2003). The use of GIS was extensive during the initial rescue and relief operations undertaken immediately after the terrorist events of September 11, 2001. Applications of GIS in this event ranged from the positioning of logistical support and resources to the

Table 2.3 Potential use of GIS and other geo-technologies

1. Mapping of disaster-affected areas and assessing disaster loss and damage
2. Risk mapping and modeling
3. Mitigation and spatial planning for resource mobilization
4. Spatial decision support system
5. Alternative transport routes planning and searching
6. Target risk communication planning, implementation, and monitoring
7. Spatial monitoring of rescue operations by using real time communication and geographic visualization
8. Real-time monitoring

production of public maps of the damage by both print and electronic media as well as on the Internet (Cutter, 2003). This technology is increasingly used in hazard vulnerability studies as well as modeling, predicting, and monitoring hazard risk (Cutter et al., 1999; Montz, 1994). Susan Cutter and her colleagues (2001) used GIS as a tool for understanding both biophysical and social vulnerability at specific place. Emani et al. (1993) used it to investigate vulnerability to extreme storm events and sea-level rise, and Lowry et al. (1995) applied GIS to examine vulnerability to hazardous chemical releases.

Among all the disasters that have occurred thus far during the twenty-first century, GIS has been most widely used in studies dealing with Hurricane Katrina. For example, Boyd (2010) estimated and mapped the fatality rate for deaths that occurred due to direct exposure to floodwaters that impacted the Orleans and St. Bernard parishes of Louisiana during and after Hurricane Katrina. After developing geo-referenced datasets on both flood deaths and the flood-exposed population, he prepared flood fatality maps. Using a GIS (Arc Map 9.3) for 15 neighborhoods of New Orleans, Curtis (2010) examined chronic disease as an evacuation impediment prior to landfall of Hurricane Katrina.

Showalter and Lu (2010) recently published an edited volume which contains papers that applied a wide variety of geospatial and geo-information technologies in examining different types of urban hazards, such as sea-level rise/flooding, earthquakes/tsunamis, hurricanes, air pollution, and climatic events. The communication of hazard risk to the public through the development of early warning systems based on use of GIT is a recurring theme throughout the contribution to this volume.

The spatial overlay capability of GIS makes it an ideal research tool to identify areas exposed to multiple hazards. Several studies have applied this tool to assess hazard risk by combining basic features related to the same area with or without assigning a "weight" (Valpreda, 2004). Mapping exposure to multiple hazards is identified by entering all of the data on primary and secondary hazards into a GIS that creates separate layers. These layers are then overlaid to produce composite maps displaying areas subject to multiple hazards. GIS has contributed to the investigation of chemical hazards (Sorensen *et al.*, 1992) and toxic releases (Chakraborty and Armstrong, 1997; McMaster *et al.*, 1997; Scott and Cutter, 1997). It has also been used in modeling airborne exposure pathways (Hepner and Finco, 1995; Scott, 1999).

For estimating potential and actual losses from disasters, a GIS-based disaster model called HAZUS (HAZards US) was developed in 1997 by the Federal Emergency Management Agency (FEMA) (Cutter, 2003). Over the years HAZUS has been improved to estimate first-order damage caused by three disasters: floods, earthquakes, and hurricanes/cyclones. HAZUS also helps to improve response times to disasters by conveying the best places that resources should be placed in hazard-prone areas. HAZUS-MH (HAZards US-Multi Hazard) is currently the most up-to-date HAZUS software; it has the ability to analyze multiple hazards at once, suggesting that if a hurricane brings with it massive flooding, this version could assess both hazards at the same time. However, the efficient application of either HAZUS model is not possible without updating and providing data on local building inventories, geology, and critical infrastructure. This is a very data-intensive task and emergency management agencies rarely have sufficient economic and human resources to collect the relevant data to make use of either model (Cutter, 2003).

Both HAZUS and HAZUS-MH provide general estimates of the casualties, as well as the fatal and nonfatal injuries resulting from earthquakes, hurricanes, floods, and wind-related hazards. HAZUS estimates injuries by location, time of day, and population group. It characterizes injuries using a four-point severity scale, where $1 =$ minor or moderate injury; $2 =$ serious injury; $3 =$ severe or critical injuries; and $4 =$ fatal injury (Cropper and Sahin, 2008). Both models incorporate current scientific and engineering knowledge as well as the latest GIS technology to produce estimates of hazard-related impact before or after a disaster occurs.

Along with GIS, the use of remote sensing in hazard research has also grown since the 1990s. Both of these research tools have been used to assess fire and other hazards (Helfert and Lulla, 1990; Curtis and Mills, 2009), calculate areas under flood and the depth of flood water (Lougeay *et al.*, 1994), and map oil spills (Jensen *et al.*, 1993). Hazard mapping or disaster cartography, which has a long tradition in hazard geography, has benefited most from these two contemporary research tools. Disaster cartography has long been an important part of describing and understanding extreme events. Disasters maps are prepared to help understand the different aspects of a disaster across the globe – what has happened, what might happen, and even what is happening (Curtis and Mills, 2009). GIS technology has also been applied to interactive mapping on the Internet, addressing both natural and technological hazards (Hodgson and Cutter, 2001).

Sanyal and Lu (2006) compiled a GIS-based regional flood hazard map for Gangetic West Bengal, India. Using GIS, they developed a composite hazard index with the help of several variables: flood frequency, population density, transportation networks, access to potable water, and availability of high ground. The data for the last variable came from a digital elevation model which was derived from high-resolution imagery (also see Osti and Tokioka, 2008). Similarly, Penton and Overton (2007) used satellite imagery and elevated models to model the extent and depth of flooding in a complex river system in Australia.

Gillespie et al. (2007) claim that since 2000 there have been a number of spaceborne satellites carrying sensors that have changed the way researchers assess and predict natural disasters. The technology used by satellite sensors are able to quantify physical geographic phenomena associated with movements of the earth's surface (earthquakes and mass movements), water and wind (floods, tsunamis, and storms), and wildfires. Most of these satellites contain active and/or passive sensors that have been used successfully to assess flood and earthquake damage. In addition to remotely sensed images, aerial photography also provides an excellent data source for disaster cartography.

Hazard research has also benefited from other technological innovation of recent decades such as the Internet-based World Wide Web, Google Earth, digital cameras, and global positioning systems (GPS). Many websites offer disaster information at various scales with varying abilities for manipulation; such sites are useful for research and public awareness purposes (Tobin and Montz, 2004). Images and photographs of a particular disaster are often posted on the World Wide Web seeking emergency assistance across nations. For the same purpose, disaster videos are also released through YouTube. Further, the use of Google Earth in hazard research began with the San Diego, California wildfires of 2007. At that time Google Earth was used to track the spread of the fire, and the direction fires were spreading. It was also used to relay updates as to the status of the fires, which roads had been closed for traffic, and where mandatory evacuation orders were in place (Curtis and Mills, 2009). Google Earth has since been used in several other disaster situations, providing excellent geographic details and a means to disseminate information smoothly, quickly, and easily to anyone with Internet access.

In addition to the modern technologies mentioned above, other technological advancements have improved our identification of hazard events. For example, development of Doppler radar has greatly improved weather monitoring and forecasting. As indicated, satellite-based remote sensing has provided enhanced images of such phenomena as hurricanes and wildfires as they develop and spread. The Spatial Video Acquisition System (SVAS) has been used to capture damage and recovery data after Hurricane Katrina and the "Super Tuesday" 2008 tornadoes that struck the Mid-South part of the United States from central Arkansas to central Tennessee and from Kentucky to Mississippi (Curtis and Mills, 2009; Mills et al., 2009). SVAS was developed through the National Center for Geocomputation (NCG) in Maynooth, Ireland. In brief, it consists of digital video cameras linked to a GPS receiver through their audio channels. These cameras are mounted on the windows of the field vehicle, and as the vehicle drives through the disaster-impacted areas, each frame of the video is tagged with geographic coordinates. These coordinates provide geo-referenced data in GIS, which is able to create spatial patterns of damage (Mills *et al.*, 2009).

There are many potential uses for the application of GIS and new technologies in hazard studies. Using this tool it is possible to identify vulnerability hotspots – the geographic areas occupied by demographic segments and social groups that are most vulnerable to disaster impacts. This can be accomplished by using a GIS to either overlay or mathematically combine data on hazard exposure. Still, there is great need for better integration of physical processes and social models to enhance the prediction of disaster impact. Lack of relevant, reliable, and up-to-date data is a major problem in using GIS in hazard research. Spatial and temporal data acquisition and integration and sharing of data are needed to expand the use of GIS in disaster and emergency management. Closer collaborations among GIS users, and making this technology more accessible will also help expansion of its utilization among hazard researchers and practitioners.

Changes in the sources of data and data collection procedures have also occurred in hazard geography over the last two decades. Hazard geographers have traditionally used primary data collected directly from respondents. However, they have been increasingly using secondary data, which are available for areal units of different sizes. Some of these secondary data sources are discussed in Chapter 1. Secondary data are also available below the national level. With widespread availability of the Internet in the recent decades, hazard geographers, like other hazard researchers, have been using this technology as an important source of information for their studies. In addition to data, many published materials are available on the World Wide Web, which facilitates research and collaboration on hazard-related and other topics. Along with technological advancement, advances in data collection procedures and protocols have also greatly facilitated disaster research.

With GIS, use of spatial and secondary data in hazard research has increased in recent years. In addition, a change is also evident in data collection procedures as just mentioned. Although participatory research methodologies have been used as a tool to collect accurate information since the 1970s, hazard researchers, including geographers, have recently started to use this data collection technique (Ozerdem and Bowd, 2010; Paul and Dutt, 2010). Participatory methods include the Rapid Rural Apprisal (RRA), focus group discussions, semi-structured interviews with local people, open group meetings, and careful field observations. Some of these methods have been used to collect both qualitative and quantitative information regarding various aspects of disasters and these methods have been applied to the study of several types of disasters, such as river floods (e.g., Thompson and Sultana, 1996) and tropical cyclones (e.g., Ikeda, 1995; Paul, 2010; Paul *et al.*, 2011).

2.3 Conclusion

Available evidence suggests that the number of disasters and the scale of monetary losses caused by these extreme events are increasing, while the number of disaster-induced deaths is declining. These circumstances require a shift of research interests more toward examining ways to reduce property losses. Although a drastic change in research focus is not anticipated, hazards research continues to analyze the physical and human use systems when confronted by extreme natural events. Continuation of this type of research is needed to better understand each component as well as their interactions. More attention, however, will be given toward emerging hazards such climate change, the human dimensions of global change, technological hazards, and all types of terrorism. One of the major research objectives for hazard geographers in the near future should focus on reducing vulnerability to all types of threats. Slow-onset disasters, particularly drought, also deserve more attention among hazard geographers.

Like the past decade, much hazard research in the coming years will be hybrid in nature. Cross-disciplinary collaborations of the past will not only continue, but will likely intensify in the future. Unfortunately, collaboration among hazard geographers in different countries is only slowly developing. However, more than a decade ago, Mitchell (1999) stated that urban disasters, particularly in developing countries, have not received adequate attention from hazard researchers. This is still true. More research is also needed on local knowledge and indigenous approaches to hazard mitigation, response, adjustments, and preparedness – particularly in developing countries.

As in the past, applied hazard research will probably dominate the discipline of hazard geography. With the advancement of technology and tools, hazards research in geography has the potential to become more popular among hazard researchers. GIS, remote sensing, GPS, computer modeling and simulation, interactive mapping, and hot spot analysis will facilitate a tremendous growth of hazard research in geography in the near future. Further developments in geo-information technologies and their integration in future may help refining existing models/approaches for prediction of extreme natural events and their management.

Past and contemporary hazard research in geography has often been accused of lacking a sound theoretical foundation, particularly those studies addressing practical problems. By adapting an integrative approach, future geographical hazard studies should reduce this tension between theory and practice. Montz and Tobin (2011) rightly observe that the thrust of hazard research is no longer seeking simple theoretical frameworks, but rather research is now multifaceted, covering topics with more practical implications. Hazard geographers need to undertake ground-breaking research of both basic and applied nature. Few studies have attempted to establish the extent to which political characteristics of a nation or a community and public response to disaster are related, more studies are certainly needed to fill this important gap. Like public emergency managers at various levels, NGOs have been playing an important role from disaster response to mitigation since the 1980s, particularly in developing countries. Their activities and experiences deserve the close attention of hazard geographers. It is abundantly clear that hazard research in geography will continue to thrive and thus will reduce the suffering of people from the impacts of disasters whenever and wherever they occur.

References

Ackerman, E.A. (1963) Where is a research frontier? *Annals of the Association of American Geographers* **53**: 429–440.

Alexander, D. (1991) Natural disasters: a framework for research and teaching. *Disasters* **15**: 209–226.

Alexander, D. (1997) The study of natural disasters, 1977–1997. Some reflections on a changing field of knowledge. *Disasters* **21**: 284–304.

Alexander, D. (2000) *Confronting Catastrophe*. Oxford: Oxford University Press.

Barrows, H.H. (1923) Geography as human ecology. *Annals of the Association of American Geographers* **13** (1): 1–14.

Blaikie, P. and Brookfield, H.C. (1987) *Land Degradation and Society*. London: Methuen.

Blaikie, P. et al. (1994) *At Risk: Natural Hazards, People's Vulnerability and Disaster*. London: Routledge.

Bolin, R. and Stanford, L. (1999) Constructing vulnerability in the first world: the Northridge Earthquake in Southern California, 1994. In *The Angry Earth: Disaster in Anthropological Perspective* (ed. A. Oliver-Smith and S.M. Hoffman). New York: Routledge, pp. 89–112.

Boyd, E.C. (2010) Estimating and mapping the direct flood fatality rate for flooding in Greater New Orleans due to Hurricane Katrina. *Risk, Hazards & Crisis in Public Policy* **1** (3): 91–114.

Burton, I. (1962) Types of agricultural occupance of flood plains in the United States. Department of Geography Research Paper No. 75. University of Chicago, Chicago.

Chakraborty, J. and Armstrong, M.P. (1997) Exploring the use of buffer analysis for the identification of impacted areas in environmental equity assessment. *Cartography and Geographic Information Systems* **24** (3): 145–157.

Collins, A.E. (2009) *Disaster and Development*. London: Routledge.

Coppola, D.P. (2007) *Introduction to International Disaster Management*. Boston: Elsevier.

Corson, M.W. (1999) Hazardscapes in reunified Germany. *Environmental Hazards* **1** (1): 57–68.

Cropper, M.L. and Sahin, S. (2008) *Valuing Mortality and Morbidity in the Context of Disaster Risk. Background Paper for the Joint World Bank–UN Assessment on Disaster Risk Reduction*. World Bank: Washington, D.C.

Cuny, F.C. (1983) *Disaster and Development*. New York: Oxford University Press.

Curtis, A. (2010) Chronic disease as an evacuation impediment: using a geographic information system and 911 call data after Katrina to determine neighborhood scale health vulnerability. *Risk, Hazards & Crisis in Public Policy* **1** (3): 63–89.

Curtis, A. and Mills, J.W. (2009) *GIS, Human Geography, and Disasters*. San Diego: University Readers.

Cutter, S.L. (1996) Vulnerability to natural hazards. *Progress in Human Geography* 20 (4): 529–539.

Cutter, S.L. (2001) The changing nature of risks and hazards. In *American Hazardscapes: The Regionalization of Hazards and Disaster* (ed. S.L. Cutter). Washington, D.C.: Joseph Henry Press, pp. 1–12.

Cutter, S.L. (2003) GI science, disasters, and emergency management. *Transactions in GIS* 7 (4): 439–445.

Cutter, S.L. and Solecki, W.D. (1989) The national pattern of airborne toxic releases. *Professional Geographer* 41 (2): 149–161.

Cutter, S.L. et al. (1999) *South Carolina Atlas of Environmental Risks and Hazards*. Columbia: University of South Carolina Press.

Cutter, S.L. et al. (2000) Revealing the vulnerability of people and places: a case study of Georgetown County, South Carolina. *Annals of the Association of American Geographers* 90 (4): 713–737.

Cutter, S.L. et al. (2001) Subsidized inequities: the spatial patterning of environmental risks and federally assisted housing. *Urban Geography* 22: 29–53.

Cutter, S.L. et al. (2003) *The Geographical Dimension of Terrorism*. New York: Routledge.

Dash, N. (1997) The use of geographic information systems in disaster research. *International Journal of Mass Emergencies and Disasters* 15 (1): 135–146.

Emani, S. et al. (1993) Assessing vulnerability to extreme storm events and sea-level rise using geographical information systems (GIS). In *GIS/LIS '93 Proceedings*. Bethesda, MD: ACSM, ASPRS, AM/FM, AAG, and URISA, pp. 201–209.

Emel, J. and Peet, R. (1989) Resource management and natural hazards. In *New Models in Geography* (ed. R. Peet and N. Thrift). London: Unwin Hyman, pp. 49–76.

Flint, C. (2003) Terrorism and counterterrorism: geographic research questions and agendas. *The Professional Geographers* 55 (2): 161–169.

Frerks, G.E. et al. (1995) A disaster continuum. *Disasters* 19 (4): 245–246.

Gillespie, T.W. et al. (2007) Assessment and prediction of natural hazards from satellite imagery. *Progress in Physical Geography* 31 (5): 459–470.

Hagman, G. (1985) *Prevention Better than Cure: Human and Environmental Disasters in the Third World*. Geneva: Swedish Red Cross and League of Red Cross and Red Crescent Societies.

Haque, C.E. (1997) *Hazards in a Fickle Environment: Bangladesh*. Boston: Kluwer Academic Publishers.

Hastings, J.V. (2008) Geography, globalization, and terrorism: the plots of Jemaah Islamiya. *Security Studies* 17: 505–530.

Helfert, M.R. and Lulla, K.P. (1990) Mapping continental-scale biomass burning and smoke pails over the Amazon Basin as observed from the space shuttle. *Programmetric Engineering and Remote Sensing* 56: 1367–1373.

Hepner, G.F. and Finco, M.V. (1995) Modeling dense gaseous contaminant pathways over complex terrain using a geographic information system. *Journal of Hazardous Materials* 42: 187–199.

Hewitt, K. (ed.) (1983) *Interpretations of Calamity*. Winchester, MA: Allen & Unwin.

Hewitt, K. (1987) The social space of terror: towards a civil interpretation of total war. *Society and Space, Environment and Planning* D5: 445–474.

Hewitt, K. (1997) *Regions of Risk: A Geographical Introduction*. Harlow: Longman.

Hewitt, K. and Burton, I. (1971) *The Hazardousness of Place: A Regional Ecology of Damaging Events*. Research Publication 6. Toronto: University of Toronto, Department of Geography.

Hodgson, E. and Cutter, S.L. (2001) Mapping and the spatial analysis of hazardscapes. In *American Hazardscapes: the Regionalization of Hazards and Disasters* (ed. S.L. Cutter). Washington, D.C.: Joseph Henry Press, pp. 37–60.

Hodgson, E. and Palm, R. (1992) Attitude and response to earthquake hazards: a GIS design for analyzing risk assessment. *GeoInfoSystem* 2 (7): 40–51.

Ikeda, K. (1995) Gender differences in human loss and vulnerability in natural disasters: a case study from Bangladesh. *Indian Journal of Gender Studies* 2 (2): 171–193.

Jensen, J.R. et al. (1993) Coastal environmental sensitivity mapping for oil spills in the United Arab Emirates using remote sensing and GIS technology. *Geocarto International* 8 (2): 5–13.

Kapur, A. (2010) *Vulnerable India: A Geographical Study of Disasters*. New Delhi: Sage.

Kasperson, R.E. et al. (1988) The social amplification of risk: conceptual framework. *Risk Analysis* 8 (2): 177–187.

Kates, R.W. (1962) Hazard and choice perception in flood plain management. Research Paper No. 78. University of Chicago, Chicago.

Khan, S. and Crozier, M.J. (2009) "Hazardscape": a holistic approach to assess tipping points in humanitarian crises. Submitted to Annual Summer Academy on Social Vulnerability: "Tipping Points in Humanitarian Crises" held in Hohenkammer, Munich, Germany, July 26–August 1, 2009.

Khan, F. and Mustafa, D. (2007) Navigating the contours of the Pakistani hazardscapes: disaster experience versus policy. In *Working with the Winds of Change: Towards Strategies for Responding to the Risks Associated with Climate Change and Other Hazards* (ed. M. Moench and A. Dixit). Kathmandu: ProVention Consortium, pp. 193–234.

Leach, B. (1978) Geography, behavior, and Marxist philosophy. *Antipode* 10 (2): 33–37.

Lewis, J. (1997) Development, vulnerability and disaster reduction: Bangladesh cyclone shelter projects and their implications. In *Reconstruction after Disaster: Issues and Practices* (ed. A. Awotona). Aldershot: Ashgate, pp. 45–56.

Liverman, D. (1990) Drought impacts in Mexico: climate, agriculture, technology, and land tenure in Sonora and Puebla. *Annals of the Association of American Geographers* 80 (1): 49–72.

Lougeay, R. et al. (1994) Two digital approachs for calculating the area of regions affected by the Great American Flood of 1993. *Geocarto International* 9 (4): 53–59.

Lowry Jr, J.H et al. (1995) A GIS-based sensitivity analysis of community vulnerability to hazardous contaminants on the US/Mexico Border. *Photogrammetric Engineering and Remote Sensing* 61: 1347–1359.

McEntire, D.A. (2007) *Disaster Response and Recovery*. Hoboken, NJ: Wiley.

McMaster, R.B. et al. (1997) GIS-based environmental equity and risk assessment: methodological problems and prospects. *Cartography and Geographic Information System* 24 (3): 172–189.

Middleton, N. and O'Keefe, P. (1998) *Disaster and Development: The Politics of Humanitarian Aid*. London: Pluto Press.

Mills, J.W. et al. (2009) *The Spatial Video Acquisition System as an Approach to Capturing Damage and Recovery Data After a Disaster: A Case Study from the Super Tuesday Tornadoes*. Boulder, CO: The Natural Hazards Center, University of Colorado at Boulder.

Mitchell, B. (1979) *Geography and Resource Analysis*. New York: Longman.
Mitchell, J.K. (1979) Social violence in Northern Ireland. *Geographical Review* 69 (2): 179–201.
Mitchell, J.K. (1989) Hazard research. In *Geography in America* (ed. G.L. Gaile and C.J. Willmott). Columbus, OH: Morrill, pp. 410–424.
Mitchell, J.K. (ed.) (1999) *Crucibles of Hazards: Mega-Cities and Disasters in Transition*. Tokyo: United Nations Press.
Mitchell, J.K. et al. (2001) *Field Observation of Lower Manhattan in the Aftermath of the World Trade Center Disaster: September 30, 2001*. Quick Response Report no. 139. Boulder, CO: The Hazards Center, University of Colorado at Boulder.
Montz, B.E. (1994) Methodologies for analysis of multiple hazard probabilities: an application in Rotorua, New Zealand. Unpublished Research Report, Center for Environmental and Resource Studies, University of Waikato, Hamilton, New Zealand.
Montz, B.E. and Tobin, G.A. (2011) Natural hazards: an evolving tradition in applied geography. *Applied Geography* 31: 1–4.
Montz, B.E. et al. (2003) Hazards. In *Geography in America: At the Dawn of the 21st Century* (ed. G.L. Gaile and C.J. Wilmott). Oxford: Oxford University Press, pp. 479–491.
Mustafa, D. (2005) The production of urban hazardscape in Pakistan: modernity, vulnerability and the range of choice. *Annals Association of American Geographers* 95 (3): 566–586.
Mustafa, D. et al. (2011) Pinning down vulnerability: from narratives to numbers. *Disasters* 35 (1): 62–86.
Oliver-Smith, A. (1986) *The Martyred City: Death and Rebirth in the Andes*. Albuquerque: University of New Mexico Press.
Osti, R.T. and Tokioka, T. (2008) Flood hazard mapping in developing countries: a prospect and problems. *Disaster Prevention and Management* 17 (1): 104–113.
Ozerdem, A. (2003) Disaster as manifestation of unresolved development challenges: the Marmara Earthquake, Turkey. In *Natural Disasters and Development in a Globalizing World* (ed. M. Pelling). London: Routledge, pp. 199–213.
Ozerdem, A. and Bowd, R. (2010) *Participatory Research Methodologies: Development and Post-Disaster/Conflict Reconstruction*. Burlington, VT: Ashgate.
Palm, R. (1990) *Natural Hazards: An Integrative Framework for Research and Planning*. Baltimore: Johns Hopkins University Press.
Palm, R. and Hodgson, M.E. (1992) Earthquake insurance: mandated disclosure and homeowners response in California. *Annals of the Association of American Geographers* 82 (2): 207–222.
Parker, D. et al. (1997) Reducing vulnerability following flood disasters: issues and practices. In *Reconstruction after Disaster: Issues and Practices* (ed. A. Awotona). Aldershot: Ashgate, pp. 23–44.
Pattison, W.D. (1964) The four traditions of geography. *Journal of Geography* 63: 211–216.
Paul, B.K. (2006) Disaster relief efforts: an update. *Progress in Development Studies* 6 (3): 211–223.
Paul, B.K. (2010) Human injuries caused by Bangladesh's Cyclone Sidr: an empirical study. *Natural Hazards* 54: 483–495.
Paul, B.K. and Dutt, S. (2010) Applications of participatory research methods in post-disaster environment: the case of Cyclone Sidr, Bangladesh. In *Participatory Research Methodologies: Development and Post-Disaster/Conflict Reconstruction* (ed. A. Ozerdem and R. Bowd). Burlington, VT: Ashgate, pp. 85–98.

Paul, B.K. *et al.* (2011) Post-Cyclone Sidr illness patterns in coastal Bangladesh: an empirical study. *Natural Hazards* 56 (3): 841–852.

Penton, D. and Overton, I. (2007) Spatial modeling for floodplain inundation combining satellite imagery and elevation models. MODSIM 2007 International Congress on Modeling and Simulation. Modeling and Simulation Society of Australia and New Zealand.

Platt, R. (1986) Floods and man: a geographer's agenda. In *Geography Resources and Eenvironment: Themes from the Work of Gilbert F. White*, Vol. 2 (ed. R. Kates and I. Burton). Chicago: University of Chicago Press, pp. 28–68.

Rahman, M. (1991) Vulnerability syndrome and the question of peasants' adjustment to riverbank erosion and flood in Bangladesh. In *Riverbank Erosion, Flood and Population Displacement in Bangladesh* (ed. K.M. Elahi *et al.*). Savar, Dhaka, Bangladesh: Jahangirnagar University, pp. 170–187.

Rodrigue, C.M. (2002) *Patterns of Media Coverage of the Terrorist Attacks on the United States in September of 2001*. Quick Response Report no. 146. Boulder, CO: The Hazards Center, University of Colorado at Boulder.

Saarinen, T.F. (1966) Perception of drought hazard on the great plains. Research Paper No. 106. University of Chicago, Chicago.

Sanyal, J. and Lu, X.X. (2006) GIS-based flood hazard mapping at different administrative scales: a case study of Gangetic West Bengal, India. *Singapore Journal of Tropical Geography* 27: 207–220.

Scott, M.S. (1999) The exploration of an air pollution hazard scenario using dispersion modeling and a volumetric geographic information system. PhD dissertation, University of South Carolina. Dissertation microfiche: Ann Arbor, Michigan.

Scott, M.S. and Cutter, S.L. (1997) Using relative risk indicators to disclose toxic hazard information to communities. *Cartography and Geographic Information Systems* 24 (3): 158–171.

Showalter, P.S. and Lu, Y. (eds) (2010) Geospatial Techniques in Urban Hazard and Disaster Analysis. New York: Springer.

Sollis, P. (1994) The relief–development continuum: some notes on rethinking assistance for civilian victims of conflict. *Journal of International Affairs* 47: 451–471.

Sorensen, J.H. *et al.* (1992) An approach for driving emergency planning zones for chemical munitions emergencies. *Journal of Hazardous Materials* 30: 223–242.

Streeten, P. (1999) Globalization and its impact on development co-operation. *IDS Bulletin* 42: 9–15.

Sultana, F. (2010) Living in hazardous waterscapes: gendered vulnerabilities and experiences of floods and disasters. *Environmental Hazards* 9 (1): 43–54.

Susman, P. *et al.* (1983) Global disasters: a radical interpretation. In *Interpretations of Calamity* (ed. K. Hewitt). Boston: Allen & Unwin, pp. 263–283.

Thomas, D.S.K. *et al.* (2002) *Use of Spatial Data and Geographic Technologies in Response to the September 11 Terrorist Attack*. Quick Response Report no. 153. Boulder, CO: The Hazards Center, University of Colorado at Boulder.

Thompson, P.M. and Sultana, P. (1996) Distributional and social impacts of flood control in Bangladesh. *The Geographical Journal* 162 (1): 1–13.

Tobin, G.A. and Montz, B.L. (2004) Natural hazards and technology: vulnerability, risk, and community response in hazardous environments. In *Geography and Technology* (ed. S.D. Brunn *et al.*). Boston: Kluwer Academic Publishers, pp. 547–570.

UN (United Nations) (2008) *The Millennium Development Goals Report*. New York: UN.

UNDP (United Nations Development Program) (2004) *Reducing Disaster Risk: A Challenge for Development. Bureau of Crisis Prevention and Recovery.* New York: UNDP.

UNDP-DHA (United Nations Development Program–Department of Humanitarian Affairs) (1994) *Disasters and Development*, 2nd edn. New York: UNDP, Disaster Management Training Programme.

UNDRO (United Nations Disaster Relief Organization) (1992) *An Overview of Disaster Management.* New York: UNDRO.

Valpreda, E. (2004) GIS and natural hazards. In *Natural Disasters and Sustainable Development* (ed. R. Casaler and C. Margottini). Berlin: Springer, pp. 373–386.

Varley, A. (1993) *Disasters, Development and Environment.* New York: John Wiley & Sons, Inc.

Watts, M. (1983) On the poverty of theory: natural hazards research in context. In *Interpretations of Calamity from the Viewpoint of Human Ecology* (ed. K. Hewitt). London: Allen and Unwin, pp. 231–262.

White, G.F. (1945) *Human Adjustments to Floods.* Chicago: University of Chicago Press.

White, G.F. (1964) Choice of adjustment to floods. Department of Geography Research Papers No. 3. University of Chicago, Chicago.

Whyte, A.V. (1986) From hazard perception to human ecology. In *Geography, Resources, and Environment*, Vol. II (ed. R.W. Kates and I. Burton). Chicago: The University of Chicago Press, pp. 240–271.

Wisner, B. et al. (2004) *At Risk: Natural Hazards, People's Vulnerability and Disaster.* London: Routledge.

Zaman, M.Q. (1989) The social and political context of adjustment to riverbank erosion hazard and population resettlement in Bangladesh. *Human Organization* 48: 196–205.

3
Vulnerability, Resiliency, and Risk

In the course of the International Decade for Natural Disaster Reduction (IDNDR), 1990–1999, and of many other initiatives spawned over the last decade, disaster reduction has gained a lot of momentum and attention. As a result, vulnerability, resiliency, and risk have received a more prominent role in disaster research. These three concepts are increasingly used as an approach for understanding the dynamics of natural disasters and they will continue to dominate hazard research for long time. These concepts have been articulated as a powerful analytical tool for describing susceptibility to harm, powerlessness, and the marginality of both physical and social systems. This chapter provides an overview of these three important and interrelated concepts by presenting their essential components.

3.1 Vulnerability

The concept of vulnerability is central to contemporary hazard research and to the creation of practices and strategies to mitigate against the impacts of disasters (Cutter, 1996). It has dominated hazard research at least for the last two decades, if not more. During this period, the focus of researchers has shifted from the hazard itself to the combination and interaction of people's vulnerability to such events as well as the diversity of conditions that contribute to this vulnerability (Montz et al., 2003). It has evolved as a critique of mainstream technocratic hazard studies and has now become a standard vocabulary in the multidisciplinary tradition of hazard literature and research. While some scholars consider this a paradigm shift within hazard discipline, others foresee development of a new field of study called "vulnerability science." For example,

Cutter (2003, p. 6) suggests creation of vulnerability science to contribute to understanding of "those circumstances that put people and places at risk and those that reduce the ability of people and places to respond to environmental threats."

As a first step to understanding the concept of vulnerability, it is necessary to provide definitions of this term, which is accomplished in the next subsection. This subsection is followed by discussions on issues related to human vulnerability, ways to reduce vulnerability to disasters, and types of vulnerability. The final subsection deals with various measures of vulnerability with special emphasis on measures of social vulnerability.

3.1.1 Defining vulnerability

In context of vulnerability, it is necessary to understand why an individual, household, group, or community may suffer more or less than another in the wake of a natural disaster (what economists call shock, or what ecologists and resilience researchers would consider a major disturbance to the system). Hazard researchers explain the level to which these may be affected in terms of vulnerability. The etymology of the word vulnerability indicates harm, as it means "wounds," from the Latin *vulnerabilis* (Coppola, 2007). Merriam Webster's Collegiate dictionary defines vulnerable as "capable of being physically wounded" or "open to attack or damage." Vulnerability, then, is the state or condition of being capable of being wounded, attacked, or damaged. It is a measure of an entity's inability to deal with natural disasters.

An examination of the literature penned on vulnerability reveals a wide range of definitions. In her progress report on vulnerability to environmental hazards, Cutter (1996) listed 18 different definitions of vulnerability in the hazard literature. Since the publication of Cutter's 1996 paper many other definitions of vulnerability have emerged. For example, Thywissen provided 36 definitions in 2006 (Thywissen, 2006). Vulnerability definitions of only three hazard researchers are cited by both Cutter (1996) and Thywissen (2006). The existence of a large number of definitions reflects that vulnerability is a topic of interest to hazard scholars of many different disciplines, who have different ideological positions and different end purposes. As a result, none of the available definitions of vulnerability is accepted by all hazard researchers. In general, physical scientists and engineers have typically defined vulnerability in contexts of physical exposure to extreme events and adverse outcomes, while social scientists have emphasized the role of social structures and differential access to economic, political, and psychological resources in making certain groups more disadvantaged in the face of disasters (Adger, 2006; Mustafa *et al.*, 2011). Some (e.g., Cutter, 1996; Cutter *et al.*, 2000) have attempted to bridge the gap between these two perspectives. Thus, a vast majority of definitions of vulnerability fall into at least one of these three categories: (1) focus on physical factors, (2) focus on social/human factors, and (3) a combination of physical and social/human factors.

A review of the plethora of definitions is not necessary here, but it will be helpful to begin with a few of them, at least. Blaikie and his colleagues (1994, p. 9), who are pioneers in vulnerability research, define vulnerability as "the characteristics of a person or group in terms of their capacity to anticipate, cope with, resist, and recover from the impact of a natural hazard." Impacts include damage to private property, infrastructure, economic vitality, habitat, and productive ecosystem, as well as human death, injury, and illness. Blaikie and his colleagues take a rigorous approach to this term and choose to apply it only to people and not to the human-built environment. Cutter *et al.* (2003), on the other hand, maintain that the quality of the built environment influences human vulnerability. As one seismologist and disaster expert from Columbia University's Earth Institute was quoted as saying, "Earthquakes don't kill people. Bad buildings kill them" (Walsh, 2010).

Human vulnerability is inextricably tied to the level and quality of development of the human-built environment in any place. This environment includes the physical infrastructure necessary for human habitation and activities, such as homes, office buildings, schools, hotels, roads, railways, water and sewage lines, and energy transmission lines, as well as any modifications that people make to the natural environment, such as the removal of mountaintops, cutting of trees, and transformation of wetlands to grow crops or for residential use. While studying flood vulnerability, Brody *et al.* (2008) and Zahran *et al.* (2008) focused on the attributes of built environment and examined the influence of the total number of dams in a county, percentage of a county area covered by impervious surfaces, and extent of wetland alteration on vulnerability at the county level.

However, another definition of vulnerability incorporates the notion of systems: "vulnerability is the degree to which a system acts adversely to the occurrence of a hazardous event" and the capacity of the system to absorb and recover from the hazardous event (Timmerman, 1981). In a similar way, Sarewitz *et al.* (2003) state that vulnerability "describe[s] inherent characteristics of a system that create the potential for harm but [is] independent of the probabilistic risk of the occurrence ("event risk") of any particular hazard or extreme event." The risk factors in the Sarewitz *et al.* (2003) definition indicates there are elements of a disaster that lay beyond the control of humans in any regard; these are generally accepted as the physical characteristics of the event itself, not the vulnerability to any particular event. Liverman (1990) differentiates between vulnerability as defined by social, economic, and political conditions and vulnerability as a biophysical condition. Vulnerability, however, also extends to places, as Susan Cutter and her colleagues (2000) have theorized. Their work examines how the array of factors contributing to biophysical and social vulnerability together creates place vulnerability.

The most useful element of the concept of vulnerability is the idea that a disaster does not translate directly into risk, but rather is measured by the degree of vulnerability of the individuals or system in relation to that disaster. The underlying factors causing vulnerability, thus, define risk. Therefore, vulnerability and risk cannot be considered the same. Risk is generally considered to be primarily just a physical factor, and does not take into account the social characteristics that can determine a population is more vulnerable to a hazard. This opens up the chance to use the term "vulnerability" to identify, classify, and level people (Lein, 2009). Seemingly, government agencies often use such definitions to attract and obtain funds from external sources (Wisner, 2003; Wisner et al., 2004). This has encouraged a negative meaning of weak, passive people, unable to cope.

However, vulnerability is also defined in terms of three physical and socio-economic variables: exposure, sensitivity, and adaptive capacity. Exposure is the frequency of an event, which is considered as a physical factor. It is a risk measure directly related to proximity and the environmental characteristics at a particular place. Sensitivity is the magnitude of an event in relation to human response, which can be considered a combination of physical and socio-economic factors. It reflects the degree to which people and place can be harmed. Adaptive capacity is the ability of the people or a system to easily and quickly cope with the impact of a disaster. Adaptive capacity is considered to be primarily a socio-economic factor. All these factors are vital variables that determine the level of vulnerability of a given population or a place (Cutter, 1996; Adger, 2006; Polsky et al., 2007).

The vulnerability of a person, a group, or a system is not simply a product of the intensity or magnitude of a disaster rather it evolves over a long period of time and involves a combination of physical as well as socio-economic, demographic, and other factors, including attributes of the built environment. In development literature, vulnerability covers a multitude of issues like food insecurity, market situation, production relations, and social and economic justice (Sen, 2002). Vulnerability is determined by social systems and power, not by natural forces. It needs to be understood in the context of social, economic, and political systems that operate at national and even international scales (Wisner et al., 2004). The vulnerability perspective considers most natural disasters as unnatural. A concept that is useful in refining the concept of vulnerability is that of "resilience," often characterized as the opposite of vulnerability (Coppola, 2007). Cardona (2004) maintains that vulnerability is related to physical and socio-economic fragility, susceptibility or lack of resilience of the exposed elements. He points out that population growth, rapid urbanization, environmental degradation, global warming, and war have all increased vulnerability.

Anthony Oliver-Smith (2004) argues that vulnerability is fundamentally a political ecological concept: "Vulnerability is the nexus that links the relationship that people have with their environment to social forces and institutions and the cultural values that sustain or contest them" (p. 10). According to Bolin and Stanford (1999), vulnerability concerns the complex of social, economic, and political circumstances in which people's lives are embedded and these factors structure the choices and opinions they have in coping with environmental hazards. They claim that: "The most vulnerable are typically those with the fewest choices, those whose lives are constrained, for example, by discrimination, political powerlessness, physical disability, lack of education and employment, illness, the absence of legal rights, and other historically grounded practices of domination and marginalization" (Bolin and Stanford, 1999, pp. 9–10).

In their Pressure and Release (PAR) model, Blaikie and his colleagues (1994) and Wisner *et al.* (2004) posit that a disaster occurs because people are vulnerable – that for physical or economic or social reasons they are exposed and will suffer damaging losses if a hazard strikes. This vulnerability is the result of a set of unsafe conditions (e.g., being unable to afford safe housing, having to engage in dangerous livelihoods, and living in a location with high incidence of hazard events), which are then nested in explanatory fashion within dynamic pressures (e.g., lack of education, training, and appropriate skills, rapid urbanization, and perhaps even a decline in soil productivity), and those, in turn, within what are termed root causes (Figure 3.1). The root causes are implanted historically and structurally in the society by limited access to power, structures and resources, and ideologies of political and economic systems. The PAR model is shown diagrammatically in Figure 3.1, which illustrates that the disaster is crunched between the hazards on the one side, and the "progression of vulnerability" on the other.

Blaikie *et al.* (1994) consider that vulnerability is socially constructed. They also emphasize that vulnerability should be considered in applied disaster practises to release vulnerable people from the dynamic pressures. Thus, the applied practitioner can identify which variables lead to increased vulnerability, and thereby identify and develop those variables in order to reduce vulnerability. The PAR model has been criticized in failing to provide a systematic view of the mechanisms and processes of vulnerability. It has been argued that operationalizing this model necessarily involves typologies of causes and categorical data on hazard types, limiting the analysis in terms of quantifiable or predictive relationships (Adger, 2006). However, the PAR model, complementary to the access model, insists more on structural aspects and it is the only framework of disaster risk and vulnerability that clearly distinguishes between different causal levels (Nathan, 2011). The great merit of this model is to recognize that the deepest causal factors may often be quite remote from the disaster event itself.

3.1.2 Human vulnerability and related issues

Vulnerability as a concept has proven useful as means of assessing disasters within their broader socio-economic, political, and environmental contexts and it has provided a helpful guide in the formulation of approaches and policies towards hazard preparedness and relief provision (Bankoff, 2001). Vulnerability perspectives are concerned with how social structures and discourses made

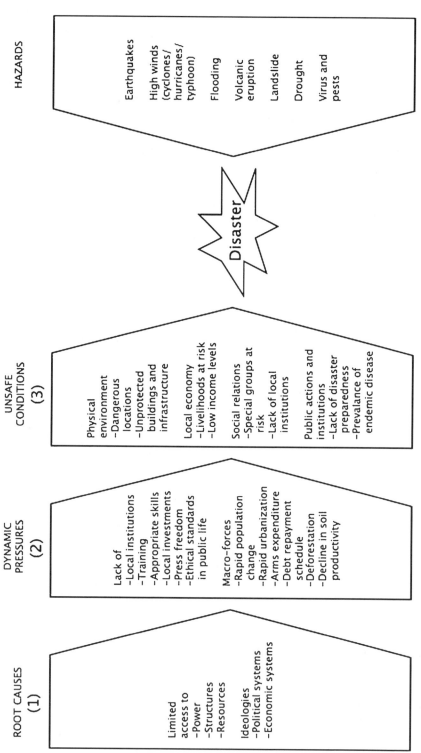

Figure 3.1 Pressure and release (PAR) model: the progression of vulnerability. *Source*: After Wisner *et al.* (2004), p. 51.

various groups differentially vulnerable to the ill-effects of natural hazards. These perspectives maintain that human actions are more responsible than nature for creating extreme events, and vulnerability results from exploitative capitalist economic development that degrades the conditions of the poor. More specifically, vulnerability is associated with a lack of power and accordingly, groups marginalized through poverty, illiteracy, race, ethnicity, or immigration and minority status (O'Riordon, 1986).

Vulnerability is associated with a lack of power and, accordingly, groups marginalized through poverty, illiteracy, race, ethnicity, or immigration and minority status. Vulnerability perspectives further claim that members of vulnerable or disadvantageous groups, such as the poor, females, the elderly, children, ethnic or religious minorities, and recent migrants, suffer more from disasters because public sources systematically discriminate against them in the provision of disaster assistance (Blaikie et al., 1994; Bolin and Stanford, 1999; Cutter et al., 2000). There is some evidence that a high density of minority population actually acts as a buffer in a disaster, through a common bonding through adversity – therefore more minority translates into less social vulnerability. But this varies by location (Curtis and Mills, 2009).

However, tragically, systematic discrimination tends to promote dominant-class interests. Political ecologists have preferred to examine issues of vulnerability and equity (Liverman, 1990; Taylor, 1990; Mitchell, 1999). They suggest targeting short-term relief assistance to the most vulnerable populations, and also linking that assistance with long-term development; they also champion empowerment by gender and class.

Although poverty is the leading cause of vulnerability, poverty is not synonymous with vulnerability (Wisner et al., 2004). There is a need to deconstruct poverty into its component parts in order to understand why certain groups are more vulnerable to a particular extreme event (Watts and Bohle, 1993). For example, Swift (1989) points out that not all poor people are equally vulnerable to starvation, and it may not be the poorest who die during a famine. Vulnerability is a multilayered and multidimensional social space defined by the determinate political, economic, and institutional capabilities of people in specific places at specific times.

Similarly, all ethnic or immigrants groups are not equally vulnerable to disasters. For example, Asian Indian immigrants in the United States are generally more affluent and well-educated than many other US immigrant groups. Asian Indians in the US have a median annual income at least US$15 000 higher than the household median income of these groups. Following the logic, it can be assumed that all elderly people are not equally vulnerable to natural hazards. For example an elderly person of a poor family has higher vulnerability than an elderly person of a wealthy family.

Existing economic literature often defines and measures household vulnerability in a different way. In this literature a household is regarded as vulnerable in proportion to which income shocks translate into consumption shocks. If two households have nearly the same consumption pattern in each state, but the second household has a more variability in income, then the second household is regarded as less vulnerable (Glewwe and Hall, 1995, 1998). Thus vulnerability is defined as the ability to smooth consumption in response to shocks, measured by observed changes in consumption over time. As a result, people who are very poor may not be considered vulnerable as they may not experience a large change in their consumption in response to shocks. At the same time, non-poor people who face large adverse shocks resulting in large consumption changes may be considered vulnerable, even though they are so well-off as not to become poor after such shocks (Dercon and Krishnan, 2000).

As noted, vulnerability is a function of exposure, sensitivity, and adaptation capacity of a person, group, or system. Sensitivity and adaptive capacity of a particular system depends on a range of socio-politico-economic characteristics, such as gender, age, political affiliation, livelihood, access to resources, and influence on decision-making and wealth (entitlements). Nonstructural hazard mitigation and preparedness measures, such as a warning system, an evacuation program, and land use control, which are more associated with the term "resilience," have gradually tended to grow in prominence in recent years (Few, 2003).

As indicated, the concept that is useful in refining the concept of vulnerability is that of resilience, which originates in agro-ecology and natural resource management literature (Berkes *et al.*, 2003). The ability of a livelihood to be able to cope with and recover from impacts of disasters is central to the theme of livelihood resilience. Those households not resilient enough to cope with short- and/or long-term adversities eventually fail to make their livelihoods sustainable. Broadly speaking, the ownership of and access to productive assets and linkages with higher power source institutions determine one's level of vulnerability and subsequent entry into or exit from the poverty level.

3.1.3 Reducing human vulnerability to disasters

Susman *et al.* (1983) applied a "theory of marginalization" to examine why some groups, communities, and countries suffer more than others when a natural hazard event occurs. This theory is very relevant in reducing human vulnerability to hazards and disasters. They (Susman *et al.*, 1983, p. 264) wrote, "Vulnerability is the degree to which different classes in society are differently at risk, both in terms of the probability of occurrence of an extreme physical event and the degree to which the community absorbs the effects of extreme physical events and helps different classes to recover." They recognized that poor people were "generally" the most vulnerable to disasters. Cuny (1983) also maintains that the poor suffer the most from disasters. Thus, one way to reduce vulnerability is to reduce poverty and place disaster response in the context of development. Poverty alleviation will contribute in improving the economic conditions of the poor. Although difficult to achieve, alleviation of poverty is the number one objective of the United Nations Millennium Development Goals. As many as 189 countries of the world adopted these development goals in 2000 some of which are to be achieved by the year 2015.

One of the ways proposed by Cuny (1983) to reduce vulnerability to disasters is to reduce the physical vulnerability to hazard events. He proposed mapping the risks of the probability that specific hazard events would occur and then to identify communities that "are particularly susceptible to damage or destruction" based on their proximity to the zone of the probable hazard risk. Addressing the root causes of inequality and acknowledging human rights for all people will help in reducing human vulnerability. Cuny (1983) also proposed to reduce vulnerability through building social capital in communities, which would include the development of formal institutions to help increase the number of ways in which individuals and communities can cope with hazard risk and hazard events (Cuny, 1983).

Social capital theory, which is defined by Pierre Bourdieu (1985) as the resources that can be derived through one's social network, is also relevant in the present context. It refers to the collective value of all "social networks" and the inclinations that arise from these networks to assist each other. A social network consists of people of different socio-economic backgrounds. Generally, poor members of such a network have one or more friends or relatives who have some degree of wealth and/or political power. These friends and relatives with greater means often provide aid to the poor members of their social network during a post-disaster period. The wealthy members of a social network also often take responsibility for distributing disaster aid among victims of extreme natural events. In such cases, vulnerable members of a social network may receive their proper share, or in some instances, an even larger amount of relief.

There are risks to social capital, however, namely overembeddedness. Although the sharing of resources, information, and helping fellow members are key aspects of social capital, it may also lead to the stretching of limited resources so thinly that the group cannot take effective action. Group decision-making, with its inherent inefficiencies, may replace that of individuals who have power and position in society. This decentralization of decision-making inhibits effective and efficient decisions.

In addition to the issue of inefficient decision-making, another problem of an overembedded social network is that of the individual relying on his or her network for information and direction, rather than listening to external sources, such as the government and/or authority figures. Mark Granovetter (1973) termed this "The Strength of Weak Ties." He categorized social ties as either "strong" in the case of family and close friends, or "weak," in the case of causal acquaintances. Cordasco and Johnson (2006) claim that presence of strong ties in a society often results in a paucity of weak ties. Having few weak ties result in limiting one's exposure to the breadth of information that can be provided via a broader social network (Granovetter, 1983).

It is important, however, to incorporate the view of social capital in disaster research and management. Cordasco and Johnson (2006, p. 6) further claim that: "By engaging community-based organizations, such as religious institutions and grass roots social groups, we will establish the vitally needed 'bridging' ties that will help us reach into the community and exchange information that is needed to prepare for and respond to future disasters."

3.1.4 Types of vulnerability

There are several types of vulnerability. Different sets of variables are used to define different types of vulnerability. A discussion of types of vulnerability is essential to examine how it might be reduced in the post-disaster phase. Such a discussion will also help in selecting appropriate variables to develop a composite index of vulnerability.

3.1.4.1 Individual/household vulnerability

In terms of scale, individual/household vulnerability is the smallest form of vulnerability because it identifies personal/household factors influencing the vulnerability of a single person/household. Individual/household vulnerability can vary greatly from person to person, or household to household, depending on such characteristics as level of education, income, race, gender, and age as well as past disaster experience. When individual vulnerability is concerned, a well-educated male with a moderate-to-high income level would be considered less vulnerable to a hazard than a male who is not educated. The elderly and the very young are more vulnerable simply because of their lessened mobility and, in most cases, their reliance on another person for support and protection. Another factor which makes some elderly vulnerable to disasters is isolation (Pelling, 2003). While studying the 1995 heat-wave disaster in Chicago, Klinenberg (2002) found that isolation contributed to the disproportionate number of deaths of sick and single elderly men.

In addition to the above characteristics, individual vulnerability also differs due to personal choices and decisions. Such decisions can affect both short-term and long-term vulnerability to hazards and other aspects of lives of individuals.

At the household level household assets, such as landholding size, quality of land, ownership of farm animals, and money saved, determine household vulnerability. Households with fewer assets suffer greater impacts to natural hazard events and have greater difficulty recovering afterward. Wisner and his colleagues (2004) stress that a major explanatory factor in the creation (and distribution of impacts) of disasters is the pattern of wealth and power, because these act as major determinants of the level of vulnerability across a range of people. Understanding the context-specific structural features and constraints of wealth and power are important in any study of individual or household level hazard vulnerability. This type of vulnerability is closely related to social vulnerability, which is discussed below.

3.1.4.2 Social vulnerability

Towards the end of the twentieth century hazard researchers began to question the unequal distribution of disaster effects within a population, with some population subgroups and localities affected disproportionately by disaster outcomes. These studies led to the development of social vulnerability that focused on the social and economic forces that shape disaster outcomes (Zahran et al., 2008). People are seen as being more or less socially vulnerable primarily because of differential access to resources and political power. Social vulnerability is defined by the possession of social attributes that increase susceptibility to disasters (Blaikie et al., 1994). The primary difference between individual and social vulnerability is the scale. Social vulnerability is caused by both internal and external forces, and usually these forces are associated with human characteristics rather than physical characteristics.

Among other things, factors that generally contribute to increasing levels of social vulnerability are poverty, race, and isolation. Poor African Americans in New Orleans, for example, did not have the financial resources to afford an evacuation prior to Hurricane Katrina's arrival. Socially isolated people may not have the information or the assistance they need to evacuate themselves from harm's way or to recover in the aftermath of an extreme event. These factors have been shown to negatively impact behavioral responses and outcomes across the disaster cycle from preparedness through recovery (Fothergill et al., 1999). Much evidence supports the notion that social structure may limit response options available to victims of disaster.

As indicated, social vulnerability is increased for low income and low status persons, homeless people, females, the elderly and young children, migrant workers, ethnic minorities, large family, single-parent families, and special-needs populations (e.g., people with pre-existing chronic conditions and people who rely on home care). As a group, these people often disproportionately occupy the most hazardous geographical areas and thus they are frequently exposed to natural disasters. Similarly, these people also disproportionately occupy the oldest and most poorly maintained buildings. Thus, those who are most socially vulnerable are also likely to experience the greatest physical impacts from natural disasters.

Vulnerability also increases in societies with rapid population growth, high rates of unemployment/underemployment, high dependency ratio, weak housing, and lack of nearby medical services. Rapid population growth displaces the population segments with the fewest social, economic, and political resources from safer to hazard-prone areas. Social and political instability, lower status for women, and racial/religious conflicts all contribute to social vulnerability. Social vulnerability may be reduced in a variety of ways. It can be reduced through increasing levels of social capital, which refers to the collective value of all "social networks" and the inclinations that arise from these networks to assist each other (Bourdieu, 1985).

All members of a social network are not equally poor. For example, NGOs in Bangladesh, whose members are generally poor with no or little land, provide their members with the social capital they need to recover quickly after a natural disaster (Paul, 2003). Such support is not available to the poor who have some land and thus are not eligible to become members of many NGOs. Buckland and Rahman (1999) reported that in the case of the 1997 Canadian Red River flood, flood-impacted communities with strong social capital were better prepared and more effective responders to the disaster. Social vulnerability may also be reduced by integrating formal and informal means of reducing hazards.

Social vulnerability is not only a complex concept, it is also dynamic. Social vulnerability of an individual or a group can change over time if the conditions contributing this vulnerability improve. For example, increased employment opportunities, better housing, and increased institutional and structural system put in place may potentially reduce the level of vulnerability of individuals and groups (Cutter *et al.*, 2003; Wisner *et al.*, 2004). The temporal and spatial scales are important to consider when undertaking vulnerability assessment. Data analysis performed at the national level may not accurately reflect a social profile at the subnational level, since there may exist unique sets of circumstances in localized areas that could yield different results at different scales (Schmidtlein *et al.*, 2008).

3.1.4.3 Institutional vulnerability

Institutional vulnerability has to do with government policy, and public and private institutions, and how effective such policies and organizations are in mitigating vulnerability and/or recovering from disasters. It also refers to the ability of relevant organizations to properly response to hazard events. When referring directly to government institutions, this type of vulnerability refers to governmental agency's ability to make policies and enact them in an effective manner. Thus, the institutional vulnerability does not necessarily mean that institutions are vulnerable, rather the institution is causing others to become more vulnerable to hazards and disasters. If an institution, such as a local government, fails to enforce building codes for a certain hazard that is frequent in the area, this could be considered an example of institutional vulnerability. Another source for institutional vulnerability is when relevant organizations do not fully appreciate and commit to the threat of disasters; this can leave a community completely unprepared.

Two things that help determine public and private institutions' ability to respond to and prepare for disasters are their capacity and commitment. The type and age of organizations and institutions have positive impact in reducing individual and household level vulnerability. Without enough capacity and commitment relevant institutions will be ineffective in time of crisis. The level of capacity and commitment of hazard-related organizations and institutions are directly related to the size of the institution. In other words, a large city with 100 000 people will likely have less institutional vulnerability than a small city of 4000 people, primarily because the larger city has more resources to spend on hazard mitigation and preparedness programs compared to the smaller city (Paul and Huang, 2004). "Institution-building" is an important tool in all communities and countries in order to maximize the potential for well-designed policies that are less prone to vulnerability during a disaster.

3.1.4.4 Economic vulnerability

This type of vulnerability is generally concerned with the financial ability of a community, city, or country to protect its inhabitants from the devastating impacts of disasters. Within societies, level of vulnerability may differ by economic class. As noted, the poor, as a group, are much more likely to suffer from the consequences of disasters because they are not financially able to adopt hazard mitigation measures. They are forced to live on more dangerous land and their houses are more likely to be constructed of materials that are unable to withstand extreme events. Some of the economic factors that affect vulnerability at the national level include: gross domestic product (GDP), debt, sources of national income, and funds reserved for disasters. Low income, poverty, unemployment, landlessness, and unequal distribution of land are some of the factors that represent economic vulnerability at individual level.

3.1.4.5 Physical vulnerability

Physical vulnerability generally involves what in the built environment is physically at risk of being affected. Some of the components of this vulnerability include: human settlements in hazard-prone areas, rapid urbanization, rapid population growth and land use change, poor quality of housing, infrastructure, and inadequate physical protection. Indicators of physical vulnerability are: deaths and injuries, damage to crops, buildings, properties, roads, and disruption of normal life. Hazard researchers often use structural vulnerability to mean damage to all types of buildings and infrastructure by natural disasters. These structures are vulnerable to wind, seismic, and water forces because of inadequate designs, inadequate construction materials, or both. The focus of physical vulnerability studies also includes the physical dimensions of disaster events such as duration, frequency, seasonality, and diurnal factors.

It is, however, worth noting that moving into risky areas does not necessarily imply that vulnerability has been increased. For example, a building may be placed in a flood-prone zone, but raising the structure onto stilts reduces its physical vulnerability. Other flood-proofing techniques, such as river channel improvements, institution-building, improved maintenance and management of watercourses, and improved flood forecasting and warning systems, are able to reduce the physical vulnerability of an area to flood hazards.

3.1.4.6 Environmental vulnerability

Environmental vulnerability refers to the conditions of natural environment that either contribute to or reduce the sufferings of disaster victims from impacts of extreme events. This type of vulnerability is caused by environmental degradation, which, in turn, is caused by poor environmental practices, such as overgrazing and deforestation. Instead of reducing risks or threat of harm, technology often increases vulnerability. For example, a dam protects people from flooding, but floods often occur due to dam failure. Indicators of environmental vulnerability include: rapid population and urban growth, and migration to risk-prone areas.

3.1.4.7 System vulnerability

System vulnerability is loosely associated with institutional vulnerability, which relates to how well a certain system can cope with disasters, whether that involves physical destruction to buildings and roads or loss of inputs from outside sources. It refers to the capacity of a community to diversify its dependence for production or services in order to prevent a complete loss of a particular service in a disaster event. This refers especially to water, electricity, and waste disposal, as well as emergency health care and other vital services. System vulnerability is possibly the most easily corrected of all types of vulnerability.

Individuals depend for obtaining certain services on their communities or place of residences. Their vulnerability to a particular disaster often increases or decreases depending on the nature of these services. For example, instead of cities relying upon a single large water supply system, it is preferable to have a series of small and self-contained water supply plants. In the case of flooding, for example, it is unlikely that all the plants will be destroyed or damaged (Parker et al., 1997). In such a case, non-affected plants will supply necessary water to residents of the flood-impacted city. System vulnerability is more of a problem for non-natural disasters, such as Bhopal disaster which caused following gas leakage in December 1984.

In the context of compliance with cyclone evacuation orders in Bangladesh, several studies (e.g., Chowdhury et al., 1993; Ikeda, 1995; Paul, 2009a) reported that coastal residents are less willing to travel more than 1 mile (1.6 km) to take safer refuge in public cyclone shelters during an emergency. Before the landfall of Cyclone Sidr on November 15, 2007, in many impacted areas the shelters were located 2–3 miles (3.2–4.8 km) apart. Realizing the inadequacy of the number of existing shelters, the Bangladesh government has a plan to construct a significant number of new cyclone shelters (Paul, 2009a). To be most effective, experts urge that these shelters need to be constructed within close proximity to each other. A denser network of smaller public shelters would be preferable to less numerous larger shelters, because such a network would reduce the house-to-shelter distance, allowing for greater utilization of such facilities.

Similarly, reliance upon simple, low-cost intermediate technologies is preferable to more modern methods. In order to reduce flood susceptibility, where possible, treatment plants, related installations, and roads should be sited on relatively high ground. Parker et al. (1997) claim that transferability may be increased to raise resilience in a number of ways, such as by building-in a degree of system "redundancy." For example, in order to reduce the possibility of food shortages in flood-affected areas, it is necessary to plan to store food on high ground. Thus the system vulnerability can be reduced by reducing dependence on a single source of production or service; increasing the level of transferability in order to maximize available resources; and the innovation of efficient and resilient systems that can withstand major disasters and reduce the need for reconstruction costs.

It is evident from the above that discussions of system vulnerability would most likely revolve around resources. The placement, distribution, replacement value, and dependency factor of resources all heavily affect system vulnerability. The warning against putting "all your eggs in one basket" applies nicely here, in that a system is less vulnerable when its resources are varied and widespread. Reliance on more than one type of resource is important, because while one type may be destroyed in a disaster, the other may come away unscathed. Likewise if resources are widely distributed, some may go unharmed simply because they were not within the physical range of the disaster's effects.

As noted, the complexity of a system must also be taken into consideration; one good example is technology. It is often tempting for private companies and government organizations to convert completely to the newest and latest technologies in order to make their system run fastest and longest. This, however, often gets in the way when disaster strikes. In many instances the first means of operation to be struck are those involving electricity, but even more significant for many government operations is interference with satellite operations. Those systems that are slightly less technologically advanced may in fact be better prepared for a disaster, and thus less vulnerable. However, this factor is case specific, and certainly in some instances technology is not a hindrance but a saving grace.

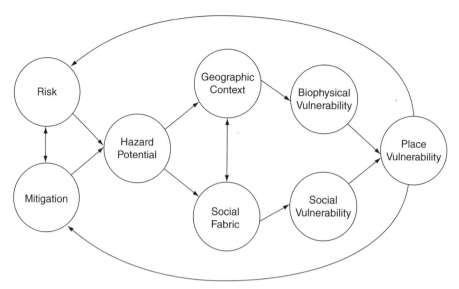

Figure 3.2 The hazards-of-place model of vulnerability. *Source:* Cutter (1996), p. 536.

3.1.4.8 Place vulnerability

Cutter and her colleagues (2000) in their hazards-of-place model of vulnerability introduced this type of vulnerability. This model intends to organize and combine both a traditional view of vulnerability (biophysical risk) with more recent ideas on social vulnerability. They combine risk, which is the probability of a hazard, with mitigation, efforts to reduce risks, to generate the potential for a hazard (Figure 3.2). The hazard potential is then "filtered through" (1) the social fabric to create social vulnerability and (2) the geographic context to produce biophysical vulnerability. The geophysical context deals with physical location of the place, and the level to which that place is at risk from a hazard. The social fabric context concerns socio-economic and other factors rooted in the social landscape of a place. The geographical context, in turn, creates the biophysical vulnerability, and the social fabric creates the social vulnerability. The interaction between these two creates the place vulnerability. The knowledge about place vulnerability aids in building community resilience to disaster impacts.

The types of vulnerability discussed above are not mutually exclusive and they are interconnected. For example, income, gender, age, ethnic minority, and family size are all characteristics of both individual/household level vulnerability as well as social vulnerability. Similarly, economic vulnerability can lead to social vulnerability, or physical vulnerability can cause environmental vulnerability. Social vulnerability is intrinsically linked to place vulnerability. This is because a person's economic status is directly related to where they work and live. Thus, an indicator (e.g., rapid population growth) may represent more than one type of vulnerability. In developing a composite index of vulnerability at any scale, caution needs to be exercised in order to avoid double count.

Moreover, all of the different types of vulnerability are dynamic and continue to change based on the location, timing, and type of any hazard event. The frequency of an event, the magnitude in relation to human response, and people's ability to adjust in the past and in the future to minimize damage caused by hazards are all factors that contribute to vulnerability. The typical response to vulnerability is to reduce the exposure to the hazard, minimize destructive consequences, improve the capacity to manage in a hazard event and reinforce the potential for recovery.

3.1.5 Measures of vulnerability

In measuring human vulnerability at different scales, hazard researchers have used a large number of variables. Because many of these variables are strongly correlated, most of these researchers have developed a composite measure of vulnerability in general and social vulnerability in particular. They have examined a variety of hazard and disaster contexts identifying various dimensions of vulnerability. An examination of variables associated with vulnerability and composite indices will help in further developing appropriate indicators of vulnerability to extreme natural events.

Cutter *et al.* (2003) developed the Social Vulnerability Index (SoVI) for counties of the United States. Using the hazards-of-place model described earlier, they developed an index for social vulnerability. They collected relevant data for all US counties and used 42 variables (reduced by testing for multicollinearity from an original set of 250, and further reduced from a subset of 85 variables through application of conversion to percentages, per capita variables, and density function) as independents. Principal component analysis (PCA) was then applied in order to identify what variables describe the most variance within the data set. The PCA produced 11 factors that explained 76% of the variance across all US counties. The results showed that personal wealth (per capita income being the dominant variable) was the highest rated factor, explaining 12% of the variance in the dataset, followed by age (11.9%).

In Cutter *et al.*'s SoVI, levels and qualities of development figured importantly in two of their factors. The factor "density of the built environment" (density of commercial and manufacturing establishments, housing units, and new housing permits) explained 11.2% of the variance in the data set. The other factor titled "housing stock and tenancy" included variables such as the nature of the ownership of houses (whether rented or owned), the location of their residence (whether urban or not), and the nature of housing stock (whether or not it was a mobile home). In the United States, urban areas are more at risk for damaged dwellings while rural areas are at increased risk for destruction of mobile homes (Cutter *et al.*, 2003).

In order to produce the SoVI, the authors then placed the factor scores for each county into an additive model that resulted in a composite index score of vulnerability for each county in the United States. The work of Cutter and her colleagues (2003) illustrates a collection of a wide range of variables narrowed down through intense statistical calculation, with the final result, the SoVI, producing one single number describing the potential for counties in the United States to be harmed by environmental disasters. The SoVI, however, indicates higher vulnerabilities based in part on population size. They also found that the most socially vulnerable counties were those with the most densely built environments. Of the 3141 US counties, 393 were classified as being most vulnerable. The utility of the index is in revealing the geographic distribution of vulnerability. If mitigation and prevention are the goal of the policymaker, they can apply this index to help them look more closely at parts of the United States that need assistance in building community resilience to hazard impacts.

In 2006 Cutter and Emrich applied the SoVI to an analysis of the counties and parishes along the Gulf Coast that were affected by Hurricane Katrina. They found the model to be a consistent and robust indicator of social vulnerability (Cutter and Emrich, 2006). In particular the SoVI was able to reveal how resiliency within a single county or parish had declined from one point in time to another (they used data from the decadal US Census). One policy application of the SoVI would be to reduce "overall social vulnerability" at the local level "by focusing mitigation and planning on the most important component for each community, rather than implementing broad-brush approaches that might miss the more intricate place-based differences in social vulnerability" (Cutter and Emrich, 2006, pp. 111–112).

Burton (2010) also developed a social vulnerability index for a selected census tract of Mississippi's Gulf Coast impacted by Hurricane Katrina. The SoVI provided the comparative framework for Burton's study.

As noted, Cutter *et al.* (2000) examined the vulnerability of Georgetown County, South Carolina, in terms of exposure to harm from technological and natural disasters. They used the term "place vulnerability" and claim that the interaction between biophysical and social vulnerability creates place vulnerability (Figure 3.2). Their place vulnerability was based on eight social (total population, total housing units, number of females, number of nonwhite residents, number of people under age 18, number of people over age 65, mean house value, and number of mobile homes) and 16 environmental factors (the frequency of 16 different environmental hazards). Both indices were developed for multiple hazards and they were not weighted.

Another application of vulnerability indices deals with coastal vulnerability, usually focused on the impact of hurricanes or storm surge associated with them. Dixon and Fitzsimons (2001) examined the vulnerability of Texas Gulf Coast communities to hurricanes. Using Saffir–Simpson intensity categories for historical hurricanes as well as population and property value data for each county, the authors developed an additive model resulting in five categories of risk scores and five exposure scores. Their Hurricane Vulnerability Index (HVI) is then derived by adding the two scores. This method illustrates an attempt to combine not only data on events that have already occurred, but to couple those data with the potential for harm to a county given the population and assumed worth of the property in that community, creating an index value that, while time-dependent, serves to assign a measure of potential loss in the event of a future hurricane.

In a similar work, Pethick and Crooks (2000) developed a Coastal Vulnerability Index (CVI) where they examined geomorphological factors such as exposure of cliffs, sand dunes, spits, and marshes in terms of their contribution to vulnerability of the coastal system as a whole. The CVI is relatively simple: vulnerability is considered as relaxation time divided by return interval of an erosion event. While the CVI is not as rigorous as the indices of Cutter *et al.* (2003) and Dixon and Fitzsimons (2001), it still offers insight into the different windows through which vulnerability can be viewed.

Since the terrorist attacks on September 11, 2001, attempts have been made to determine spatial vulnerability to agricultural bioterrorism. For example, the US Department of Homeland Security's Office for Domestic Preparedness (USDHS-ODP) issued both jurisdictional and state level security assessment handbooks to attempt to update threat, vulnerability, and response capabilities to terrorist activities on agricultural vulnerability in 2003 (USDHS-ODP, 2003a,b). Through a standardized form, this organization collected extensive data and subsequently calculated the vulnerability scores at both local and state levels to determine a vulnerability rating for the entire jurisdiction. Madden and Wheelis (2003) determined the threat of deliberately introduced plant pathogens as weapons against US crops. Stephen's (2004) analysis of vulnerability in Delanta, Ethiopia, focused on food security and global foreign aid. She utilized 12 variables (e.g., grazing land, crop land, dairy types available, livestock holdings, crop sales prices, total income, and expenditure) to gauge the vulnerability of the rural population of Delanta.

Recently Mustafa *et al.* (2011) have developed a theoretically driven and empirically tested quantitative Vulnerability and Capacity Index (VCI) for use at the local scale. They used 12 indicators to represent material, institutional and attitudinal aspects of differential vulnerability and capacities to develop this index. One of the strengths of this VCI is that it looks at common core drivers of vulnerability across household and community scales and across rural and urban divides without changing drastically. The authors claim that the VCI is an attempt at a practical compromise that will help bridge the gap between research and fruitful policy.

3.2 Resilience

The terms vulnerability and resilience are highly complementary since greater resilience is what is achieved when vulnerability is reduced (Foster, 1995). Resilience and vulnerability reduction are compatible with the goals of sustainable development, broadly defined by the World Commission on Environment and Development (WCED) as: "development that meets the needs of the present without compromising the ability of future generations to meet their own needs" (WCED, 1987, p. 188). Communities that are sustainable and resilient are able to minimize the effects of disasters, and at the same time, possess the ability to recover rapidly from these extreme events (Tobin, 1999).

The term resilience is derived from the Latin *resilio*, meaning "to jump back" (Klein et al., 2003). From an engineering point of view, it is the ability of a material to recoil or spring back into shape after bending. The concept of resilience was introduced in 1973 by ecologist C.S. (Buzz) Holling (Walker et al., 2006). He has drawn this widely contested concept from agro-ecology and natural resource management literature. Although resilience as a topic has been discussed among hazard researchers for more than three decades, this term gained prominence in the disaster literature after Hurricane Katrina in 2005, supplementing the concept of disaster resistance (Tierney and Bruneau, 2007). The 2004 World Disaster Report places resilience at the center of calls for stronger approaches to risk reduction that require local knowledge and implies a paradigm shift towards people-centered approaches as well as understanding of how communities and individuals survive and cope with disasters (IFRC, 2004).

Drawing on diverse literature from ecology, social science, the human–environment system, and natural hazards, this part of Chapter 3 provides an overview of the important elements of resilience.

3.2.1 Defining resilience

As noted, resilience has become an essential concept in natural hazards research in recent years and is central to the development of disaster reduction at the local, national, regional, and global levels. Resilience may be considered as the capacity of physical and human systems to respond to and recover from extreme events (Tierney and Bruneau, 2007). Resilience is often described as a buffer, or a shock absorber, promoting sustainable livelihoods by allowing individuals and/or systems an opportunity to cope during an extreme event and not depleting all resources or options for recovery in the following period. Like vulnerability, there are many definitions of resilience. Zhou et al. (2010) provided at least 29 definitions of resilience and Thywissen (2006) provided 16 definitions. These divergent definitions of resilience implies that this term means different things to people in different fields, such as ecology, economics, political science, mathematics, and archaeology. Zhou et al. (2010) maintain that the varied meanings of resilience arise from different epistemological orientations and subsequent methodological practises by researchers in different disciplines.

A careful review of relevant literature reveals that resilience is defined in the context of: hazard mitigation, recovery from the impact of disaster and coping capacity, and a reduction in vulnerability. Bruneau et al. (2003, p. 733) define resiliency as: "the ability of social units (e.g., organizations, communities) to mitigate hazards, contain the effects of disasters when they occur, and carry out recovery activities in ways that minimize social disruption and mitigate the effects of future disasters." Adger (2000) defines resilience more specifically in relation to a community's ability to withstand external social, economic, and political shocks. Unlike Zhou *et al.* (2010, p. 29), who define disaster resilience "as the capacity of hazard-affected bodies (HABs) to resist loss during disaster and to regenerate and reorganize after disaster in a specific area in a given period," no distinction is made here between resilience and disaster resilience. Both the terms are characterized by a reduced probability of system failure, reduced consequences resulting from system failure, and a reduction in time to system restoration (return to a pre-disaster level of normalcy).

Resilient systems tend to reduce physical damage including the probabilities of deaths and injuries, negative economic and social effects, and the time for recovery following an extreme event. Resiliency reflects a concern for improving the capacity of both physical and human systems to respond to and recover from extreme natural events. Correira *et al.* (1987) consider resilience a measure of the system recovery time. Similarly, Pimm (1984) defines resilience as the speed with which a system returns to its original state following a disturbance. Both definitions are now known as "engineering resilience" and highlight the difference between resilience and resistance; the latter is the extent to which a disturbance is actually translated into an impact (Adger, 2000).

By contrast, Holling (1973) defines resilience as a measure of how far the system could be perturbed without shifting to a different regime. Holling's definition is now often termed "ecological resilience," and many ecologists argue that such resilience is the key to biodiversity conservation and that diversity itself enhances resilience, stability, and ecosystem functioning (Adger, 2000). However, there is debate regarding whether stability is always a desirable characteristic of resilience. Resilience may refer to the extent to which a system is able to absorb adverse effects of a disaster, or it may refer to the recovery time for returning after a disaster. Thus, a highly resilient system can be characterized by its capacity to endure despite high stress, or its ability to bounce back quickly.

The IPCC (2001) treats resilience essentially as the opposite of vulnerability. But is resilience really the opposite of vulnerability? Is resilience a determinant of vulnerability, or is it the other way around? Answering to these questions is a difficult task because the relationship between vulnerability and resilience is not easy to identify. Zhou *et al.* (2010) identify some overlap between these two concepts. Mitigation measures, such as the construction of a dyke around a flood-prone community, will reduce the flood vulnerability of that community. Since the emphasis of disaster resilience is in the process of enhancing the capacity to resist, the dyke will increase flood resiliency of the community. Thus, overlap exists between vulnerability and resiliency because of construction of the dyke.

The ability or capacity to cope with and recover from impacts of disasters is central to the theme of resilience. The Victorian (Australia) Department of Human Services (2000) claims that the higher the resilience, the less likely there is to be damage, and the faster and more effective recovery is likely to be.

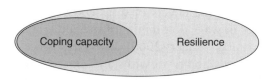

Figure 3.3 Relationship between coping capacity and resilience. *Source:* Thywissen (2006), p. 38.

Literature review shows a large overlap between coping capacity and resilience, and these two terms are often used synonymously (Thywissen, 2006). However, as shown in Figure 3.3, resilience is a more encompassing term than coping capacity. The latter includes those strategies and measures that act directly upon damages sustained during an extreme event by containing the impact and/or by bringing about emergency assistance. It also encompasses adaptive strategies that modify behavior or activities in order to lessen damage caused by the event. Resilience includes all these things, plus the capacity to remain functional during an event and to completely recover from it as quickly as possible.

The time it will take for a disaster-impacted community to return to its pre-disaster level for a given extent of damage can be measured with the help of a resilience triangle (Figure 3.4). For example, if a tornado causes an equal amount of physical damage (say 50%) to two communities of equal size, the resilience triangle of tornado resilient community will be smaller than the triangle of the lesser resilient community (Figure 3.4). Also, the time required to full recovery from the impact of tornado for the first community will be shorter than for the second community.

3.2.2 Framework of resilience

Examining the attributes and determinants of resilience, investigators of the Multidisciplinary Center for Earthquake Engineering Research (MCEER),

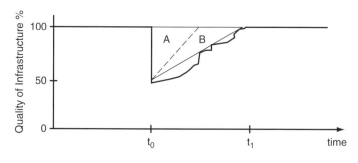

Figure 3.4 The resilience triangles. *Source:* Modified after Tierney and Bruneau (2007), p. 15.

University of Buffalo, USA, developed the R4 framework of resilience: robustness, redundancy, resourcefulness, and rapidity (Tierney and Bruneau, 2007). The first component of this framework refers to the ability of systems, system elements, and other units of analysis to withstand a given magnitude of disaster without significant degradation or loss of function. Robustness then, reflects the inherent strength of a system. Lack of robustness can cause break down of the system as happened with the breaking of levees in New Orleans in 2005 following Hurricane Katrina.

The second R4 component – redundancy – refers to the extent to which elements in systems are sustainable (i.e., capable of satisfying functional requirements if significant damage of functionality occurs). Redundancy allows for alternative options, choices, and substitutions. Lack of this component hinders proper response to an extreme event. A considerable number of people in New Orleans were not able to comply with the mandatory evacuation order prior to the landfall of Hurricane Katrina because public transportation was unavailable (Harrington et al., 2006).

The resourcefulness component of the R4 resilience framework is concerned with the capacity to diagnose problems, establish priorities, and mobilize adequate resources for quick recovery from the impact of an extreme event. Lack of resources delay the recovery process – as was the case for remote areas of Pakistan devastated by the 2005 earthquake. The last R4 component is rapidity, which refers to the capacity to meet priorities and achieve goals in a timely manner (i.e. the speed with which disruption can be overcome, and services and livelihoods restored).

Understanding of the R4 framework of resilience provides helpful guide to define and achieve acceptable levels of loss, damage, and system performance. This framework highlights the multiple paths of resilience and each one of these components of resilience can be improved through investments in hazard mitigation measures (Tierney and Bruneau, 2007).

3.2.3 Themes in resilience studies

Hazard researchers have come up with four distinct themes in resilience studies: resilience as a biophysical attribute, a social attribute, a social-ecological system (SES) attribute, and an attribute of specific areas (Zhou et al., 2010).

3.2.3.1 Resilience as a biophysical attribute

Many hazard researchers (e.g., Holling et al., 1995; Chapin et al., 1997; Holling, 2001; Folke et al., 2004) have examined the resilience of biophysical or technological systems. These researchers are concerned with the physical properties of systems, including the ability to resist damage and loss of function, and to fail gracefully. Biophysical attribute includes physical components that add redundancy. Studies dealing with resilience as a biophysical attribute primarily focus on the key components of systems, such as biodiversity and functional diversity, which provide a system with variety of responses to varied perturbations.

3.2.3.2 Resilience as a social attribute

The second theme of resilience studies deals with social resilience, which is broadly defined as the capacity of a group or a community to bounce back or respond positively to adversity (Maguire and Hagan, 2007). Social resilience is an important component of the circumstances under which individuals and social groups adapt to environmental change. Social resilience studies focus on describing the behavioral response of social entities, including economic, demographic, and institutional factors in both temporal and spatial contexts. Key economic indicators of social resilience are the nature of economic growth and the stability, and distribution of income. Population displacement, permanent or circular migration, and mobility are key demographic factors, which either enhance or reduce social resilience at various levels (Adger, 2000). For example, several million people from South and Southeast Asian nations work in the Middle Eastern countries as guest workers. These workers have to return to their home countries after their contracts expire. However, these workers help enhance the resilience of their families as well as their home countries by sending most of their wages homes, which is called remittance.

Variables mentioned above determine whether a given social group is either more vulnerable or more adaptable to extreme events. As mentioned, social vulnerability indicators include poverty, low levels of education, and the lack of access to resources for protective actions such as evacuation. Social capital, including trust and social networks, and social memory, including experience in dealing with change, are essential for the groups or communities to adapt to and shape change (Olick and Robbins, 1998; Enemarck, 2006).

Adger (2000) and Kimhi and Shamai (2004) claim that social resilience has three properties: resistance, recovery, and creativity. The first property – resistance – relates to a social group's efforts to withstand the impacts of a disaster and can be understood in terms of the degree of disruption that can be accommodated without the social group itself undergoing long-term change. Recovery relates to a social group's ability to pull through the disaster, and can be understood in terms of the time taken for the social group to recover from negative impacts of a disaster. Finally, creativity is represented by a gain in resilience achieved as part of the recovery process, and it can be attained by adapting to new circumstances and learning from the disaster experience.

Adger (2000) maintains that social resilience could be measured through proxies of institutional change and economic structure, property rights, access to resources, and demographic change. Social resilience can be increased through institutional development, diversification, land reform, marketing, and human capability building (Rockstrom, 2003).

3.2.3.3 Resilience as a social-ecological system (SES) attribute

The third theme of resilience studies focuses on the resilience of social-ecological systems, which incorporate diverse mechanisms for living with and learning from, change and crisis (Adger *et al.*, 2005). Learning to live with the change and uncertainty is one of the critical factors in building an SES (Folke *et al.*, 2003). Focusing on diversity is another factor in building an SES, because it helps increase options for coping with external stresses and disturbances. Combining different types of knowledge for learning is an effective strategy for bridging scales to stimulate innovation. Finally, creation of opportunity for self-organization and cross-scale linkages is the last critical factor in building any SES.

3.2.3.4 Resilience as an attribute of specific area

A fourth direction is emerging where resilience is conceived as a biophysical, social, or social-ecological attribute, but within a specific area. This concept suggests that the area impacted by an extreme event is able to withstand the event without external assistance. Like place vulnerability, Cutter *et al.* (2008) developed a disaster resilience of place (DROP) model to present the relationship between vulnerability and resilience; this can be applied to address real problems in real places.

Given the fact that rural areas are now confronted with a spectrum of changes, some researchers (Schouten *et al.*, 2009) advocate differentiating the area attribute of resilience into rural and urban resilience. The former "refers to the capacity of a rural area to adapt to changing external circumstance in such a way that a satisfactory standard of living in maintained, while coping with its inherent ecological, economic, and social vulnerability" (Schouten *et al.*, 2009, p. 3). Heijman *et al.* (2007) introduced the concept of rural resilience, which describes how rural areas are affected by external shocks and how they influence system dynamics. The rural environment and its natural resources are conditioned by actions of the populations living there and therefore rural areas can be treated as a social-ecological systems (Ambrosio-Albala and Delgado, 2008).

3.2.4 Disaster resilience of "loss–response" of location (DRLRL)

Borrowing from the DROP model, Zhou *et al.* (2010) proposed the "disaster resilience of 'loss–response' of location" (DRLRL) model, which is very relevant,

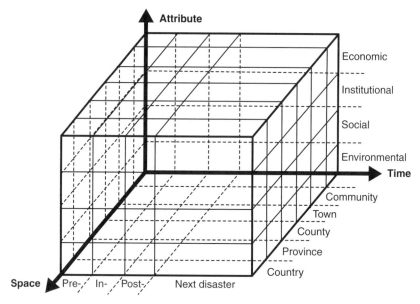

Figure 3.5 Three dimensions of disaster resilience of "loss–response" of location model. *Source:* Zhou *et al.* (2010), p. 29.

helping to understand the concept of resilience in a holistic way. They defined disaster resilience from a three-dimensional mode: spatial, temporal, and the attributes of hazard-affected bodies (Figure 3.5). Determining which localities are more resilient to an extreme event forms the basis for a spatial pattern of disaster resilience. That is, location-specific attributes help in understanding what makes some localities more resilient to disasters and how resilience can be enhanced within these areas. As shown in Figure 3.5, disaster resilience differs across spatial scale. Disasters may often occur at multiple spatial scales, with feedback across scale. For this reason, natural disasters cannot be fully understood at only one level.

As just indicated, whether localities are more disaster resilient depends on the attributes of locality, and these attributes are disaggregated into economic, institutional, social, and environmental categories (Figure 3.4). The environmental attribute, along with disaster magnitude, largely determines the degree of disaster loss. Other attributes of a particular locality determine the capacity of response to loss or loss potential from extreme natural events. In general, localities that have low loss potential and strong response capacity are more resilient than those have high loss potential and weak response (Zhou *et al.*, 2010). This results in the former locality type recovering from the impacts of disasters more quickly than the latter locality type.

The DRLRL model has identified time as an important factor of resilience, because resilience can fluctuate over time due to the changing characteristics of hazard-affected bodies (Kulig and Hanson, 1996), and effective resilience management requires a clear understanding of the temporal stages of change. The time dimension of the DRLRL model is divided into three periods: before (pre-), during (in-), and after (post-) disaster (Figure 3.5). It also can be divided into several periods according to the number of disasters, and the post-disaster period in one disaster may overlap the pre-disaster one in the next disaster. Further, disaster resilience of a locality may vary from one period to the other.

It is important to note that resilience also needs to be seen as a constantly evolving process that adjusts to changing reality and its social constructions. For example, the introduction of hand-pumped shallow tube wells (HPSTWs) in the 1970s increased resilience among Bangladeshi people, particularly children, by protecting them against waterborne diseases. Prior to the widespread introduction of HPSTWs, most people in rural Bangladesh drank bacteria-infested surface water from dug wells, ponds, rivers, and lakes. Ironically, this eventually exposed rural inhabitants of Bangladesh greater risk of arsenic-related conditions, due to contamination of these sources within two decades. Now, people in arsenic-impacted areas are using stored rainwater, filtered water, and other sources to obtain safer drinking water. In addition, various governmental and nongovernment agencies are using neighboring community cohesiveness to allow residents to share water from safe tube wells (Paul, 2009b).

It is clear from the above discussion that the concept of resilience is widely used, but its specific meaning is contested. There is no precise definition of resilience because scholars in many disciplines use and explain this concept from their own professional perspective. However, most scholars do agree that the concept itself is a dynamic one and that resilience can be enhanced by adopting mitigation measures, developing robust organizational and community capacity to respond to disasters, and by improving the coping capabilities of individual households and businesses. In addition, resilience can be enhanced by adjusting, and adapting to hazards, and learning from disasters.

3.3 Risk

Closely related to both vulnerability and resilience is the term "risk," which has multiple conceptions and meanings. Like vulnerability, much has been written about the concept of risk, to the extent that it has crystallized into a full-fledged field of study (Alexander, 2000). As with both vulnerability and resilience, risk has also been a topic of interest to researchers in many disciplines for the last 40 years, and therefore many definitions of risk exist. For the same reason, the literature on this topic/concept is extensive and diverse. No wonder the concept of risk has varying meanings to researchers in different disciplines. Even among risk managers, there is no single accepted definition for the term (Coppola, 2007).

The objectives of this part of Chapter 3 are to cover essential facets of risk that are crucial to understanding this concept. In the context of natural hazards, this concept is very important because it shapes both individual and collective perceptions and actions. It is also important for individual households and hazard-prone communities to take measures to reduce risk associated with extreme events. It is generally held that risk reduction by prior mitigation is cheaper than money spent on recovery and reconstruction after the occurrence of an extreme event. Burton and Pushchak (1984) divided the field of risk into three distinct parts: risk perception and communication, risk estimate/assessment, and risk management. For this discussion, risk perception and risk communication are treated separately. However, before covering these distinct aspects of risk, several definitions of risk are presented.

3.3.1 Defining risk

With reference to natural hazards, most available definitions of risk are expressed as a product of two or more components. In general, risk is the likelihood or probability of hazard occurrence of a certain magnitude. One of the simplest and common definitions of risk, the one preferred by many risk managers, is the likelihood of an event occurring multiplied by the consequences of that event:

$$\text{Risk} = (\text{Likelihood of Hazard Occurrence})(\text{Consequence})$$

(Ansel and Wharton, 1992), where likelihood is expressed either as a probability (e.g., 20%) or a frequency (e.g., 1 in 5 years) of occurrence of a disaster, whichever is appropriate for the analysis being considered. Consequence is a measure of the effect of the hazard on humans and on built and natural environments. It is expressed as a combination of three factors: expected loss of lives, expected number of people injured, and expected damage. The last factor is reported in currency, generally US dollars for international comparison (UNDP, 2004).

In addition to a quantitative representation of the likelihood of hazards in terms of probability and frequency, it is possible to use a qualitative representation of likelihood. For instance, event likelihood could be expressed as follows (Coppola, 2007):

- Certain: >99% chance of occurring in a given year (1 or more occurrences per year)

- Likely: 50–99% chance of occurring in a given year (1 occurrence every 1–2 years)

- Possible: 5–49% chance of occurring in a given year (1 occurrence every 2–20 years)

- Unlikely: 2–5% chance of occurring in a given year (1 occurrence every 20–50 years)

- Rare: 1–2% chance of occurring in a given year (1 occurrence every 50–100 years)
- Extremely rare: <1% chance of occurring in a given year (1 occurrence every 100 or more years).

Note that this is just one of a limitless range of qualitative terms and values that could be used to describe the likelihood component of risk.

Van Dissen and McVerry (1994) and Twigg (1998) generally agree with the above definition of risk. However, they replace consequence with vulnerability:

$$\text{Risk} = (\text{Hazard Probability})(\text{Vulnerability})$$

However, analysts are uncertain whether vulnerability adds to or multiplies the hazard probability. Therefore, an alternate expression has been suggested by CARE (2003):

$$\text{Risk} = (\text{Hazard Probability}) + (\text{Vulnerability})$$

In evaluating earthquake risk in New Zealand, Van Dissen and McVerry (1994) defined probability as the likelihood of an earthquake occurring (based on results of a seismicity model) and vulnerability as the damage potential for property, using a damage ratio. However, Van Dissen and McVery fail to incorporate geographic differences in population size and density (or what might be termed exposure) as well as adjustments undertaken at various levels to minimize potential loss (Tobin and Montz, 1997).

Instead of either consequence or vulnerability, Alexander (2000, p. 10) uses loss and defines risk "as the likelihood, or more formally the probability, that a particular level of loss will be sustained by a given series of elements as a result of a given level of hazard impact." According to him, the elements at risk consist of populations, communities, the built environment, the natural environment, as well as economic activities and services, which are under threat of disaster in a given area.

Lowrance's (1976) definition of risk is similar to three definitions already provided. He defines risk as "a measure of probability and severity of harm" (Lowrance, 1976, p. 8). By severity, he means expected loss, vulnerability, or consequences of occurrence of an extreme event. He substitutes consequence by severity or magnitude of an event:

$$\text{Risk} = (\text{Probability of occurrence of an extreme event})(\text{Magnitude})$$

In all the preceding definitions, risk is divided into two components and they are considered "technological risk" because these definitions combine two elements of risk in a logical, mathematically sound manner, thus yielding a means of comparison (Tobin and Montz, 1997). Hazard risk is, therefore, often viewed as existing on a two-dimensional plane: probability of occurrence and the extent of probable consequences, each of these dimensions can be dichotomized as high and low (Tobin and Montz, 1997). For example, a category 5 hurricane in Texas can be considered a low probability–high consequence event, while a thunderstorm in the same state can be considered a high probability–low consequence event.

However, the difficulty in expressing risk of a hazard using probability and consequence dimensions lies in the fact that identical values may represent either high probability–low consequence or low probability–high consequence risks. For example, a category 4 hurricane may have a return period of approximately 100 in a particular coast (i.e., 0.01 probability of occurring in any given year). According to the formula, the risk could be described as $(0.01 \times 4) = 0.04$. A category 2 hurricane with a 0.02 probability (i.e., a 50-year event) of occurrence has exactly the same risk as the category 4 hurricane mentioned above.

Despite identical risk, these two events would have significantly different outcomes. Tobin and Montz (1997) maintain that if we are unconcerned about attitudes toward these different outcomes, technical risk is an appropriate measure. In reality, we are not unconcerned with people's views and perceptions of risk, which differ from person to person. Such differences are important because they influence attitudes, actions, and vulnerability. Hence, Whyte (1982) suggested altering the (technical) risk formula from:

$$\text{Risk} = (\text{Probability of occurrence of an extreme event})(\text{Magnitude})$$

to

$$\text{Risk} = (\text{Probability of occurrence of an extreme event})(\text{Magnitude})^n$$

where n is social values. It has been argued that when n is sufficiently high, mitigation measures are sought in order to lower the ultimate value of n. It is not easy, however, to quantify the value of n.

Sometimes, risk probability is equated with hazard probability. For example, Alwang et al. (2001), in their definition, use these two terms as synonyms. They maintain that risk is characterized by a known or unknown probability distribution of events. These events are themselves characterized by their magnitude, including size and speed, their frequency and duration, and their history. However, risk is usually considered a component of a hazard event and thus these two terms are not synonymous. The United Nations Disaster Relief Organization (UNDRO) defines risk as an overlapping part between hazard and vulnerability (Figure 3.6) (UNDRO, 1979). Mitchell (1990) conceptualizes hazards as a multiplicative function of risk, exposure, vulnerability, and response:

$$H = f\{(\text{Risk})(\text{Exposure})(\text{Vulnerability})(\text{Response})\}$$

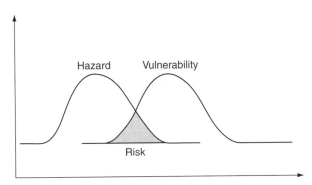

Figure 3.6 United Nations Disaster Relief Organization (UNDRO) definition of risk. *Source:* Menoni (2004), p. 173.

where risk is the probability of an adverse effect, exposure is the size and characteristics of the at-risk population, vulnerability is the potential for loss, and response is the extent to which mitigation measures are in place.

In contrast to Mitchell, Crichton (1999) considers risk as the probability of a loss, which, according to him, depends on three elements: the hazard event itself, vulnerability, and exposure. If any of these three elements in risk increases or decreases, then the risk increases or decreases respectively. Similarly, risk is also considered as a function of hazard occurrence probability, population, and vulnerability (UNDP, 2004):

$$\text{Risk} = f\{(\text{Probability of hazard occurrence})(\text{Population})(\text{Vulnerability})\}$$

Thywissen (2006, p. 39) defines risk using four components "as a function of hazard, vulnerability, exposure, and resilience" (Figure 3.7):

$$\text{Risk} = f\{(\text{Hazard})(\text{Vulnerability})(\text{Exposure})(\text{Resilience})\}$$

One problem with expressing risk using two or more components is that if the value of any component is zero, then the risk is zero too.

As noted, most definitions of risk are expressed only in multiplicative form. One exception is the definition provided by Fournier d'Albe (1979), who conceptualized risk as:

$$R = \{(\text{Hazard probability})(\text{Expected loss})\}/\text{Preparedness (loss mitigation)}$$

To calculate the frequency or probability of occurrence of a hazard requires reliable data for a period as long as 100 years or so. Usually, such data do not exist, particularly in developing countries. Furthermore, data representing shorter time periods may provide unreliable probability for particular hazards and regions. For newer technological hazards, the historical database may be quite

Figure 3.7 Components of risk. *Source:* Thywissen (2006), p. 39.

inadequate to support a reliable statistical assessment of risk. However, even if historical data are available, other problems may arise. For example, human fatalities can be expressed in many different ways (Table 3.1) and the selection of a particular fatality indicator is a complex task. In addition, it is difficult to assign a dollar value for intangible losses, including the loss of human lives.

With reference to expressing fatality risks, each way of summarizing deaths embodies its own set of values. For example, "reduction in life expectancy" treats deaths among the young as more important than among older people, who have less life expectancy to lose. In contrast, deaths per facility assign equal mortality probability for people who work in the facility and those who do not; the probability is also same for children and adults. One way to overcome this problem

Table 3.1 Some ways of expressing fatality risks

Deaths per million people in the population

Deaths per million people within x miles of the source of exposure

Deaths per unit of concentration

Deaths per facility

Deaths per ton of air toxin released

Deaths per ton of air toxin absorbed by people

Deaths per ton of chemical produced

Deaths per million dollars of product produced

Loss of life expectancy associated with exposure to the hazard

---

Source: Slovic and Weber (2002), p. 5.

is by assigning unequal weightings for different kinds of deaths. However, to arrive at any selection requires a value judgment (Slovic and Weber, 2002).

Another problem lies in the fact that the degree of risk varies according to the perspective. For example, the probability of a person being struck by lightning is low and the risk can be viewed as small for the population as a whole. However, the vast majority of lightning strikes are fatal. From the viewpoint of the victim, the risk of death would be very high. Finally, the risk and ultimately the threat (often defined as hazardousness) changes over time as human use and environmental processes change. For example, seismic risk in Dhaka, Bangladesh increased over time because of the rapid growth in population and noncompliance with building codes (Paul and Bhuiyan, 2010). Thus, hazard risk is both dynamic and complex.

3.3.2 Risk perception

Risk perception, which is commonly used in reference to natural hazards, refers to the subjective judgment that individuals make about the characteristics and severity of a risk (Slovic, 1992, 1999). It reflects the decision-maker's own interpretation of the likelihood of being exposed to the conditions propagating risk. Hazard research in social science disciplines clearly suggests that people make decisions and take actions regarding hazards based on their personal perception of risk, rather than on some externally derived measure of threat. Risk perception differs between technical experts and members of the general public, across age, gender, and culture.

In risk perception literature, men are often described as "risk-takers" and women as "risk-avoiders" (e.g., Cutter *et al.*, 1992). Females place higher value on personal health and well-being, and therefore generally feel less secure about all types of hazards than do males. However, the notion that women perceive higher risk associated with hazards is not unanimously established. Flynn *et al.* (1994) concluded that the level of perceived risk was more or less similar for men and women. However, Arcury *et al.* (1987) maintained that males were more concerned and knowledgeable regarding problems associated with hazards than were females.

Risk perception nevertheless is regarded as a valid, even essential component of risk management along with scientific assessments. How a specific environmental problem is perceived by individuals or groups is essential to prepare an effective plan for risk communication. It also plays an essential role in developing effective and socially acceptable public emergency preparedness measures at the household, community, regional, and national levels (e.g., Hewitt, 1983, 1997). Increased preparedness along with hazard awareness significantly reduces individual and community risk to environmental hazards. Furthermore, individuals and communities with a low level of perceived risk are likely to adjust poorly to a hazard event, whereas those with higher level of perceived risk often behave in ways that reduce their risk (Burton *et al.*, 1993). Risk perception is vital to understand people's decision-making and adjustments before, during, and after a hazard event. With the growing importance of public involvement in hazard planning, risk perception now plays a key role in shaping disaster policies at all government levels (Brody *et al.*, 2003).

Three paradigms have been proposed by social scientists to explain why different people make different estimates of the dangerousness of hazard risks (Weber, 2001). These paradigms are: the axiomatic measurement paradigm, the socio-cultural paradigm, and the psychometric paradigm. Studies within the first paradigm have focused on the way in which people subjectively transform objective risk information. The second paradigm focuses on culture rather than individual psychology as an explanation for differences in risk judgments. Anthropologists and sociologists claim that risk perception has its root in cultural and social factors. They maintain that the idea of culture is critical to explaining differences in risk perception (Palm, 1998).

Although the first studies on risk perception date back to the early 1960s (e.g., Bauer, 1960; Slovic, 1962), the concept of "perceived risk" became prominent only since the late-1970s with the seminal work of Slovic, Fishhhoff, and Lichtenstein (Fischhoff *et al.*, 1978; Slovic *et al.*, 1980). In reaction to a paper by Starr (1969), these researchers introduced a new methodological approach to investigate risk perception, one which is now referred to as the psychometric paradigm. Studies using the psychometric paradigm have shown that people's emotional reactions to risky situations that affect judgment of the "riskiness" go beyond their objective consequences. Researchers of this paradigm claim that for most people, risk is more than some combination of "size of damage" and "probability of damage," which are the core parameters of technical risk (Kaplan and Garrick, 1981). They reported that both experts and lay people in general held incorrect perceptions regarding several aspects of hazards. For example, people often underestimate the probability of harm from a given hazard compared to statistical estimates, or may even deny that there is any risk whatsoever (Slovic *et al.*, 1974).

Researchers have identified four systematic biases that could account for these misconceptions regarding risk. These biases result from a set of general inferential rules that individuals seem to use in common situations. These judgment rules are technically known as heuristics and serve the purpose of making difficult mental tasks simpler (Slovic *et al.*, 1980). The four biases noted above are: availability, overconfidence, desire for certainty, and the notion that "it won't happen to me or us," which has also been termed "optimistic bias" (Witte *et al.*, 2001). Combined with other factors/misconceptions, these biases contribute to the acceptance of certain risks while others are deemed unacceptable.

One common misconception includes what Burton and his colleagues (1993) termed the "gambler fallacy" or the idea that once a disaster has occurred, it is unlikely to occur again for quite some time. Gardner *et al.* (1987) reported that many residents of a community in southern California were no longer concerned with fire risk because they had experienced a wildfire within recent memory.

Slovic and his co-investigators (e.g., Slovic *et al.*, 1980) also identified determinants of perceived risk. These are: dread/severity, exposure, and familiarity (new vs. experienced risk). Dread here refers to extreme fear, which is a clear example of what people think about a risk in terms of their intuitive feelings, a process which is called the heuristic effect. Exposure is how many people are endangered by the hazard, and familiarity is whether the risk is routine and something that the respondent is knowledgeable about. Of these three factors, dread is found to have the most influence on human perception of risk, while exposure and familiarity play secondary and tertiary roles, respectively. Several researchers (e.g., McCaffrey, 2004) have made an attempt to examine the connection between these determinants of perceived risk.

Studies (e.g., Davis, 1989; Paul and Bhuiyan, 2010) also found that an individual's risk perception is based on past experience and memory. Their past experience is highly correlated with familiarity mentioned above. Disaster experience alters personal perceptions of hazards, and changes individual attitudes and behavior concerning hazard preparedness. Experience also determines whether a person attaches lesser/greater importance to a particular risk than to other statistically significant ones. Memory is similar to experience and it is affected by variables such as age, time since experience, and the actual severity of the experienced disaster (Paul and Bhuiyan, 2010). A memorable accident usually evokes an elevated perception of risk. An examination of individual perceptions allows one to determine how people view the threat of extreme events, how such attitudes are influenced, and how such views relate to the options people consider in coping with hazard impacts (Slovic, 1999). Other determinants are: type of risk (voluntary vs. involuntary), individual characteristics (e.g., age, educational level, income, and gender), nature of consequence (immediate vs. delayed), and controllability of risk.

3.3.3 Measures of perceived risk

Research shows that perceived risk is quantifiable and predictable. Various statistical techniques, such as psychophysical scaling and multivariate techniques, can be used to produce quantitative representations of risk perception. However, some researchers (e.g., Tobin and Montz, 1997) maintain that quantitative measures are less important than the qualitative attributes of a risk. People tend to evaluate risk in a multidimensional, but subjective manner. As a result, some risks become "socially amplified," while others are "socially attenuated."

Several researchers (e.g., McClelland *et al.*, 1990) qualitatively measured perceived risk using three components: the "risk ladder," probability of exposure, and relative severity (dread). The first component is a numerical assessment comparing risks in terms of their annual mortality probabilities. To formulate a risk ladder, exposure levels and associated risk estimates are arranged with low levels at the bottom of the ladder and high ones at the top. The rationale behind usage

Table 3.2 One way to measure earthquake risk perception

Question	Likert scale				
	Not probable			Highly probable	
	1	2	3	4	5

1. In your view how probable is it that a severe earthquake will occur in your town?
2. In your view, how probable is it that a severe earthquake will cause death in your family?
3. In your view, how probable is it that a severe earthquake will cause injury in your family?
4. In your view, how probable is it that a severe earthquake will cause damage of your property?
5. In your view, how imminent is it that a severe earthquake will occur in your town?
    This year
    Within the next three years
    Within the next five years

of this component is that people are easily able to distinguish risks they have experienced from the ones they have not (Underwood, 1971).

Probability of exposure or perceived exposure to risk involves individual views regarding the likelihood that their community will be affected by an extreme event of interest. Relative severity focuses on the individual's perceived severity of risk compared to their community's risk – as the individual perceives it. It seems that the first and third components of perceived risk overlap each other. For this reason, Ozdemir and Kruse (2000) measured perceived risk on the basis of just two components (perceived exposure and perceived severity).

Holtgrave and Weber (1993) reported that a hybrid model of risk perception that incorporates both affective variables (dread) and cognitive variables (outcomes and probabilities) provides the best fit for risk perception of situations with particular outcomes. Mulilis and Duval (1995) in their Person-Relative-to-Event (PrE) theory also measured risk perception in terms of hazard probability, imminence, and severity. They disaggregated severity into concern and fear of death, injury, and property damage. In the context of seismic risk perception, this can be measured in terms of earthquake probability, severity, and imminence (Table 3.2).

In Table 3.2 earthquake severity is reflected by expected deaths, injuries, and property damage. Five questions are included in the table to help understand the seismic risk perception of the respondents. A 1–5 Likert scale, where 1 signifies the lowest probability/severity/imminence and 5 the highest, is used to record responses. Based on these responses, a composite risk perception score can be calculated for each respondent by adding Likert scale value of earthquake probability, severity, and average imminence value (i.e., ([{this year} + {next 3 years} + {next 5 years}]/3). It is important to note that often preventability is used as a component of risk perception. Differential weights are also assigned to each selected component.

3.3.4 Risk analysis

In the context of extreme events, risk analysis is the process of defining and analyzing the dangers to individuals and communities posed by potential natural hazards. Starr's (1969) work on social benefits versus technological risk marked the development of the risk analysis paradigm, which gained wide acceptance in 1983 (Mileti, 1999). Risk analysis provides the foundation for the entire recovery planning effort. It encompasses risk assessment and risk management.

3.3.5 Risk assessment

Risk assessment refers to the determination of risk or a formal method for establishing the degree of risk an individual or a community faces from single or multiple hazards. A risk assessment can be either quantitative or qualitative. In the former case, an attempt is made to numerically determine probabilities of occurrence for extreme events and the likely extent of losses should a particular event takes place. Qualitative risk analysis, which is used more often, involves neither estimating probabilities nor prediction of losses. Instead, qualitative methods involve defining the various threats, determining the extent of vulnerabilities, and devising countermeasures should an event occur.

Risk assessment represents the first step toward risk management and contains two components: a probability statement of an extreme event and its potential consequences or magnitude of the potential loss. Risk assessment for hazards is usually carried out by trained experts with the aim of producing reliable and repeatable results. Unfortunately, even experts may fail to reach to a consensus on the nature of risk. Differences in inference may result when experts measure risk based on risk perception and subjective judgments. Lack of data and/or differential data quality also creates this inferential problem. One of the key issues in risk assessment accounting for the differing views people hold regarding the importance of various risks.

In estimating the degree of risk, another challenge relates to determining acceptable risk, which is not easy to define. There is a need to clearly differentiate between acceptable risk, accepted risk, necessary risk, and tolerable risk. Acceptable risk is essentially the degree of human and material loss that is perceived as bearable for actions taken to reduce disaster risk (Blanchard, 2005). UNISDR (2004) defines it as: "The level of loss a society or community considers acceptable given existing social, economic, political, cultural, technical, and environmental conditions."

As noted, tolerable risk is different from acceptable risk. The former represents temporarily acceptable risk. An individual may be willing to tolerate a risk because it is confined to a brief time period, or associated with a short-term activity (Tobin and Montz, 1997). However, some risks are part of a lifestyle and thus are considered accepted risks. These risks are viewed as the consequence of living in hazardous environments, such as floodplains, coastal areas, and mountain slopes. Given the context, necessary risks are also accepted risks, but in a different context necessary risks may not acceptable. One example of a response to a necessary risk might be to construct dykes to protect a community from flooding, rather than moving the community off the floodplain (Tobin and Montz, 1997).

Given that the provision of absolute safety is impossible, there is merit in trying to determine the level of risk which is acceptable to a society or a group for common activities or probable situations. One must always specify "acceptable to whom" and that implies a conscious decision based on all available information. In engineering terms, acceptable risk is used to assess structural and nonstructural measures undertaken to reduce possible damage to a level which does not harm people and property (UNISDR, 2004). In engineering, a popular way to define acceptable risk is to specify a recurrence interval, which expresses the failure rate of the systems. For example, the appropriate protection level for agricultural levees is determined by cost–benefit analysis, specially maximizing net benefits and considering incremental crop damage versus the added cost of raising a levee (Mileti, 1999).

Acceptable risk is a more policy-driven issue than a scientific one because many political, social, and/or economic factors influence the collective determination of what risks are acceptable and what risks are not. Two factors confounding the acceptability of risks are the benefits associated with certain risks, and the creation of new risks by eliminating existing ones. For example, dismantling of all nuclear power plants is needed to completely eliminate the risk from such facilities. However, the resulting shortage of power might require using more energy derived from coal or gas plants. This, in turn, would create increased carbon-based pollution, which would create increased health and environmental risks (Coppola, 2007).

There are three steps in risk assessment: type of hazard, probability of occurrence of the hazard, and estimate losses in terms of dollars of damage, fatalities, injuries, illnesses, or some other unit of analysis, such as loss of assets, loss of stockholder confidence, and loss of income. Items to consider in determining the probability of a specific disaster should include but not be limited to: geographic location, topography of the area, frequency of the disaster, and proximity to major highways. By its nature, risk assessment inherently incorporates uncertainty associated with determining the probability of an event occurring and the resulting magnitude of losses. It is difficult to quantify both components. The chance of error in the measurement of these two components is large. Fortunately, sensitivity analysis may help resolve some of the uncertainties associated with probability estimates.

Risk assessment should be thought of as a dynamic, ongoing process, not as a one-time project. The process is described as a set of steps that are continually repeated. At the outset, however, there is a startup process that usually is not repeated. There are several shortcomings of risk assessment. Such assessments tend to be overly quantitative and reductive. Risk assessment also usually ignores qualitative differences among risks. Some maintain that risk assessments may omit important nonquantifiable and/or inaccessible information, such as variations among classes of people exposed to hazards. Furthermore, quantitative approaches divert attention from precautionary or preventative measures. Others consider risk managers little more than "blind users" of statistical tools and methods.

3.3.6 Risk management

In the context of natural hazards, risk management refers to both mitigation – actions taken to reduce the threats to life, property, and environment posed by extreme events – and preparedness – ensuring the readiness of individuals and communities to forecast, take precautionary measures and respond to an impending disaster (Christoplos et al., 2001). According to Plate (2002, p. 211):

> Risk management is a methodology for giving rational considerations to all factors affecting the safety or the operation of large structures or systems of structures. It identifies, evaluates, and executes, in conformity with other social sectors, all aspects of the management of a system, from identification of loads to the planning of emergency scenarios for the case of operational failure, and of relief and rehabilitation for the case of structural failure.

In essence, risk management seeks to minimize, distribute, or share the potentially adverse consequences caused by disasters. This means that the relevant public or private entities need to apply resources to reduce risk by reducing the occurrence of disasters, particularly of high magnitude ones, as well as the negative effects of these extreme events. Prior to implementing risk reduction measures, these entities should prioritize such measures, and their strategy should be to minimize spending and maximize outcomes. Disaster risk reduction measures include mitigation and preparedness aspects of the emergency cycle discussed in Chapter 5. These measures are undertaken to make individuals, households, and communities or society as a whole more resilient to disasters. The main purpose of risk management is to reduce all impacts of natural disasters by implementing and integrating all risk reduction measures.

Similar to risk assessment, uncertainty is a serious problem in risk management. Tobin and Montz (1997) list three broad types of uncertainty in risk management. First, there is considerable uncertainty with respect to important parameters of extreme events, such as frequency, magnitude, and extent of damage. Second, gathering the necessary data adds considerably to the uncertainty and further contributes to management difficulties. Uncertainty also results when mitigation measures are implemented because the magnitude and significance of uncertainty are woven into policy. However, any potential errors emanating from uncertainty in policy are subsumed and, by default, are no longer considered important.

Tobin and Montz (1997) identified four essential processes in risk management. This process begins with identifying exposure – an inventory of those areas, people, and property that are subject to the possibility of loss from a hazard event. Exposure is of two types: physical and financial. Physical exposure refers to identification of areas that are at risk from natural hazards of a given magnitude. As noted, all areas are not equally prone to all types of hazards. Thus it is important to identify which types of events pose risks to an area at what level of severity. In this instance, exposure is defined in the context of frequency and magnitude of hazards. Financial exposure entails a risk management strategy to reduce all types of disaster-induced losses (e.g., property damage, damage to infrastructure, and income loss). Therefore, it is important to know the details of potential losses.

The second process in risk management is identification of options available for reducing potential losses associated with a particular hazard. These options are traditionally known in hazard literature as adjustments, which are defined as those human activities intended to reduce or minimize any negative impacts of an extreme event (White, 1974). Adoption of appropriate adjustments for reducing risk is one of the main goals of every hazard management endeavor. Hazard loss reduction options or adjustments are often grouped as: (a) those that affect the cause, (b) those that modify the hazard, (c) adjustments that modify loss potential, and (d) adjustments to losses by spreading the losses, planning for losses, and bearing the losses (Burton et al., 1993). As indicated, options for reducing losses are hazard-specific, and the outcomes of these options differ depending on the risk threshold under consideration. Understandably, not all options are necessarily available at a given place at a given time. There may be several reasons for this, including lack of resources (including technology), lack of interest by outside investors, and a population that has no power to induce change (Tobin and Montz, 1997).

One way of reducing loss is to affect the causes of a hazard event. For example, flood flows can be reduced by land use change and watershed treatment, and by cloud seeding. The former can be achieved by conservation measures such as afforestation, terracing, and contour ploughing over large areas of the drainage basin. The goal of cloud seeding is to stop flood-producing rains, or to reduce flood flows after flood-producing rains. Similarly, river dredging is helpful in reducing the intensity of floods. Likewise, risk of climate change associated with global warming can be reduced by releasing less greenhouse gases in the atmosphere. This would be another example of affecting the cause of hazards. For some hazards, such as earthquakes, there is no known way of altering or affecting the earthquake mechanism.

Exposure can be reduced by modifying the physical processes of hazards. The aim here is to reduce the damage potential associated with a particular hazard by some degree of physical control over the processes involved. This strategy may also be termed "environmental control." It is based on the premise that prevention is better than cure and can be accomplished through both large- and small-scale environmental control. Building dams to store water in the upper part of basin or embankments and levees to contain the flood flows further downstream are examples of controlling floods by modifying the hazard events. At a smaller scale, hazard-resistant design can be applied to individual buildings through structural adaptations, such as raising the floor level, making them and their contents less susceptible to flooding. Apart from structural measures, hazard-resistant design can also be achieved through implementing strict building codes, limiting activities in hazardous areas, and other local regulations which imply a high degree of community acceptance and support (Smith, 1992).

One important way to reduce hazard exposure is by altering the vulnerability of society to hazardous events by implementing nonstructural measures, such as introducing hazard forecasting, establishing warning systems, disseminating warnings, evacuating people, and implementing other preparedness and mitigation measures. Some of these measures require advanced technology, and even structural devices but, in contrast to event modification, this approach is rooted in social science rather than engineering science. The range of responses is much wider here than for the event modification approach. Also, unlike the previous two groups of options, potential loss-modifying measures can be applied to all hazards. Unfortunately, many of these measures fail to operate in a fully effective manner primarily because of poor hazard perception and/or education on the part of potential victims of extreme natural events.

Finally, hazard exposure can be reduced by adjusting to losses in three different ways. The most common means of adjusting losses is through insurance, which provides some certainty that funds will be available to cover at least part of the losses incurred from an event. However, there are two main problems with relying on insurance as a risk management option: insurance may not be available for all hazards and receiving compensation from insurance companies involves a lot of paper work. Provision of emergency aid by both domestic and/or international sources can also spread losses incurred by natural disasters. This topic is discussed in greater detail in Chapter 7. It is important to mention that some options cannot be implemented at the individual level; they need to be carried out at the public or government level.

It cannot be stressed enough that people in hazard-prone areas need to plan for losses before the occurrence of an extreme event. For example, they need to buy specific hazard insurance and to create reserve funds for such events because they will probably need to bear some of the losses. Disaster relief and insurance rarely accounts for total property damage caused by an extreme event. Risk may be transferred, to some degree, beyond which it is assumed by the disaster victim. For individuals or households, assuming risk involves potentially drawing on personal resources to cover losses when an event occurs. Both individuals and households need to mobilize assets to respond to a disaster. They can use their savings, sell their assets, and borrow money from their kin, friends, and relatives to cope with the impacts of disasters. The options already discussed are not mutually exclusive and individual options often work best in some combination.

The last two essential processes in risk management are to evaluate the efficacy of available options for reducing hazard losses, and finally, choosing and implementing "best" options for a particular hazard in a given area. The selection of an appropriate mix of options on the part of individuals, government, and the private sector is based on several factors, including exposure, the frequency and severity of events, cost-effectiveness, and the acceptance and adoptability of such options by residents of hazardous areas.

The risk management process is a complex one involving economic, social, political, technical, and perceptual factors. It must be recognized that a decision to do nothing to reduce hazard losses is not risk management, but denial or avoidance. Avoidance, however, often may be seen as the most cost-effective option. Increase or decrease of losses incurred by disasters in an area depends on use of sound risk management principles (hazard identification, risk analysis, and impact analysis). Risk managers should, therefore, ensure unity of effort and collaboration among all levels of government and all elements of a community. They need to value knowledge-based approaches, based on education, training, experience, ethical practice, public stewardship, and continuous improvement.

3.3.7 Risk communication

People in hazardous areas are not always able to properly identify risks associated with natural hazards and undertake appropriate mitigation measures. This is primarily because they do not have the necessary and/or accurate information for both approaching and potential dangers, and they are also not usually completely aware of the options available in confronting such dangers. Therefore, there is a need for disseminating scientific information in an easy-to-understand format which will help them take appropriate action to reduce risks from natural hazards. Risk communication seeks to supply people at risk with the information they need to make informed, independent judgments about the risks associated with extreme events. Experts and the media are the two main sources from which lay people derive their knowledge of risk. Other sources include risk managers, risk message preparers, and risk analysts.

It has been identified in the existing literature that there remains a significant gap between the general public's perception of risk and the objective risks as identified by expert groups. One reason for such a gap is that scientists usually define risk in terms of effects on populations, while the ordinary citizen is primarily concerned with effects on individuals. Experts, however, often either overestimate or underestimate the public's ability to evaluate risk and options. In either case, necessary and/or sufficient information is not communicated to the public. As a consequence, the public often fail to properly respond to risk. This is particularly the case if hazard prediction fails repeatedly. In such case, people lose their faith in the ability of experts to accurately predict extreme natural events. It is not only necessary to provide accurate information, but also the information should be complete, simple, straightforward, and should be provided as quickly as possible. Further, information should be delivered with brevity, clarity, and effectiveness.

The media also create problems. Although media coverage acts as an influential force in disseminating information and in encouraging emergency aid, media people often deliberately exaggerate the risk in order to make it sensational (Heeger, 2007). In addition, journalists may have a limited understanding of risk and options, which makes their reporting prone to error (Tobin and Montz, 1997). The media often present formidable challenges to responders and emergency managers. Media correspondents often suffer a lack of sensitivity and frequently provide inaccurate reporting in an effort to get the news out first (Payne, 1994). For example, it was erroneously reported during the early days of Hurricane Katrina that there were widespread rapes and murders at the New Orleans Superdome and Convention Center. There were not always witnesses to these crimes, and many of these reports appear to have been inaccurate (McEntire, 2007). There are other problems with the media in times of disasters. Some of these are: convergence at the scene, additional demands placed on emergency managers and responders, interference with emergency response operations, creation of additional safety problems, lack of technical understanding, misrepresentations and perpetuation of myths, overstatement of impact, and damage to individual and/or agency reputation (McEntire, 2007).

However, it needs to be remembered that the media also help to: educate the public on preventive and preparedness measures, warn the public about an impending hazard, provide information regarding evacuation and sheltering, and relay what the government and others are doing and how they are responding to the disaster (McEntire, 2007). In the case of 1984 Bhopal disaster, for example, a series of newspaper articles warned the people living around the Union Carbide Plant of the chemical hazard for at least a couple of years (Varma and Varma, 2005). Unfortunately, these warnings were largely ignored by the company authorities, government officials, and the victims alike. People associated with disaster management have to deal with media personnel in both emergency and non-emergency situations. Therefore, it is important that they should strive to maintain good relations with the media.

Risk communication is an interactive process of exchange of information and opinion among individuals, groups, and risk managers. It also involves discussion about risk types and levels, and about methods for managing risks. Risk communication benefits include improved decision-making, both individually and collectively. Other benefits are: a better educated public, increased coordination between various levels of government, and the development of working relationships between diverse interest groups. Successful risk communication can assist risk managers in preventing ineffective and potentially damaging public response to natural disasters.

Successful risk communication at individual and community levels can promote changes in people's risk reduction behavior and their beliefs about hazards. This, in turn, may lead to the adoption of appropriate actions to mitigate risk associated with hazards (Mileti, 1999). However, deciding to take such actions is a complex process. A large body of literature on hazard warning response exists to understand decision-making processes, barriers, and factors influencing adoption of appropriate response with evacuation mandates. This literature, which is discussed in Chapter 5, is very useful in helping to make risk communication successful.

References

Arcury, T.A. *et al.* (1987) Sex differences in environmental concern and knowledge: the case of acid rain. *Sex Roles* **16** (9): 463–472.

Adger, W.N. (2000) Social and ecological resilience: are they related? *Progress in Human Geography* **24** (3): 347–364.

Adger, W.N. (2006) Vulnerability. *Global Environmental Change* **16** (3): 268–281.

Adger, W.N. *et al.* (2005) Social-ecological resilience to coastal disasters. *Science* **309** (5737): 1036–1039.

Alexander, D. (2000) *Confronting Catastrophe*. Oxford: Oxford University Press.

Alwang, J. *et al.* (2001) *Vulnerability: A View from Different Disciplines*. Social Protection Discussion Paper Series, No. 0115. World Bank: Washington, D.C.

Ambrosio-Albala, M. and Delgado, M. (2008) Understanding rural areas dynamics from a complex perspective: an application of prospective structural analysis. 2008 EAAE International Congress, August 26–29, 2008. Gent, Belgium.

Ansel, J. and Wharton, F. (1992) *Risk Analysis, Assessment, and Management*. Chichester: John Wiley & Sons Ltd.

Bankoff, G. (2001) Rendering the world unsafe: "vulnerability" as western discourse. *Disasters* **25** (1): 19–35.

Bauer, R.A. (1960) Consumer behavior as risk taking. In *Dynamic Marketing for a Changing World* (ed. R.S. Hancock). Chicago: American Marketing Association, pp. 389–398.

Berkes, F. *et al.* (2003) *Navigating Social-Ecological Systems. Management Practices and Social Mechanisms for Building Resilience*. Cambridge: Cambridge University Press.

Blaikie, P. et al. (1994) *At Risk: Natural Hazards, People's Vulnerability, and Disasters*. New York: Routledge.

Blanchard, W. (2005) Select Emergency Management-Related Terms and Definitions. Vulnerability Assessment Techniques and Applications (VATA). http://www.csc.noaa.gov/vata/glossary.html (accessed January 24, 2006).

Bolin, R. and Stanford, L. (1999) Constructing vulnerability in the first world: the Northridge Earthquake in Southern California. In *The Angry Earth: Disaster in Anthropological Perspective* (eds A. Oliver-Smith and S.M. Hoffman). London: Routledge, pp. 89–112.

Bourdieu, P. (1985) The Social Space and the Genesis of Groups. *Theory and Society* **14** (6): 723–744.

Brody, S.D. et al. (2003) Mandating citizen participation in plan making: six strategic planning choices. *Journal of the American Planning Association* **69** (3): 343–351.

Brody, S.D. et al. (2008) Identifying the impact of the built environment on flood damage in Texas. *Disasters* **32** (1): 1–18.

Bruneau, M. et al. (2003) A framework to quantitatively assess and enhance the seismic resilience of communities. *Earthquake Spectra* **19** (4): 733–752.

Buckland, J. and Rahman, M. (1999) Community-based disaster management during the 1997 Red River Flood in Canada. *Disasters* **23** (2): 174–191.

Burton, C.G. (2010) Social Vulnerability and hurricane impact modeling. *Natural Hazards Review* **11** (2): 58–68.

Burton, I. and Pushchak, R. (1984) The status and prospect of risk assessment. *Geoforum* **15** (3): 463–475.

Burton, I. et al. (1993) *The Environment as Hazard*. New York: Guilford Press.

Cardona, O.D. (2004) The need for rethinking the concepts of vulnerability and risk from a holistic perspective: a necessary review and criticism for effective risk management. In *Mapping Vulnerability. Disasters, Development & People* (eds Bankoff, G. et al.). Wallingford, Oxfordshire: CABI/IWMI/IDRC, pp. 41–52.

CARE. (2003) *Managing Risks, Improving Livelihoods: Programme Guidelines for Conditions of Chronic Vulnerability*. Nairobi, Kenya: East and Central Africa Regional Management Unit.

Chapin, F.S. III et al. (1997) Biotic control over the functioning of ecosystems. *Science* **277**: 500–504.

Chowdhury, A.M.R. et al. (1993) The Bangladesh Cyclone of 1991: why so many people died. *Disasters* **17** (4): 291–303.

Christoplos, I. et al. (2001) Re-framing risks: the changing context of disaster mitigation and preparedness. *Disasters* **23** (3): 185–198.

Coppola, D.P. (2007) *Introduction to International Disaster Management*. Amsterdam: Elsevier.

Cordasco, K.M. and Johnson, R.W. (2006) The paradox of social capital as a liability in disaster management: understanding the evacuation failure of Hurricane Katrina. *Natural Hazards Observer* **30** (3): 5–6.

Correira, S. et al. (1987) Engineering risk in regional drought studies. In *Engineering Reliability and Risk in Water Resources* (ed. P. Duckstein). Boston: Martinus Nijhoff Publishers, p. 588.

Crichton, D. (1999) The risk triangle. In *Natural Disaster Management* (ed. J. Ingleton). London: Tudor Rose, pp. 102–103.

Cuny, F.C. (1983) *Disasters and Development*. New York: Oxford University Press.

Curtis, A. and Mills, J.W. (2009) *GIS, Human Geography, and Disasters*. San Diego, CA: University Readers.

Cutter, S.L. (1996) Vulnerability to environmental hazards. *Progress in Human Geography* 20 (4): 529–539.

Cutter, S.L. (2003) The vulnerability of science and the science of vulnerability. *Annals of the Association of American Geographers* 93 (1): 1–12.

Cutter, S.L. and Emrich, C.T. (2006) Moral hazards, social catastrophe: the changing face of vulnerability along the hurricane coasts. *The Annals of the American Academy of Political Science* 604: 102–112.

Cutter, S. L. et al. (1992) En-gendered fears: feminity and technological risk perception. *Industrial Crisis Quarterly* 6 (1): 5–22.

Cutter, S.L. et al. (2000) Revealing the vulnerability of people and places: case study of Georgetown County, South Carolina. *Annals of the Association of American Geographers* 90 (4): 713–737.

Cutter, S.L. et al. (2003) Social vulnerability to environmental hazards. *Social Science Quarterly* 84 (2): 242–261.

Cutter, S.L. et al. (2008) A place-based model for understanding community resilience to natural disasters. *Global Environmental Change* 18 (4): 598–606.

d'Albe, F. (1979) Objectives of volcanic monitoring and prediction. *Journal of Geological Society* 136: 321–326.

Davis, M. (1989) Living along the fault line: an update on earthquake awareness and preparedness in Southern California. *Urban Resources* 5 (4): 8–14.

Department of Human Services (2000) *Assessing Resilience and Vulnerability in the Context of Emergencies: Guidelines*. Melbourne: Victorian Government Publishing Service.

Dercon, S. and Krishnan, P. (2000) Vulnerability, seasonality and poverty in Ethiopia. *Journal of Development Studies* 36 (6): 25–53.

Dixon, R.W. and Fitzsimons, D.E. (2001) Toward a quantified hurricane vulnerability assessment for Texas Coastal Counties. *Texas Journal of Science* 53 (4): 345–352.

Enemarck, C. (2006) Pandemic pending. *Australian Journal of International Affairs* 60 (1): 43–46.

Few, R. (2003) Flooding, vulnerability and coping strategies: local response to a global threat. *Progress in Development Studies* 3 (1): 43–58.

Fischhoff, B. et al. (1978) How safe is safe enough? A psychometric study of attitudes towards technological risks and benefits. *Policy Sciences* 9 (1): 127–152.

Flynn, J. et al. (1994) Gender, race, and perception of environmental health risks. *Risk Analysis* 14 (6): 1101–1108.

Folke, C. et al. (2003) Synthesis: building resilience and adaptive capacity in social-ecological systems. In *Navigating Social-Ecological Systems: Building Resilience for Complexity and Change* (eds F. Berkes et al.). Cambridge: Cambridge University Press, pp. 352–387.

Folke, C. et al. (2004) Regime shifts, resilience and biodiversity in ecosystem management. *Annual Review of Evolution Systems* 35: 557–581.

Foster, H. (1995) Disaster mitigation: the role of resilience. In *Proceedings of a Trilateral Workshop on Natural Hazards* (ed. D. Etkin). Ontario, Canada: Merrickville, pp. 93–108.

Fothergill, A. et al. (1999) Race, ethnicity, and disasters in the United States: a review of the literature. *Disasters* 23 (2): 156–173.

Gardner, P.D. *et al.* (1987) The risk perceptions and policy response toward wild land fire hazards by urban home owners. *Landscape and Urban Planning* **14** (1): 163–172.

Glewwe, P. and Hall, G. (1995) Who is most vulnerable to macroeconomic shocks? Hypothesis tests using panel data from Peru. Living Standards Measurement Study Working Paper No. 117. Washington, D.C.: The World Bank.

Glewwe, P. and Hall, G. (1998) Are some groups more vulnerable to macroeconomic shocks than others? Hypothesis tests based on panel data from Peru. *Journal of Development Economics* **56** (1): 181–206.

Granovetter, M. (1973) The strength of weak ties. *American Journal of Sociology* **78** (6): 1360–1380.

Granovetter, M. (1983) The strength of weak ties: a network theory revisited. *Sociological Theory* **1**: 201–233.

Harrington, L. *et al.* (2006) Southeastern Louisiana evacuation/non-evacuation for Hurricane Katrina. In *Learning from Catastrophe: Quick Response Research in the Wake of Hurricane Katrina*. Boulder, CO: Hazards Research Center, University of Colorado, pp. 327–352.

Heeger, B. (2007) Natural disasters and CNN: the importance of TV news coverage for provoking private donations for disaster relief. *Natural Hazards Observer* **31** (5): 1–12.

Heijman, W. *et al.* (2007) Rural resilience as a new development concept. Paper presented at the EAAE Seminar held in Novi Sad, Serbia.

Hewitt, K. (1983) *Interpretations of Calamity*. Boston: Allen and Unwin.

Hewitt, K. (1997) *Regions of Risk: A Geographical Introduction to Disaster*. Essex: Longman.

Holling, C.S. (1973) Resilience and stability of ecological systems. *Annual Review of Ecological System* **4**: 1–23.

Holling, C.S. (2001) Understanding the complexity of economic, ecological and social system. *Ecosystems* **4**: 390–405.

Holling, C.S. *et al.* (1995) Biodiversity in the functioning of ecosystems: an ecological synthesis. In *Biodiversity Loss: Economic and Ecological Issues* (eds C. Perrings *et al.*). Cambridge: Cambridge University Press, pp. 44–83.

Holtgrave, D. and Weber, E.U. (1993) Dimensions of risk perception for financial and health risks. *Risk Analysis* **13**: 553–558.

Ikeda, K. (1995) Gender differences in human loss and vulnerability in natural disasters: a case study from Bangladesh. *Indian Journal of Gender Studies* **2** (2): 171–193.

IFRC and RCS (International Federation of Red Cross and Red Crescent Societies) (2004) *The World Disaster Report: Focus on Community Resilience*. Geneva: IFRC.

IPCC (International Panel on Climate Change) (2001) *Climate Change 2001*. Cambridge: Cambridge University Press.

Kaplan, S. and Garrick, B.J. (1981) On the quantitative definition of risk. *Risk Analysis* **1** (1): 11–27.

Kimhi, S. and Shamai, M. (2004) Community resilience and the impact of stress: adult response to Israel's withdrawal from Lebanon. *Journal of Community Psychology* **32** (4): 439–451.

Klein, R.J.T. *et al.* (2003) Resilience to natural hazards: how useful is this concept? *Environmental Hazards* **5**: 35–45.

Klinenberg, E. (2002) *Heat Wave: A Social Autopsy of Disaster in Chicago*. Chicago: The University of Chicago Press.

Kulig, J. and Hanson, L. (1996) *Discussion and Expansion of the Concept of Resiliency: Summary of a Think Tank*. Lethbridge, Canada: Regional Center for Health Promotion and Community Studies, University of Lethbridge.

Lein, H. (2009) The poorest and most vulnerable? On hazards, livelihoods and labeling of riverine communities in Bangladesh. *Singapore Journal of Tropical Geography* 30 (1): 98–113.

Liverman, D.M. (1990) Drought and agriculture in Mexico: the case of Sonora and Puebla in 1970. *Annals of the Association of American Geographers* 80 (1): 49–72.

Lowrance, W.W. (1976) *Of Acceptable Risk: Science and the Determination of Safety*. Los Altos, CA: William Kaufmann.

Madden, L.V. and Wheelis, M. (2003) The threat of plant pathogens as weapons against U.S. crops. *Annual Review of Phytopathology* 41: 155–176.

Maguire, B. and Hagan, P. (2007) Disasters and communities: understanding social resilience. *Australian Journal of Emergency Management* 22 (2): 16–20.

McCaffrey, S. (2004) Thinking of wildfire as a natural hazard. *Society and Natural Resources* 17 (6): 509–516.

McClelland, G.H. *et al*. (1990) The effect of risk beliefs on property values: a case study of hazardous waste site. *Risk Analysis* 10 (4): 485–497.

McEntire, D.A. (2007) *Disaster Response and Recovery: Strategies and Tactics for Resilience*. Hoboken, NJ: Wiley.

Menoni, A. (2004) Land use planning in hazard mitigation: intervening in social and systematic vulnerabilities – an application to seismic risk prevention. In *Natural Disasters and Sustainable Development* (eds R. Casale and C. Margottini). Berlin: Springer, pp. 165–182.

Mileti, D.S. (1999) *Disaster by Design: A Reassessment of Natural Hazards in the United States*. Washington, D.C.: Joseph Henry Press.

Mitchell, J.K. (1990) Human dimensions of environmental hazards, complexity, disparity, and the search for guidance. In *Nothing to Fear* (ed. A. Kirby). Tucson: University of Arizona Press, pp. 131–175.

Mitchell, J.K. (1999) Natural disasters in the context of mega-cities. In *Crucibles of Hazard: Mega-Cities and Disasters in Transition* (ed. J.K. Mitchell). Tokyo: United Nations University Press, pp. 15–55.

Montz, B.E. *et al*. (2003) Hazards. In *Geography in America: At the Dawn of the 21st Century* (eds G.L. Gaile and C.J. Willmott). Oxford: Oxford University Press, pp. 479–491.

Mulilis, J-P. and Duval, T.S. (1995) Negative threat appeals and earthquake preparedness: a person-relative-to-event PrE model of coping with threat. *Journal of Applied Social Psychology* 39 (1): 1319–1339.

Mustafa, D. *et al*. (2011) Pinning down vulnerability: from narratives to numbers. *Disasters* 35 (1): 62–86.

Nathan, F. (2010) Vulnerability to natural hazards: case study on landslide risks in La Paz. In Coping with Global Environmental Change, Disasters, and Security: Threats, Challenges and Risks (eds H.G. Brauch *et al*.). New York: Springer, pp. 539–574.

Olick, J.K. and Robbins, J. (1998) Social memory studies: from "collective memory" to historical sociology of mnemonic practices. *Annual Review of Sociology* 24: 105–140.

Oliver-Smith, A. (2004) Theorizing vulnerability in a globalized world: a political ecological perspective. In *Mapping Vulnerability: Disasters, Development and People* (eds G. Bankoff *et al*.). Sterling, VA: Earthscan, pp. 10–24.

O'Riordon, T. (1986) Coping with environmental hazards. In *Geography, Resource and Environment*, Vol. 2 (eds R. Kates and I. Burton). Chicago: University of Chicago Press, pp. 272–309.

Ozdemir, O. and Kruse, J.B. (2000) Relationship between risk perception and willingness-to-pay for low probability, high consequence risk: a survey method. *Risk Analysis* 26 (4): 945–954.

Palm, R. (1998) Urban earthquake hazards: the impacts of culture on perceived risk and response in the USA and Japan. *Applied Geography* 18 (1): 35–46.

Parker, D. et al. (1997) Reducing vulnerability following flood and disasters: issues and practices. In *Reconstruction after Disaster: Issues and Practices* (ed. A. Awotona). Brookfield, USA: Ashgate, pp. 23–44.

Paul, B.K. (2003) Relief assistance to 1998 flood victims: a comparison of the performance of the government and NGOs. *The Geographical Journal* 169 (1): 75–89.

Paul, B.K. (2009a) Why relatively fewer people died? the case of Bangladesh's Cyclone Sidr. *Natural Hazards* 50: 289–304.

Paul, B.K. (2009b) Attitudes toward arsenicosis victims in rural Bangladesh: an empirical study. *Papers and Proceedings of the Applied Geography Conferences* 32: 115–123.

Paul, B.K. and Bhuiyan, R.H. (2010) Urban earthquake hazard: perceived seismic risk and preparedness in Dhaka city, Bangladesh. *Disasters* 34 (2): 337–359.

Paul, B.K. and Huang, B. (2004) Predictors for public response to tornado warnings: the May 4, 2003 tornadoes in Kansas, Missouri, and Tennessee. *Papers and Proceedings of the Applied Geography Conferences* 27: 51–57.

Payne, C.F. (1994) Handling the press. *Disaster Prevention and Management* 3 (1): 2–32.

Pelling, M. (2003) *The Vulnerability of Cities: Natural Disasters and Social Resilience*. London: Earthscan.

Pethick, J.S. and Crooks, S. (2000) Development of a coastal vulnerability index: a geomorphological perspective. *Environmental Conservation* 27: 359–367.

Pimm, S.L. (1984) The complexity and stability of ecosystems. *Nature* 307: 321–326.

Plate, E.J. (2002) Risk management for hydraulic systems under hydrological loads. In *Risk, Reliability, Uncertainty, and Robustness of Water Resources Systems* (eds J.J. Bogardi and Z.W. Kundewicz). Cambridge: Cambridge University Press, pp. 209–220.

Polsky, C., Neff, R., and Yarnal, B. (2007) Building comparable global change vulnerability assessments: the vulnerability scoping diagram. *Global Environmental Change* 17: 472–485.

Rockstrom, J. (2003) Resilience building and water demand management for drought mitigation. *Physical Chemistry of Earth* 28: 869–877.

Sarewitz, D. et al. (2003) Vulnerability and risk: some thoughts from a political and policy perspective. *Risk Analysis* 23 (4): 805–810.

Schmidtlein, M.C. et al. (2008) A sensitivity analysis of the social vulnerability index. *Risk Analysis* 28 (4): 1099–1114.

Schouten, M. et al. (2009) Resilience of social-ecological systems in European rural areas: theory and prospects. Paper presented at the 113th EAAE Seminar held in Belgrade, Republic of Serbia, December 9–11, 2009.

Sen A.K. (2002) *Rationality and Freedom*. Cambridge, MA: Harvard University Press.

Slovic, P. (1962) Convergent validation of risk-taking measures. *Journal of Abnormal and Social Psychology* 65 (1): 68–71.

Slovic, P. (1992) Perception of risk: reflections on the psychometric paradigm. In *Social Theory of Risk* (eds S. Krimsky and D. Golding). New York: Praeger, pp. 117–152.

Slovic, P. (1999) Trust, emotion, sex, politics, and science: surveying the risk-assessment battlefield. *Risk Analysis* **9** (4): 689–701.

Slovic, P. and Weber, E.U. (2002) Perception of risk posed by extreme events. A paper prepared for discussion at the conference "Risk Management Strategies in an Uncertain World." Palisades, New York, April 12–13.

Slovic, P. *et al.* (1974) Decision processes, rationality and adjustments to natural hazards. In *Natural Hazards: Local, National and Global* (ed. G.F. White). New York: Oxford University Press, pp. 187–204.

Slovic, P. *et al.* (1980) Facts and fears: understanding perceived risk. In *Societal Risk Assessments: How Safe is Safe Enough?* (eds R.C. Schwing and W. Albers). New York: Plenum Press.

Smith, K. (1992) *Environmental Hazards: Assessing Risk and Reducing Disaster*. London: Routledge.

Starr, C. (1969) Social benefits versus technological risk: what is our society willing to pay for safety? *Science* **165**: 1232–1236.

Stephen, L. (2004) Vulnerability regions versus vulnerable people: an Ethiopian case study. In *Mapping Vulnerability* (eds G. Bankoff *et al.*). London: Earthscan, pp. 99–114.

Susman, P. *et al.* (1983) Global disasters, a radical interpretation. In *Interpretations of Calamity from the Viewpoint of Human Ecology* (ed. K. Hewitt). Boston: Allen and Unwin, pp. 263–283.

Swift, J. (1989) Why are rural people vulnerable to famine? *IDS Bulletin* **20** (2): 8–15.

Taylor, A.J. (1990) A pattern of disasters and victims. *Disasters* **14** (4): 291–300.

Thywissen, K. (2006) *Components of Risk: A Comparative Glossary*. Bonn: United Nations University, Institute for Environment and Human Security (UNU-EHS).

Timmerman, P. (1981) Vulnerability, Resilience and the Collapse of Society. Environmental Monograph 1, Institute for Environmental Studies. Toronto University: Toronto.

Tierney, K. and Bruneau, M. (2007) Conceptualizing and measuring resilience: a key to disaster loss reduction. *TR News* **250**, May–June: 14–17.

Tobin, G.A. (1999) Sustainability and community resilience: the holy grail of hazards planning? *Environmental Hazards* **1**: 13–25.

Tobin, G.A. and Montz, B.E. (1997) *Natural Hazards: Explanation and Integration*. New York: The Guilford Press.

Twigg, J. (1998) Understanding vulnerability: an introduction. In *Understanding Vulnerability: South Asian Perspectives* (eds J. Twigg, J and M. Bhatt). Colombo: Intermediate Technology Publications, pp. 1–11.

Underwood, B.J. (1971) Recognition memory. In *Essays in Neobehavioralism* (eds H.H. Kendler and J.T. Spence). New York: Appleton-Century-Crofts, pp. 313–335.

UNDP (United Nations Development Program) (2004) *Reducing Disaster Risk: A Challenge for Development*. New York: Bureau for Crisis Prevention and Recovery.

UNDRO (United Nations Disaster Relief Organization) (1979) *Disaster Prevention and Mitigation*, Vol. 9. United Nations: New York.

UNISDR (United Nations International Strategy for Disaster Reduction) (2004) *Living with Risk: A Global Review of Disaster Reduction Initiatives*. Geneva: UN.

USDHS-ODP (United States Department of Homeland Security-Office for Domestic Preparedness) (2003a) State Assessment Handbook http:www.shsasresources.com (accessed August 21, 2008).

USDHS-ODP (United States Department of Homeland Security-Office for Domestic Preparedness) (2003b) Jurisdiction Assessment Handbook. http:www.shsasresources.com (accessed August 21, 2008).

Van Dissen, R. and McVerry, G. (1994) Earthquake hazard and risk in New Zealand. In *Proceedings of the Natural Hazards Management Workshop* (eds A.G. Hull and R. Coory). Lower Hutt, New Zealand: Institute of geological and Nuclear Sciences, pp. 67–71.

Varma, R. and Varma, D. (2005) The Bhopal disaster of 1984. *Bulletin of Science, Technology & Society* **25** (1): 37–45.

Walker, B.H. *et al.* (2006) Exploring resilience in social-ecological systems through comparative studies and theory development: introduction to the special issue. *Ecology and Society* **11** (1): 12.

Walsh, B. (2010) After the Destruction: What will it take to Rebuild Haiti? http://www.com/time/specials/packages/printout/0,29239,19533379_1953494_1954338.00.html# (accessed January 16, 2010).

Watts, M. and Bohle, H.G. (1993) The space of vulnerability: the causal structure of hunger and famine. *Progress in Human Geography* **17** (1): 47–67.

Weber, E.U. (2001) Decision and choice: risk, empirical studies. In *International Encyclopedia of the Social and Behavioral Sciences* (eds H.H. Kendler and N.J. Baltes). Oxford: Elsevier Science, pp. 11274–11276.

White, F. (1974) *Natural Hazards: Local, National, Global*. New York: Oxford University Press.

Whyte, A.V. (1982) Probabilities, consequences, and values in the perception of risk. In *Risk Assessment and Perception Symposium*. Toronto: Royal Society of Canada.

Wisner, B. (2003) The communities do science! Proactive and contextual assessment of capability and vulnerability in the face of hazards. In *Vulnerability: Disasters, Development and People* (eds G. Bankoff *et al.*). London: Earthscan.

Wisner, B. *et al.* (2004) *At Risk: Natural Hazards, People's Vulnerability and Disasters*. London: Routledge.

Witte, K. *et al.* (2001) *Effective Health Risk Messages*. London: Sage Publications.

WCED (World Commission on Environment and Development) (1987) *Our Common Future*. New York: Oxford University Press.

Zahran, S. *et al.* (2008) Social vulnerability and the natural and built environment: a model of flood casualties in Texas. *Disasters* **32** (4): 537–560.

Zhou, H. *et al.* (2010) Resilience to natural hazards: a geographic perspective. *Natural Hazards* **53** (1): 21–41.

4
Disaster Effects and Impacts

Environmental disasters cause loss of lives and injuries as well as damage to property. Because of their close association, the terms "disaster damage" and "losses" are frequently used interchangeably. Strictly speaking, however, there are distinctions between these two terms and these are discussed below in Section 4.1. The World Bank considers disaster effects to be the combination of total damage and total losses caused by an extreme event (Government of Bangladesh, 2008; Government of Yemen, 2009). Impacts, on the other hand, are the outcomes of the total effects of a disaster on the post-disaster physical, economic, social, health, and other environments. After presenting the key elements of disaster effects, including deaths and injuries, various impacts of disasters are discussed in some detail.

An assessment of damage and losses after disasters is essential to estimating the macro- and microeconomic impact of such events. This assessment provides a basis for defining the financial needs to achieve full recovery and reconstruction following any disaster, and formulating livelihood recovery programs. In addition, such an estimate can help in determining if public assistance is needed, and if so, what quantity and character of assistance is needed. Finally, an estimation of damage and losses is also needed to compare disaster effects with past events as well as among countries of the world (Mileti, 1999).

4.1 Disaster effects

4.1.1 Damage and losses

Damage is defined as the monetary value of fully or partially destroyed physical assets (Government of Bangladesh, 2008; Government of Yemen, 2009). It is assumed that damaged assets will be repaired or replaced to the same condition – in quantity and quality – that they had prior to the disaster. Physical assets are items

---

Environmental Hazards and Disasters: Contexts, Perspectives and Management, First Edition. Bimal Kanti Paul.
© 2011 John Wiley & Sons, Ltd. Published 2011 by John Wiley & Sons, Ltd.

of economic, commercial or exchange value that have a tangible or material existence. They can be inventories or stored, although they may go through depletion, depreciation, deterioration, or shrinkage in the storage process. Disaster damage occurs during the extreme event itself, and it is measured in physical units and valued at replacement cost. Disaster losses, on the other hand, refer to the goods and services that will not be forthcoming in the affected area until the destroyed assets are rebuilt or recovered over the span of time that elapses from the occurrence of the extreme event until the end of reconstruction and recovery period (Government of Yemen, 2009). Simply, losses mean changes in economic flows resulting from temporary absence of assets.

Disaster losses accrue from the time of the disaster until full economic recovery and reconstruction of assets are achieved. They are measured in monetary terms at current prices. There are three types of losses. Losses in productive sectors mean output will not be obtained and/or costs of production will be higher during the post-disaster period compared to the pre-disaster period. For example, disasters often fully or partially damage field crops. Thus, the output of crops will be less in areas affected by a disaster than the output of crops in areas not affected by the event. Depending on the cropping season, farmers might possibly re-cultivate crops during the post-disaster period, but at that time it may cost more because of increased labor cost and/or high seed prices. Service sectors will also incur loss of revenues and experience higher operational costs. Finally, provision of humanitarian assistance, demolition of damaged buildings and bridges, and debris removal will entail significant expenditures.

In the aftermath of major disasters, governments of affected countries generally request the World Bank and/or other international organizations such as the Asian Development Bank (ADB) and the United Nations (UN) to participate in damage, loss, and needs assessment (DLNA). These organizations use the Damage and Loss Assessment (DaLa) methodology, which was developed by the UN Economic Commission for Latin America and the Caribbean (UN-ECLAC) in the early 1970s (UN-ECLAC, 2003). This methodology has been continuously expanded and updated over the past three decades and in recent years has been strengthened, simplified, and customized for application in different regions of the world (Government of Yemen, 2009). DaLa bases the assessment of disaster impacts on the overall economy of the affected country as well as at the household and individual levels. It also provides a basis for estimating the negative impact on livelihoods, so that programs to expedite recovery may be designed and put into effect.

As noted in Chapter 2, beside DaLa, HAZUS and HAZUS-MH are used in the United States as powerful tools for analyzing potential first-order damage estimates from floods, hurricane, and earthquakes. HAZUS and HAZUS-MH require a large amount of data such as building stock, population, employment, household characteristics, and types of local economy. Some of these data may not always be readily available, even in developed countries. Also, both tools are quite limited in providing causal models that describe how consequences

Table 4.1 A comparison of damage and losses caused by two disasters

Disaster	Disaster effects (US$ million)		
	Damage	Loss	Total
Cyclone Sidr, Bangladesh	1158.0 (69.16)*	516.9 (30.84)	1674.9 (100.0)
Storm and floods, Yemen	874.8 (53.40)	762.9 (46.60)	1637.8 (100.00)

*Figures within parentheses represent the percentage of the total.
Sources: Compiled from Government of Bangladesh (2008); Government of Yemen (2009).

are likely to vary with changes in spatial extent and the intensity of a hazard event (French et al., 2010). Other loss estimation techniques are also available; however, their use largely depends on what types of losses one is interested in measuring (economic or environmental) and the level (individual, household, or national) of estimation (Hill and Cutter, 2001).

Between the two components, damage generally accounts for the larger share of disaster effects compared to losses. The World Disasters Report of the International Federation of Red Cross (IFRC) and Red Crescent Society (RCS) annually publish information on the total amount of estimated damage caused by disasters in millions of US dollars. This information corresponds to the damage value at the moment of the event and usually only to direct damage (e.g., damage to infrastructure, crops, and housing). According to the latest World Disasters Report, the world experienced average yearly damage of US$98.67 billion (2009 prices) during the last decade (2000–2009) (IFRC and RCS, 2010). Disaster damage shows marked yearly fluctuations. The total amount of damage (US$35.5 billion) reported in 2001 was the lowest of the decade. The year 2005 experienced the highest amount of disaster damage during the decade from 2000 to 2009 with US$237.4 billion followed by the preceding year, 2004, with US$154.6 billion (IFRC and RCS, 2010).

As noted, the World Disasters Report does not provide a breakdown of disaster effects by damage and losses. In order to gain insight regarding such a breakdown, two cases are presented below. Using the DaLa framework, the World Bank estimated the disaster effects of Cyclone Sidr and the 2008 storm and floods in Yemen. The former was a category IV cyclone, which occurred along coastal Bangladesh on November 15, 2007. Total damage and losses caused by Cyclone Sidr was estimated at US$1.7 billion, an amount equal to 2.8% of the country's gross domestic product (GDP) (Government of Bangladesh, 2008). The total value of the disaster effects caused by the October 2008 storm and floods in Yemen was estimated at US$1.6 billion (Table 4.1). This amount is equivalent to 6% of Yemen's 2008 GDP (Government of Yemen, 2009). Table 4.1 shows that damage accounted for slightly over 69% of disaster effects in the case of Sidr, while the corresponding percentage was slightly over 53% for Yemen's 2008 storms and floods. The somewhat equivalent distribution between damage

Table 4.2 Overall summary of damage and losses caused by Cyclone Sidr, Bangladesh

Sector/subsector	Disaster effects (US$ million)		
	Damage	Losses	Total
Infrastructure	1029.9	30.9	1060.8
Housing	839.3	–	839.3
Transport	116.0	25.0	141.0
Electricity	8.3	5.2	13.6
Water and sanitation	2.3	0.7	2.9
Urban and municipal	24.6	–	24.6
Water resource control	71.3	–	71.3
Social sectors	65.0	21.1	86.0
Health and nutrition	2.4	15.0	17.5
Education	62.5	6.0	68.5
Productive sectors	25.1	465.0	490.1
Agriculture	21.3	416.3	437.6
Industry	3.8	29.5	33.3
Commerce	–	18.2	18.2
Tourism	–	0.9	0.9
Cross-cutting issues	6.1	0.0	6.1
Environment	6.1	–	6.1
Total	1158.0	516.9	1674.9

Source: Government of Bangladesh (2008), xvii.

and losses reflects the relatively limited industrial development that exists in the affected area of Yemen (Government of Yemen, 2009).

Because the effects of disasters are not uniform across various sectors of economy, the World Bank customarily presents a breakdown of overall summary of disaster damage and losses by major sectors and subsectors of the economy. For example, Table 4.2 shows a summary of damage and losses due to Cyclone Sidr in Bangladesh. This table contains 13 subsectors, each of which belongs to one of four sectors of the economy. Depending on the nature of a national economy, two or more subsectors are often aggregated into one subsector. For example, the World Bank aggregated industry, commerce, and tourism into one subsector when presenting a summary of damage and losses caused by the October 2008 storms and floods in Yemen. Similarly, the classification of subsectors is not exclusive; for example, the housing subsector is included either under infrastructure or under the social sector.

While three of the four sectors listed in Table 4.2 are self-explanatory, cross-cutting sectors generally comprise environment and public buildings subsectors. The latter subsector is often included under the housing subsector of the infrastructure sector. Disasters also destroy the natural environment such as forest resources. For example, Cyclone Sidr totally destroyed almost one-third of the Sunderbans Forest, a World Heritage site (Paul, 2009). Comprising about 2316 square miles (6000 km^2), the Sundarbans is the world's largest mangrove forest. It covers southwestern coastal Bangladesh and the southeastern coast of West Bengal, India. Two-thirds of this forest lies within Bangladesh. The severe ecosystem disruption in the Sunderbans as a result of Cyclone Sidr included uprooted, broken, and twisted plants, and burnt foliage. The Forest Department estimated that about 30 000 acres of forest resources were severely affected and another 80 000 acres were partially affected (Government of Bangladesh, 2008). The total cost of damage and losses to the Sunderbans was assessed at about US$6.1 million (Table 4.2).

Among all the subsectors listed in Table 4.2, only the education subsector requires further explanation. An estimated 5927 educational institutions were fully or partially damaged by Cyclone Sidr, resulting in a total damage and losses of US$68.5 million. Thus, the effect on the education subsector is primarily expressed in a context of damage to buildings used for educational purposes. Similar to public buildings, educational buildings could be included under the housing subsector of the infrastructure sector (Table 4.2).

Table 4.2 clearly shows, however, that the housing subsector experienced the highest amount of damage among all subsectors included in the table. This subsector accounted for 72% of all damage caused by Cyclone Sidr or 50% of the total damage and losses. The number of homes destroyed by Sidr was almost fourfold more than the number of homes completely destroyed (reported at 400 000) by the 2004 Indian Ocean Tsunami (Paul, 2007). Cyclone Sidr left as many as 9 million people without shelter initially, and 3.5 million were without shelter over a significant period (Government of Bangladesh, 2008). Among the four sectors listed in Table 4.2, the infrastructure sector experienced the highest level of monetary damage. This sector accounted for nearly 89% all damage caused by Cyclone Sidr, or 63% of the total damage and losses. Most damage and losses attributable to Sidr were incurred in the private sector (Government of Bangladesh, 2008). This breakdown of damage and losses has significant implications for the strategies that must be adopted for efficient and effective recovery and reconstruction efforts.

Sector- and subsector-wide disaster damage and losses differ by the type of event as well as from one country to another because of differences in the importance of economic sectors to national economies. For example, the agriculture subsector and production sector are generally the most affected by floods. In contrast, the housing subsector and the infrastructure sector are most affected in total effects (damage and losses) by earthquakes. As noted, disaster effects differ by country. Damage and losses caused by the October 2008 storm and floods in Yemen were concentrated specifically in the agriculture subsector and general production sector. The agriculture subsector accounted for slightly over 63% of the total damage and losses incurred by the storm and floods. The production sector accounted for slightly over 76% of the total damage and losses (Government of Yemen, 2009). The share of the private sector of the total value of damage and losses was estimated at 75.4%, while the share of the public sector was 24.6% (Government of Yemen, 2009).

Among all subsectors, tourism was the most affected subsector by the 2004 Indian Ocean tsunami in Thailand. Economic losses suffered by the tourism industry exceeded an estimated US$2 billion. The impact of the tsunami on this industry included extensive damage to hotels, restaurants, and other services, as well as the loss of tourism income and productivity. In contrast to the tourism subsector, the total economic impact on livelihoods in agriculture, fisheries, and related industries was estimated to be US$350 million (asset loss and loss of productivity) (Zurick, 2011).

It is important to mention that disaster damage and losses also differ spatially, primarily because different regions are subject to different types of hazards and their frequencies also differ geographically. Although studies dealing with the spatial variability of disaster losses are limited, Thomas and Mitchell (2001) examined geographic variability in disaster losses for the United States between the 1975 and 1998 period. They claim that the central portion of the country bordering the Mississippi River and the Gulf Coast states incurred substantial damage. For the study period considered, California experienced largest total losses (US$40.1 billion) followed by Florida (US$33.0 billion) and Texas (US$22.0 billion). In California's case, a majority of the damage was from earthquakes and flooding, while the losses in Florida and Texas stemmed from hurricanes, floods, and tornadoes. With the exception of California, most western states experienced relatively little damage (Thomas and Mitchell, 2001).

Estimates of damage and losses for structures, built and physical environments, animals, and crops are prone to error. Disaster damage estimates are most accurate when trained damage assessors are involved. In the built environment, such assessors calculate the dollar value of both damage to structures and damage to contents within the structures. Early approximate estimates are obtained by conducting a "windshield survey" in which trained damage assessors visit the disaster impacted area and estimate the extent of damage that is visible from the street, or by conducting computer analyses using HAZUS (Adams and Huyck, 2005; Adams et al., 2009; Lindell and Prater, 2003). As noted earlier, the information needed for input to HAZUS is not always easily available.

4.1.2 Deaths

As noted, the Center for Research on the Epidemiology of Disasters (CRED) in Brussels has developed the Emergency Events Database (EM-DAT). This global database records disasters and deaths arising from them. Although there are some limitations to this database, it can provide useful indications on mortality

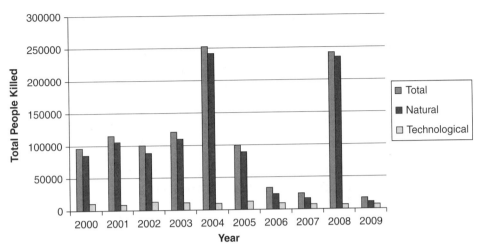

Figure 4.1 Total number of people reported killed by disasters, 2000–2009. IFRCS and RCS (2009, 2010).

rates from disasters at the country level. In addition, several national and cross-sectional studies (e.g., Borden and Cutter, 2008; Coates, 1999; Jonkman and Kelman, 2005; Nishikiori et al., 2006; Pradhan et al., 2007) analyzed death tolls either for all disasters or a specific disaster (e.g., floods, earthquakes, and tornadoes). These are reviewed along with event-specific (e.g., Hurricane Katrina and the Indian Ocean tsunami) studies in an effort to understand the extent, cause, timing, circumstance, and other salient factors associated with disaster-related mortality.

According to the CRED data, since 1960 disasters have killed nearly 5.5 million people worldwide. Droughts alone have killed over 2.5 million people, earthquakes over 1 million, and storms and hurricanes/cyclones nearly another million. The majority of these deaths have occurred in developing countries. Although the absolute number of people killed by all types of disasters is small relative to total deaths, many disaster-related deaths are preventable. Key methods of reducing deaths and injuries in the event of a natural disaster include adoption of preparedness and mitigation measures, including providing early disaster warnings and successful evacuation of at-risk populations. These measures have been successful in reducing disaster fatalities in the United States and other developed countries. For example, deaths from tornadoes in the United States have decreased dramatically over the past century (Sutter and Simmons, 2010). Brooks and Doswell (2002) estimate that the national tornado fatality rate fell from 1.8 per million in 1925 to 0.11 per million in 2000.

CRED data further suggest that the number of deaths caused by both natural and technological disasters has been declining over time. Yet, on average, 110 937 people died annually during the 2000–2009 period from these two types of disasters. Natural disasters accounted for slightly over 91% of all deaths caused by all types of disasters during this same period (IFRC and RCS, 2010). Figure 4.1 shows considerable year-to-year fluctuation in the number of deaths between 2000 and 2009. The figure also illustrates that in 2009 the number of people reported killed was the lowest of the decade, far below the decade average of 110 535. The highest number of deaths occurred in 2004. Although total fatalities caused by the 2004 Indian Ocean tsunami differ markedly by sources, this event alone accounted for more than two-thirds of the disaster fatalities in 2004. In 2008, the number of people that died was the second highest of the decade, with Cyclone Nargis in Myanmar killing 138 366 people (or 57% of all deaths) and the earthquake in Sichuan, China with 87 476 people dead or missing (36% of all deaths).

The total number of deaths caused by disasters also differs by disaster type. Among natural disasters, the number of deaths from floods was the lowest during the 2000–2009 period (IFRC and RCS, 2010). In this period 5380 people, on average, died annually from flood disasters. During this period the highest number of deaths was caused by earthquakes/tsunamis followed by droughts/food insecurity and windstorms. Earthquakes/tsunamis were responsible for an average of 45 355 deaths per year. The corresponding figures for droughts/food insecurity and windstorms were 23 018 and 17 233, respectively during the decade. Most reported deaths caused by droughts and food insecurity during the decade were attributable to the famine in the Democratic People's Republic of Korea, although the estimates provided are disputed (IFRC and RCS, 2010). It is important to mention that deaths after disasters from injuries are not generally considered deaths directly caused by such events.

Disaster deaths also significantly vary by level of economic development among countries of the world. For example, developing countries accounted for nearly 90% of all deaths caused by natural disasters during the 2000–2009 decade (IFRC and RCS, 2010). These countries experienced an average of 98 989 deaths per year, while the developed countries experienced some 11 546 fatalities per year between 2000 and 2009. The overwhelming majority of the disaster-related deaths were concentrated in Asian and African countries. This disparity, in terms of number of deaths from disasters, seems to be increasing over time because the average annual fatalities in developed countries has been declining, while the average number of deaths either remained the same or increased in developing countries during the past decade.

Irrespective of economic development level, disaster mortality significantly differs within a given country. For example, Borden and Cutter (2008) examined the spatial patterns of natural disaster mortality at the subnational level. Based on data for 1970–2004, they reported that the regions of the United States most prone to deaths from natural disasters are the South (FEMA regions IV and VI) and Intermountain West (FEMA region XIII). These regional patterns are due to the occurrence of various severe weather hazards such as floods and tornadoes. High risk of deaths in Intermountain West is likely a function of the small population size within the region (Borden and Cutter, 2008). But subregional county-level mortality patterns show even more variability. They observed a distinct urban/rural component to the county patterns as well as a coastal trend.

High disaster-related mortality tended to clusters in the lower Mississippi Valley, upper Great Plains, and Mountain West, with additional areas in west Texas, and the panhandle of Florida. In contrast to this, they found significant clusters of low disaster-related mortality in the Midwest and the urbanized Northeast (Borden and Cutter, 2008).

Using the same data set used by Borden and Cutter (2008), Thomas and Mitchell (2001) examined the state-level number of deaths caused by all disasters in the United States. They reported that Texas ranked top followed by Florida and Alabama in terms of total deaths from all disasters that occurred between 1975 and 1998. Texas alone accounted for 10% of all deaths, or more than five times the national fatality average and nearly twice the next highest state. Deaths in Texas were caused predominantly by three disaster types: floods, hurricanes, and tornadoes. California, New York, Tennessee, and Pennsylvania also stand out in the geographic distribution of disaster fatalities. Deaths in California were mostly caused by earthquakes, flooding, and winter-related storms. States with the lowest number of disaster-related deaths were in Nevada, Wyoming, and South Dakota (Thomas and Mitchell, 2001).

It is worthwhile mentioning that Borden and Cutter (2008) and Thomas and Mitchell (2001) used the Spatial Hazard Events and Losses Database for the United States (SHELDUS) as their data source. SHELDUS is a country-level hazard data set for the United States for 18 different natural hazards, such as thunderstorms, hurricanes, floods, wildfires, and tornadoes. For each event the database includes the beginning date, location (county and state), property losses, injuries, and fatalities that affected the county. These data are derived from several existing national sources such as National Climatic Data Center's monthly Storm Data publications. This database is problematic with respect to mortality associated with extreme heat events. The deaths from the 1993 Philadelphia heat wave were absent from the data set utilized by Borden and Cutter (2008) and Thomas and Mitchell (2001). Adding this single event would change the regional mortality patterns reported in these two studies (Massimo et al., 2006; Thacker et al., 2008).

However, the magnitude of disasters alone does not determine the total number of deaths from extreme natural events. For example, Cyclone Sidr, a category IV storm, killed 3406 people in Bangladesh in 2007. Despite being similar in severity, Sidr claimed far fewer lives than Cyclone Gorky, another category IV storm, which struck Bangladesh on April 29, 1991 and killed an estimated 140 000 people (Paul, 2009). International media, donor countries, and foreign aid agencies claim that the relatively low number of fatalities as well as lower than expected damage caused by Sidr was the result of the Bangladesh government's attempt to provide early cyclone warning and successful evacuation of people living on the projected path of the cyclone (Blake, 2008).

Bangladesh also experienced a devastating (unnamed) cyclone on November 12, 1970, which killed as many as 500 000 people (Paul et al., 2010). This category III cyclone is considered the deadliest cyclone in Bangladesh history and one of the worst natural disasters – in terms of human fatalities – in world history. After this disaster, the Bangladesh government introduced a cyclone warning system, the absence of which likely contributed to the large number of deaths caused by the 1970 cyclone.

Cyclone Nargis in Myanmar in 2008 killed at least 138 000 people (Fritz *et al.*, 2009). This large number of deaths has been blamed on the Myanmar military government for not issuing a cyclone warning. The IFRC and RCS (2009), however, claim that the devastation caused by Nargis was not due to a failure in early warning service – cyclone warnings were provided by the Myanmar Meteorological Service – but to a failure in other elements of hazard warning, especially those involving communications and preparedness to act.

Cyclone Nargis was highly unusual in the sense that it did not follow "normal" cyclone tracks. It hit from the west and forced water up the Irrawaddy Delta, flooding a vast, low-lying area very quickly. Even if the warnings had reached people of the delta, it was such an unusual event that many residents probably would not have believed or been prepared to act on the warnings. Furthermore, the Myanmar government's capacity to evacuate the area would likely have been severely limited by poor roads and infrastructure (IFRC and RCS, 2009).

Based on analysis of tornado data collected for the 1996–2007 period, Sutter and Simmons (2010) estimated that the relative probability of deaths from an F3 was 0.44 compared to 0.19 for F4 and 0.13 for F5 tornadoes. This pattern holds primarily because the frequency of F3 tornadoes is much higher than either F4 or F5 tornadoes. This illustrates that both frequency and magnitude should be considered, particularly for analysis of disaster-induced mortality data.

The number of deaths caused by disasters also differs according to timing of their occurrence and the distance of population concentrations from the sources of extreme events, particularly earthquakes and tsunamis. For example, tornadoes cause more deaths if they occur during the night when people are asleep compared to their occurrence during the day (Sutter and Simmons, 2010). Distance from the hazard source also determines who are more likely to die from such events. For example, the 2004 Indian Ocean tsunami caused the deaths of 186 019 people in 12 countries bordering the India Ocean (UN, 2006). However, only four affected countries – India, Indonesia, Sri Lanka, and Thailand – accounted for 99.75% of all deaths caused by this extreme natural event. The greatest number of people (129 775) died in Indonesia's Aceh province, located close to the epicenter of the 9.0 magnitude earthquake that was responsible for the tsunami. This tsunami killed 35 322 in Sri Lanka, 12 405 in India, and 8 212 in Thailand (Table 4.3). Like any disaster-induced fatalities, the numbers of deaths reported by various organizations do not correspond to each other (Table 4.3).

Relief agencies report that children account for at least a third of all deaths caused by the 2004 Indian Ocean tsunami. As many as four times more women than men were killed in some tsunami-impacted areas. Many women were waiting on the beach for their husbands to return from the sea with their catch. Such women traditionally sort the catch and take it to market. Other women were

Table 4.3 Death toll caused by the 2004 Indian Ocean tsunami by source and country

Country	Source	
	Washington Post	UN
Indonesia	131 338	129 775
Sri Lanka	31 229	35 322
India	10 749	12 405
Thailand	5 395	8 212
Somalia	290	78
Myanmar	90	61
Maldives	82	82
Malaysia	68	69
Tanzania	10	13
Bangladesh	2	–
Kenya	1	–
Seychelles	2	2
Total	216 858	186 019

Sources: Washington Post (2005) and the UN (2006).

looking after their children at home located close to the beach (Paul, 2007). It has long been established that a tsunami is no great danger for those out at sea. In the case of the Indian Ocean tsunami, the tectonic plates clashed several miles below the ocean surface, producing barely perceptible mid-ocean waves.

Oxfam, the British relief and development agency, conducted a survey a few months after the 2004 tsunami. This survey shows that only 189 of 676 survivors were female in the four villages in the Aceh Besar Province, Indonesia. This is a ratio of about 1:3. In the worst affected village, Kuala Cangkoy, for every male who died, there were four females who perished in the tsunami. In the city of Cuddalore in Tamil Nadu, India, almost three times as many women were killed as men – 391 females died compared to 146 males. In Pachaankuppam village, every single person that died was female (Gautham, 2006).

A considerable proportion of the elderly were also victims of the 2004 Indian Ocean tsunami. It was estimated that across the four hardest-hit countries – India, Indonesia, Sri Lanka, and Thailand – people over 60 accounted for almost 14% of all deaths, and nearly 93% of all displaced. In addition to the large number of local residents, at least 9000 foreign tourists (mostly European) were among the dead or missing. The European nation that suffered the most casualties was Sweden, with 428 fatalities and 116 still missing in early 2005. Other foreign countries whose citizens were dead and/or injured by the disaster included Australia, Britain, France, Germany, Japan, Nigeria, Norway, South Africa, South Korea, Tanzania, and the United States.

On October 8, 2005, a 7.6 magnitude earthquake struck the North West Frontier Province (NWFP) of Pakistan, Pakistan-administered Kashmir, and parts of India and Afghanistan. The epicenter of the earthquake was Muzaffarabad, the capital of the Pakistani-controlled portion of Kashmir. As a result, an estimated 87 350 people lost their lives in Pakistan, while this earthquake killed 1360 people in India-administered Kashmir and three people in Afghanistan. Between 50 and 60% of all fatalities in Pakistan were children. This earthquake occurred in the early morning, when many children were at school. According to UNICEF, this earthquake destroyed 10 000 schools, killed an estimated 17 000 students and over 15 000 teachers at these schools (Ozerdem, 2006).

Natural disasters also cause deaths to livestock, although the CRED does not provide this information. To gain some understanding regarding the extent of this problem, livestock deaths caused by Bangladesh's Cyclone Sidr are considered here. This event killed 242 000 livestock. In monetary terms, this accounted for 4.41% of the total damage and losses of the agriculture subsector, which includes crops, livestock, and fisheries (Government of Bangladesh, 2008). Livestock is an important economic subsector for the rural landless, as well as marginal and small farmers. Raising livestock not only provides employment, but also generates regular income from the sale of milk, egg, poultry, goats, and cattle. Most of the damage in the livestock subsector was caused by tidal surges that drowned animals and birds and by falling trees (Government of Bangladesh, 2008).

Cattle are one of the worst victims of severe winter storms or blizzards. In a blizzard, cattle try to face away from the wind and move with the storm (Cotton and Ackerman, 2007). They also herd together, creating a windbreak. Snow often piles up on cattle, eventually covering and suffocating them. Blizzards also create a number of health problems for the cattle, including hypothermia, frostbite, and trauma (CEAH, 2002). Health problems intensify for those cattle that stay in open fields in bitter cold without feed and water for days. This causes malnutrition and undernutrition, which reduces cattle weight and causes economic losses (Paul *et al.*, 2007). The wet, cold conditions in muddy corrals can also lead to conditions such as frozen feet/footrot and pneumonia. Such health problems increase veterinary expenses for ranchers. Cows that are in advanced stages of their pregnancy during blizzards often experience spontaneous abortions and still births (CEAH, 2002). Parts of five states in the Great Plains regions of the United States experienced severe winter storms in late December 2006. It is estimated that these blizzards caused slightly over 13 deaths per 1000 head of cattle (Paul *et al.*, 2007).

4.1.2.1 Disaster fatalities: causes and correlates

Understanding what causes disaster-related deaths is important from the perspective of designing measures to reduce these fatalities. Causes of death are disaster-specific. For example, Jonkman and Kelman (2005) reported that two-thirds of deaths from flooding are from drowning, one-third are from physical trauma, heart attack, electrocution, carbon monoxide poisoning, or fire. These researchers analyzed flood disaster deaths from 13 floods that occurred in six countries of Europe and North America between 1989 and 2002. Vehicle-related drowning occurs most frequently in developed countries, mainly when people try to drive across flooded bridges, roads or streams. However, Jonkman and Kelman (2005) also observed differences between flood deaths in Europe and the United States. The most striking difference to emerge from their research is that drowning in vehicles seems to be a more significant problem in the United States than in Europe. In the United States most flood deaths and damage are caused by flash floods, which likely accounts for most of the above difference between Europe and the United States in terms vehicle drowning. However, there are higher recurrence rates of floods that pose a danger to vehicles in the United States. Differing road networks (for instance, more low-water crossings than in Europe) are associated with the higher incidence of vehicle-related drowning in the United States (Jonkman and Kelman, 2005).

Causes of flood deaths also seem to differ between developed and developing countries. For example, available studies (e.g., Kuni *et al.*, 2002; Siddique *et al.*, 1991) suggest that the majority of flood deaths in Bangladesh are caused by waterborne diseases, drowning, and snake-bites. Other causes include respiratory diseases, pneumonia, and hypothemia. Diarrhea-related deaths are primarily caused by a lack of pure drinking water, ways the drinking water is stored and handled, poor hygiene practices, and the often partial and/or total deterioration of sewage and sanitation facilities which contaminate sources of drinking water in flood-affected areas (Kuni *et al.*, 2002; Ahern *et al.*, 2005).

In the case of hurricanes or tropical cyclones, nine out of 10 victims die by drowning. Hurricanes/cyclones generally occur with storm surges, which quickly flood low-lying coastal areas with anywhere from 3 feet (1 m) typical for a category 1 storm to over 19 feet (6 m) of surge for a category 5 storm. Hundreds of thousands of deaths in countries such as Bangladesh have been caused by storm surge associated with tropical cyclones (Rosenberg, 2010). Falling trees and being hit by flying debris also cause fatalities from cyclones and other wind-related disasters such as thunderstorms and tornadoes (Schmidlin, 2009). Sutter and Simmons (2010) reported that the probability of a tornado fatality in mobile homes is 10–15 times higher than for permanent homes in the United States. Between 1985 and 2007, 43.2% of all US tornado fatalities occurred in mobile homes, which comprised only 7.6% of all US housing units in 2000.

For earthquakes, the majority of deaths are caused by the collapse of buildings. Some building-related deaths may be due to asphyxia, as a result of dust released during the collapse. Earthquake-related deaths can also result from fire or stress associated with the event, the latter of which may result in deaths due to heart attacks or cardiac arrest (Noji, 1997). Similarly, blizzards may cause heart attacks and/or complicate childbirth. In fact, heart attacks suffered while shoveling

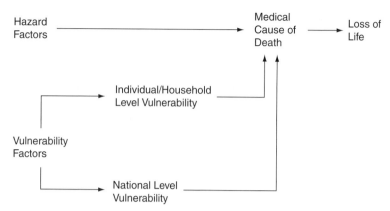

Figure 4.2 Hazard and vulnerability factors that results in loss of life due to a specific medical cause. *Source:* Modified after Jonkman and Kelman (2005).

snow are the number one cause of death during a blizzard. Blizzards also often kill people by causing traffic accidents (Paul et al., 2007). People, particularly older adults, young children, and persons with chronic medical conditions, also die resulting from exposure to extreme heat events.

Disaster literature clearly suggests that the risk of death from an extreme event is influenced by several physical characteristics of the event, such as its magnitude (intensity), scale (spatial extent), duration, and frequency (Hofer and Messerli, 2006; Mirza et al., 2001; Tobin and Montz, 1997). These characteristics are considered among hazard researchers to be directly associated with disaster mortality. For example, mortality is thought to increase with increasing flood frequency, magnitude, and duration. However, these characteristics are neither simple in themselves nor should they stand alone. For instance, a flood that inundates a large area may not cause any fatalities because of one or more of the following reasons: its short duration, the shallowness of floodwaters, and low or zero population present and exposed. In contrast, when flood duration and magnitude are substantial, deaths can increase significantly (Hofer and Messerli, 2006).

Disaster literature further reveals that deaths caused by disasters not only depend on the physical characteristics of such events, but are also a function of complex social, economic, demographic, political, and cultural factors. Jonkman and Kelman (2005) broadly group these factors as hazard and vulnerability factors. They maintain that a combination of both factors result in deaths due to a specific medical cause (Figure 4.2). Vulnerability factors exist at the household/individual and/or the national level. For example, at the household level, overall disaster mortality rates are higher among poor relative to more affluent households. Similarly, developing countries experience higher mortality rates than do developed countries. At the individual level, disaster mortality tends to decrease with higher educational level (Nishikiori et al., 2006). Because of the absence of relevant studies, the following discussion of the correlates of deaths are presented primarily by hazard types, however, this form of presentation should help in understanding the complex relationship between disasters and the deaths they cause.

As noted earlier, disaster deaths are influenced by the prevailing socio-economic and demographic characteristics of affected individuals or households. Disasters with the largest mortality impacts have usually occurred where the population at risk has very limited economic resources (Ahern et al., 2005). In examining the causal relationship between socio-economic conditions and flood-related mortality in Nepal, Pradhan et al. (2007) reported that those living in a thatched house had a 5.1-fold higher risk of experiencing a disaster-related death than those living in a cement/brick home. Further, these researchers found a wood or tin domicile was associated with 1.6 times increased risk of fatality relative to cement/brick dwellers.

Hazard researchers (e.g., Coates, 1999; Kuni et al., 2002; Jonkman and Kelman, 2005; Nishikiori et al., 2006; Pradhan et al., 2007) claim that children and the elderly are more vulnerable to deaths, particularly from flooding, than are adults. Those that die from drowning in floods are largely children and elderly individuals. Both also are at an elevated risk of mortality or morbidity from diarrhea, cholera, and respiratory symptoms (e.g., coughing, sneezing, wheezing, and sore throat) relative to adults. Similarly, the disabled experience a greater risk of death from disasters compared to able-bodied persons. Noji (1997) reports that in earthquakes people over 60 years of age, children, and the chronically ill are at greater risk of death compared to other population groups. Armenian et al. (1997) found that persons over 60 had a death rate twice that of persons below 60 in the December 1988 earthquake in Armenia – an earthquake that killed 25 000 people and injured 130 000.

Much empirical data has been produced to uphold the argument that mortality rates as a result of natural disasters are higher for women than men. Fothergill (1998), in a synthesis of over 100 papers on natural disasters, concluded that women are more likely to die from such events, particularly women in developing countries than are men. Similarly, Neumayer and Plumper (2007), in a meta-analysis of 141 countries over the period from 1981 to 2002, argue that natural disasters lower the life expectancy of women more than that of men. However, Jonkman and Kelman (2005) reported that the death rate in floods and hurricanes is higher for men than women in the United States and Europe.

Fothergill (1998) also claim that in the Bangladesh cyclone of 1991, 42% more females died than males. Using data collected from the survivors of the 1991 cyclone, Ikeda (1995) reported that the female age-cohort of 20–49 years had a mortality rate averaging four to five times higher than that of the average mortality rate for males. Sommer and Mosley (1972) suggest that similar sex- and age-group mortality patterns were observed during the devastating Bangladesh cyclone of 1970. Studies have also found that not only do more women perish from natural disasters in Bangladesh and elsewhere, but that the particular female cohort of 20–49 years appear to be the most vulnerable (Neumayer and Plumper, 2007); also that the poorer the women and/or country the more drastic this variance in gender-based mortality rates (Ahsan and Khatun, 2004); and that the stronger (i.e., greater magnitude) the disaster the larger the difference in mortality rates (Neumayer and Plumper, 2007).

In addition to individual or household level factors, national characteristics, such as the political system, level of economic development, and geography, play an important role in explaining fatalities caused by floods and other natural disasters. Using time series and cross-sectional data, Kahn (2003, 2005) empirically tested the impact of democracy, political and social institutions, and governance on the number of deaths caused by natural and anthropogenic disasters. Kahn's (2003) first study reported that one percentage higher score on the Polity Index (an index for democracy) is associated with 13% less deaths. It also shows that a one point increase in this index has the same death reduction effect as if the country's GNP per capita were US$2430 higher. In other words, a low-income country can usually reduce its fatalities if a well-functioning democracy is present and effective in mitigating the adverse effects of disasters.

Kahn's (2005) second study supports findings reported in his first study. He claims that although richer nations do not experience fewer natural disasters than poorer nations, the former countries usually do suffer less death from disasters (also see Albala-Bertrand, 1993; Toya and Skidmore, 2006). This is likely because governments of poor countries spend little or nothing on disaster prevention, mitigation, and preparedness measures. In addition, unlike richer nations, governments of poor countries are not able to provide high-quality emergency care to help protect the population against secondary disaster-related deaths, which occur after an extreme event. A 7.0 magnitude earthquake in Haiti killed 230 000 people in 2010. On the first anniversary of the earthquake the death toll from the quake was reported by Haiti government to be more than 316,000. A considerable number of these additional deaths were caused during the post-disaster period primarily because of the poor response on the part of the Haiti government (CBS News, 2011).

In contrast, more developed countries provide resources, create elaborate plans, and generally prepare themselves to meet the disaster eventuality – so casualties are reduced. While such measures do not lessen the probability that a disaster may occur, they definitely lessen its impact. A nation's geographic characteristics, such as land area, elevation above sea level, distance from the equator, and population density, are also associated with cross-national patterns in fatalities caused by natural disasters (Kahn, 2003, 2005).

Similar to Kahn (2003, 2005), Sen (1981) also claims that democracy in a country is a strong predictor of deaths due to natural disasters. Freedom of the media, in addition to the power and standing of political parties, is of considerable importance in reducing suffering from disasters. Democracy ensures people's participation in government and its accountability, assuming that the government takes proactive steps to mitigate the impacts of disasters when they do occur. Sen (1981) argues that in order to win elections, politicians experience strong pressure to propose measures to avert or lessen the impacts of disasters. Thus non-democratic countries are more susceptible to damages and deaths resulting from natural disasters than countries with a functioning democracy (Sen, 1981).

Irrespective of party affiliation, most politicians, particularly in democratic countries, respond to disaster emergencies as quickly as possible and mobilize financial resources and volunteers for their constituencies (Salkowe and Chakraborty, 2009). In contrast to this, in non-democratic countries, politicians are usually less enthusiastic about becoming involved with emergency response activities. As a consequence, such countries experience more deaths and suffer relatively more damage from natural disasters (see also Albala-Bertrand, 1993; Cohen and Werker, 2004).

4.1.3 Injuries

The CRED database suggests that the number of injuries experienced in disasters has been approximately twice the number of disaster-related deaths over the past 20 years. Injuries are defined as those requiring medical attention. It seems likely that disaster-related injuries are underreported, particularly in developing countries where medical services are not easily available. Unfortunately, the CRED database does not provide information about the nature and type of injuries suffered in disasters, nor the incidence of illness following a disaster. However, studies dealing with injuries caused by various disasters can provide insights regarding this topic. The objectives of this subsection are to present information on the extent of human injuries caused by natural disasters and report on the circumstances, type, and causes of injuries, along with risk factors associated with those injuries.

It needs to be emphasized that the extent of injuries depends on the disaster type. Flooding generally causes very few injuries, while tornadoes and earthquakes cause a significantly higher number of injuries compared with other natural disasters. However, available studies show that the death-to-injury ratios differ country to country as well as by disaster. For example, Bangladesh's Cyclone Sidr caused the deaths of 3406 people and injured more than 55 000 (Government of Bangladesh, 2008). These figures indicate the number of injuries is slightly over 16 times the number of deaths caused by this event. As noted, Bangladesh also experienced a category IV cyclone in 1991. The official death toll and injuries for this cyclone were reported as 138 868 and 460 000, respectively; providing a death-to-injury ratio of 1:3.31 (Haque, 1997). The 2004 Indian Ocean tsunami both killed and injured almost the same number and thus the death-to-injury ratio was nearly 1:1 (Paul, 2007). Hurricane Katrina killed about 2000 people and injured 5698, a death-to-injury ratio of 1:2.9 or almost three times the fatalities.

In analyzing the data collected from 13 coastal villages impacted by Cyclone Sidr, Paul (2010) claims that cyclone-related injuries were caused primarily by falling trees and wind-blown debris. Injuries caused by falling trees occurred in three different ways. First, a considerable number of trees fell on houses, causing their collapse. Second, many trees and/or branches broke due to the strong winds of this cyclone. Many people who tied themselves to these wind-ravaged trees also sustained injuries. Finally, some people fell from trees while seeking shelter there and sustained injuries. Paul (2010) further reported that 55% of all injuries associated with Cyclone Sidr were caused by falling trees and the remaining 45% were caused by wind-blown debris, particularly from corrugated tin used as roofing material in rural Bangladesh.

The proportion of injuries that occurred in houses accounted for nearly 69% of all injuries, while the remaining 31% occurred outdoors. In addition to structural collapse caused by falling trees, the walls, ceilings, and/or roofs of many houses were blown away by wind. People who took refuge in damaged houses usually experienced injuries, both minor and severe (Paul, 2010). Almost all body parts (e.g., feet, legs, arms, chests, thighs, eyes, head, back, and ribs) were injured. The top three cyclone-related injuries were blunt trauma, lacerations (from debris), and puncture wounds, with 60% of these injuries being confined to the feet, legs, and lower extremities. Less than 10% of all injuries caused by Cyclone Sidr can be considered serious (Paul, 2010).

Surprisingly, no Cyclone Sidr-related indirect injuries were reported by respondents of Paul's (2010) study. Indirect injuries are defined as those caused before or after occurrence of an extreme event. For example, people often receive injuries during evacuation prior to a cyclone's landfall or during cleanup of debris in the cyclone's aftermath. This contrasts with the situation observed in developed countries, particularly in the United States (e.g., Brenner and Noji, 1995; Brown *et al.*, 2002). Prior to a hurricane's landfall, injuries such as falls, blunt trauma, lacerations, and muscle strains generally occur because prospective victims often install plywood and metal shutters to make their homes and businesses stronger in order to withstand the physical force of the storm (Shultz *et al.*, 2005). Injuries caused during disaster aftermath cleanup efforts include puncture wounds (e.g., stepping on a nail) and muscle strains (e.g., from lifting heavy items) (Brown *et al.*, 2002). Injuries are also reported in developed countries during mass evacuations, primarily from motor vehicle accidents.

In their study of tornado-related deaths and injuries in Oklahoma due to the May 3, 1999 tornadoes, Brown *et al.* (2002) reported that 78% of all injuries were related to getting into storm shelters. They further reported that nonfatal injuries most commonly occurred when persons fell down stairs going into a storm shelter or basement; when the door of a storm shelter was closed on their body or head; and when persons fell down or ran into something as they were running to a shelter. Most of the injured persons suffered soft-tissue injuries (e.g., cuts, scrapes, and bruises) followed by fractures and/or dislocations. The most common locations of fractured bones were the upper arms and lower legs, followed by the chest/ribs, face, back, and neck (Brown et al., 2002).

Available studies (e.g., Brown et al., 2002; Schmidlin, 2009; Schmidlin and King, 1995; Sutter and Simmons, 2010) on tornado fatalities and injuries in the United States have reported that higher rates of deaths and serious injury are associated with being inside mobile homes, public buildings, and apartment complexes. Being outdoors or in motor vehicles has also often been associated with a higher risk of death or serious injury. Studies (e.g., Schmidlin and King, 1995) have found a 10-fold increased risk in both death and injury associated with being in cars and as a result of cars being lifted or rolled. The same is also true for deaths and injuries caused by high winds displacing and/or damaging mobile homes. Studies (e.g., Brown et al., 2002) also reported that the most frequent causes of tornado-induced injuries in the United States are caused by unspecified flying/falling debris, being picked up and blown by the tornado, and flying/falling wood or boards. In addition, many injuries result from concrete and bricks, nails, and screws. The most serious injuries frequently suffered in earthquake include crush injuries, fractures, and internal hemorrhaging. Noji (1997) reported that skull fractures with intracranial hemorrhage, cervical spine injuries with neurological impairment and internal injuries to the lung, liver, and spleen were the most common injuries among those hospitalized in the 1968 Iran earthquake (also see Ramirez and Peek-Asa, 2005).

A gender bias is evident with respect to injuries associated with natural disasters. Existing literature suggests that more women are injured from hurricanes/cyclones and tornadoes than are men (e.g., Brown et al., 2002; Paul, 2010). No conclusive evidence has been found regarding the nature of any association between age and injuries. However, Paul (2010) reported that the injury rate (13.84%) was highest among persons 50 years or older, followed by the 15–49 age group (11.38%). Surprisingly, the injury rate is lowest (4%) among children (0–14 age group). This is an unexpected finding and one which is difficult to explain. Possibly Cyclone Sidr survivors took extra precautions to save their children's lives. In addition, household income and educational level are considered important determinants of disaster injuries (e.g., Paul, 2010).

There remain problems with measuring physical impacts. In some cases it is difficult to determine how many deaths and injuries are actually caused by a disaster. In estimating the number of disaster casualties, deaths and injuries are often rounded to the nearest thousand or even to the nearest ten thousand (Noji, 1997). Disaster-related impacts may be a contributing factor of deaths for people with pre-existing health issues or conditions. Moreover, some deaths are an indirect consequence of a disaster event. For example, some fatalities may arise from structural fires following an earthquake.

Further research on disaster mortality and injuries are needed to broaden our knowledge regarding various aspects of disaster-induced injuries. Similarly, few studies have been conducted on the geographic variability of hazard-specific or event-specific mortalities and their determinants. In fact, only recently have hazard researchers become interested in deaths associated with different disasters types. Therefore, there is a need to fill research gaps in both areas (i.e., disaster-induced deaths and injuries).

4.2 Disaster impacts

Natural disasters produce a wide range of impacts at different levels. According to Lindell and Prater (2003), assessing these impacts, particularly at the community level, is important for three reasons: (1) impact assessments can be used by community leaders in order to make a more informed decision as to how much (if any) external assistance may be needed, (2) impact assessments can target specific sectors of the community to determine the impact on disproportionately affected certain people or businesses, and (3) impact assessments arising from previous events can be used as projection tools to better determine the effects of similar disasters on particular communities.

Impacts of natural disasters are often difficult to trace and measure accurately. This is particularly true for economic impacts because disasters affect both formal and informal economies, and tracing and measuring the effects on the latter is problematic at best. For this reason, before discussing the various impacts of disasters, a typology of disaster impacts is first presented. This typology differentiates between direct and indirect impacts, tangible and intangible impacts, and primary, secondary, and tertiary impacts. Since this typology is equally applicable for disaster damage and losses, no distinction is made among disaster impacts, damage, or losses.

Direct impacts/damage/losses are those caused by the physical contact of the extreme natural event with humans and/or with property. These impacts are attributable to the destruction of buildings, machinery, or public infrastructure. Employment losses traceable to the destruction of a workplace are considered a direct impact (Mileti, 1999; Pan *et al.*, 2008). These first-order impacts also include deaths and injuries caused directly by physical contact between humans and the extreme natural event. Indirect impacts, in contrast, are those caused by the consequences of physical contact of disasters with people and/or their property. These second-order impacts are less easily connected to the event and may manifest much later (Smith, 1992). When a cow or bull is killed by a natural disaster, this is a direct loss or impact, but when the income from cattle product sales is lost, this is an indirect loss or impact. An accurate accounting of all indirect losses or impacts is normally quite difficult.

Based on measurability, both direct and indirect impacts are often subdivided into "tangible" or "intangible" impacts (Figure 4.3). Tangible impacts are described as those that can be measured in monetary terms, such as the damage to a dwelling. In general, only tangible losses are included in the estimation of future events and the reporting of past events (Coppola, 2007). By contrast, intangible impacts are those either that cannot be expressed in financial terms

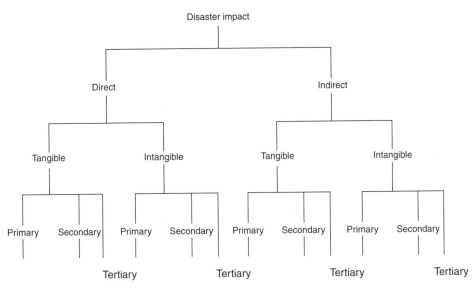

Figure 4.3 The impacts of natural disasters. *Source:* Modified after Smith and Ward (1998), p. 35.

(e.g., the loss of an archaeological site through erosion caused by flash flooding, and disaster-induced stress, fear, anxiety, pain, discomfort, and inconvenience) or for which monetary estimates are considered undesirable and unacceptable, such as loss of human life (Tobin and Montz, 1997). This latter consideration is the primary reason that deaths and injuries are assessed as a separate category from the cost measurements of disaster effects.

Although difficult, assigning a monetary value to the life of an individual human being is routinely utilized in most developed countries in order to assess the value of risk-reducing measures and to provide compensation to next-of-kin in the event of loss of life. In assigning such a value in cost–benefit analysis for hazard mitigation, tradeoffs between monetary wealth and fatal safety risks are summarized in the value of a statistical life (VSL). In 2008, the mean VSL in the United States was estimated at US$7 million (Kniesner *et al.*, 2010; Viscusi *et al.*, 2010), but can range up to US$22 million, depending on the specific type of study employed. This value is also used for quantifying loss of human life due to natural disasters. FEMA (1994) developed protocols to monetize injuries and deaths. Similarly, attempts have been made to quantify intangible disaster impacts based on some percentage of direct tangible losses. A wide range of estimates can be found in the literature, from a low of 15% to a high of 75% (Viscusi *et al.*, 2010). Chambers (1975) even suggested that intangibles might prove to be twice the tangible figure (also see Kniesner *et al.*, 2010).

Direct tangible impacts arise from damage to structures, buildings and their contents, as well as infrastructure. Indirect tangible impacts include lost production, lost wages, or time off for dealing with disaster impacts. Intangible direct impacts take a while to sort out, while intangible indirect impacts may last a relatively long time and also have repercussions on family life. These impacts include health costs, not those of injuries accruing during and immediately after the event, but rather long-term stress and depression that may be latent (Norris et al., 2010).

Parker et al. (1997) and Smith and Ward (1998) have further placed tangible and intangible impacts into primary and secondary impact categories (Figure 4.3). Primary impacts are considered to be the "first-round" impacts associated with a disaster and constitute the immediate losses resulting from the event itself. Typically, these impacts lead to further impacts, which are termed secondary impacts, or even tertiary impacts. For example, crop damage is a primary impact of a disaster, whereas the food shortage that may follow is a secondary impact. Famine caused because of local food shortage can be considered a tertiary impact. Secondary and tertiary impacts are long-term impacts that are set off as a result of a primary event, and are often termed "knock-on" effects. These include things such as the loss of habitat or permanent changes in the position of river channel caused by a flood. Secondary impacts are mainly intangible, while tertiary impacts are largely tangible.

4.2.1 Types of disaster impacts

Disaster impacts are classified in many different ways. One way to broadly classify such impacts is by making distinctions between physical and social impacts. As indicated, physical impacts include property damage and casualties (deaths and injuries). Such impacts are usually the most obvious and easily measured. Social impacts can develop over a long period of time and can be difficult to quantify, let alone observe when they occur. Study of social impacts is necessary because they can redirect the character of social institutions, result in permanent new and costly regulations for future generations, change ecosystems, and even disrupt the stability of political regimes (Mileti, 1999). In addition, a better understanding of the social impacts of disasters can provide a basis for pre-impact prediction and the development of contingency plans to prevent and/or minimize adverse consequences from future extreme events. Social impacts are often disaggregated into demographic, economic, political, institutional, psychological, and health impacts. In this section, some of these impacts are discussed independently.

4.2.1.1 Social impacts

Disasters impact a number of social units (e.g., individuals, households, and businesses) and each one of these units suffer disruption of their normal function for the either short or for the long term, or in some cases both. There are many examples of social impacts which affect people of different age, ethnicity, occupation, and gender. For example, disasters often completely destroy educational buildings. As a consequence, students may suffer academically. Students from areas impacted by Hurricane Katrina had to attend schools located in non-impacted areas. Many of these displaced students seemed to have lost their ability to concentrate on assignments and manifested symptoms of clinical depression (Picou and Marshall, 2007). In the spring of 2006, less than 33% of all public schools had reopened in Orleans, Plaquemines, and St. Bernard Parishes in Louisiana. By that time only 66% of all public schools had reopened in the New Orleans metropolitan area, and more than half of the grocery stores, restaurants, convenience stores and other retail food establishments remained closed for business.

Irrespective of personal characteristics, disasters disrupt the lives of virtually all people in the disaster-impacted area. Hurricane Katrina resulted in the massive displacement of over 348 000 K-12 students and their families throughout Louisiana, Mississippi, and Alabama (Picou and Marshall, 2007). This event not only separated people from their friends and relatives, but also separated a considerable number of them from their pets. People displaced by Hurricane Katrina also suffered both physical and mental distress. Government officials estimated that over half a million Katrina victims required mental health assistance to deal with higher rates of anxiety, depression, and anger. Disasters also frequently render people homeless and jobless, thus increasing their discomfort and suffering. In developing countries, poor people who have lost their homes to a disaster usually live under open sky for some time. Still others may live in severely damaged homes for extended periods of time. Such disruptions and discomforts are very difficult if not impossible to measure.

Increases in divorce, suicide, and crime rates, violence against women, and increased consumption of alcohol are some of additional examples of other social impacts of natural disasters. Utilizing rigorous epidemiological and statistical methods, Krug et al. (1998) concluded that suicide rates in the United States increased in the four years after floods by 13.8%, in the two years after hurricanes by 31.0%, and in the first year after earthquakes by 62.9%. These increases were attributable to post-traumatic stress disorder and depression in the wake of disasters. It is also widely believed in developed countries that crime and divorce rates increase in the aftermath of natural disasters. Alexander (1998) urges caution in accepting these findings. Several studies (e.g., Drabek, 1986; Siegel et al., 1999) reported an increased incidence in pro-social behavior such as donating emergency aid and a decreased incidence of antisocial behavior such as crimes. In the context of developing countries, available studies (e.g., Fisher, 2005; Miller, 2005) clearly suggest that alcohol consumption, substance abuse, and antisocial behavior increased among men in the aftermath of the 2004 Indian Ocean tsunami in India and Sri Lanka. Violence against women will be discussed in Chapter 8.

4.2.1.2 Economic impacts

Property damage caused by disaster impacts often has severe economic consequences. This damage is expressed in asset values that can be measured by the cost of repair or replacement. Measuring property damage is usually not difficult because most such damages are direct economic losses. This is not always true, however, particularly for uninsured damaged property. For insured property, the insurers record the amount of the deductible and reimbursed loss, but uninsured losses are not recorded (Lindell and Prater, 2003). The number of working days lost or the volume of production lost due to disasters are also examples of economic losses. The value of lost production is relatively easy to calculate, while the lost opportunities, lost competitiveness, and damage to reputation are much more difficult (Coppola, 2007).

Economic impacts are wide ranging – they affect financial ability and condition of individuals, households, the impacted community and/or region, or even a nation. For example, the overall economic impact of Hurricane Katrina was estimated to be about US$150 billion, making it the costliest natural disaster in the history of the United States. Major factors that contributed to such an extensive economic impact were reductions in oil supply, food export, tourism, and other forms of trade business. The Gulf Coast contributes about 10% of the nation's oil supply and was severely disrupted due to impacts associated with Hurricane Katrina.

As noted, economic impacts of disasters are both direct and indirect. Examples of the former include cost of repair or replacement of damaged or destroyed structures, loss of agriculture/inventory, cleanup costs, and loss of income and rental costs. Among many indirect impacts include reduction in business and personal spending. This impact often lasts longer than direct impacts. For example, damage to infrastructure continues to disrupt business activities for a relatively long time relative to damage to a specific business or even a specific retail sector (e.g., first food restaurants). When only fisheries revenues are considered, the projected indirect loss from the the 2010 Deepwater Horizon Oil Spill in the Gulf Coast of the United States is estimated at US$74, US$47, and US$22 millions in 2011, 2012, and 2013, respectively (IEM, 2010).

4.2.1.3 Demographic impacts

Out-migration of disaster survivors is one of the important demographic impacts of such extreme events. Natural disasters may trigger outward migratory flows for many reasons: lack of access to employment or other income sources, material assistance, safety and security, non-availability of natural resources, poverty, and previous migratory experience (International Organization for Migration, 2007). Migration flows as a result of natural disasters are often categorized as "distress migrations" or "forced migrations." The degree of damage to property and livelihood assets, the sufficiency and quality of the aid response and the extent of poverty largely determine the volume of these migratory flows. Paul (2005, 2003, 1998) found that in situations where the distribution of disaster aid was equitable, appropriate, and without irregularities, people do not usually move from disaster affected areas (similar to US measures taken to prevent mass out-migration in the post-disaster period). Instead, people from areas not affected by disasters often migrate to affected areas with the hope of receiving emergency assistance.

In his study of 2004 tornado survivors in north-central Bangladesh, Paul (2005) reported that nearly one-third of respondents stated they suspected that people from distant places came to tornado-affected areas in the hope of receiving disaster relief. Some claimed unsuccessfully that they were residents of a tornado-affected village and told emergency relief providers that they had lost their home. Respondents reported 15 such false claims by outsiders.

Two characteristics of disaster-induced migration deserve emphasis. First, such migration in most cases results in internal migration, and second, it usually causes only temporary displacements. If permanent migration is the result of a disaster, it may likely be a reflection of the impacted area authority's deficient response rather than the result of any disaster impact (Raleigh *et al.*, 2008). The most popular destinations for most rural disaster migrants in developing countries are large cities and such migrants contribute to the rapid growth of squatter settlements in destination cities (Paul, 2005). Numerous studies support this statement. For instance, after the major earthquakes in the Indian state of Gujarat in 2001 and in El Salvador, many poor rural victims migrated to cities like Ahmedabad, Bhuij, and San Salvador in search of work (International Organization for Migration, 2007).

Other destinations of disaster migrants are neighboring states and foreign countries. For example, the 2000 drought in Orissa, India caused the migration of nearly 60 000 people to the neighboring state of Andhra Pradesh. Hurricane Mitch in 1998 led many survivors in Latin America to migrate to relatives already living abroad, capitalizing on pre-established transnational networks (International Organization for Migration, 2007). In addition, natural disasters tend to increase the risk of sex trafficking of women and children. As economic opportunities and social support mechanisms become stretched or completely disintegrate, women and children may easily fall victim to trafficking traps in the aftermath of disasters. It is difficult, however, to estimate the volume of such trafficking.

Disasters, particularly high magnitude ones, often alter age and sex composition of the affected area and/or communities. As noted, women, children, and the elderly experience a higher risk of death and injuries from natural disasters than do men and adults. This has several demographic consequences. Large disasters such as the 2004 Indian Ocean tsunami made many children orphans. These orphans need special protection both immediately and continuing into the medium and even possibly long term. Without social and public support, disaster orphans may become victims of trafficking. It was reported that many orphans from Honduras and El Salvador were forced into brothels in Guatemala subsequent to Hurricane Mitch in 1998 (International Organization for Migration, 2007). MacDonald (2005) reported that after the 2004 tsunami, women of impacted coastal areas in India were pressured into marrying earlier than in the recent past and into having more children closer together; such changes have significant implications for their education, livelihoods, and reproductive health.

4.2.1.4 Health impacts

The potential for natural disasters to cause direct and indirect harms to human health is great. Direct health impacts of the disasters include deaths and injuries. These impacts were discussed earlier in this chapter. The indirect health impacts of natural disasters include potential for an increase in communicable, waterborne, and other diseases such as hepatitis and malaria as well as pneumonia, eye infections, and skin diseases. These health issues pose a significant threat to the lives and well-being of disaster survivors. Deaths often occur from communicable and other diseases after a disaster and for this reason these indirect health impacts are often referred to as the "second wave of death and destruction." The occurrence of communicable and other diseases are disaster- as well as country-specific.

Most of the communicable and waterborne diseases that occur, particularly in the aftermath of floods and hurricanes/cyclones in developing countries, are caused by a severe shortage of clean drinking water, nonhygienic living conditions, and lack of food. Disaster survivors in developing countries generally live in damp, dirty, and cramped conditions in their homes and/or temporary shelters. Such conditions facilitate the spread of numerous adverse health conditions from person to person within the household (Tapsell, 2009).

Post-disaster human health is also closely associated with changes in the balance of the natural environment. For example, flooding caused by overflow of river banks and/or by storm surges alters the balance of the natural environment and ecology, allowing vectors of disease and bacteria to flourish. Outbreaks of cholera and a higher incidence of malaria can result from such alterations. Noji (2005) maintains that an increase in disease transmission and the risk of epidemics in the post-flood period depends on population density and displacement and the extent to which the natural environment has been altered or disrupted. However, the Health Protection Agency (HPA) states that infection from flooding is rare in the UK as pathogens get diluted and provide low risk. There was no evidence of increased outbreaks of illness following the 2007 floods in the United Kingdom (HPA, 2008).

Illnesses are also caused by other indirect impacts of natural disasters such as damaged infrastructure, population displacement, and reduced food production as well as the release of contaminants (e.g., from storage and waster disposal

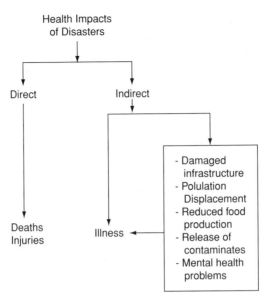

Figure 4.4 Health impacts of disasters. *Source:* Modified after Paul *et al.* (2011).

sites) into the water and air of disaster-impacted areas (Figure 4.4). In the context of health impacts, damaged infrastructure refers to all types of health care facilities such as hospitals, medical clinics, and ambulatory services, including the electricity supply on which most of these facilities depend. When either complete or partial damage to such facilities is caused by severe natural disasters, it is difficult to provide necessary care to the ill and injured. Lack of proper medical attention may also result from the absence of physicians and other health personnel as well as an insufficient supply of appropriate medicine. These indirect impacts not only prolong suffering, they also increase the probability of death from illnesses and/or injuries (Paul *et al.*, 2011).

Damage to other infrastructure, such as roads and bridges, may also impact health outcomes. Such damage can cause serious delays (or may even prevent) the provision of both emergency and regular medical supplies and personnel for treating injuries or for controlling disease outbreaks (Kuni *et al.*, 2002). Restricted access may also prevent the initiation of emergency immunization programs and other health interventions that may be required subsequent to a disaster. Release of contaminants poses serious public health risks, including cancer, for survivors of several natural disasters such as floods. Floodwaters can mix with raw sewage and thus dramatically increase the incidence of waterborne diseases. Although the release of toxic chemicals is diluted by floodwater, causing toxicity levels to decline, uncontrolled release of various chemicals – some of which may interact with each other – poses a considerable public health risk. A serious threat to public health manifests if waste storage facilities or industrial plants are flooded. In addition, sediments of flood-affected areas are often contaminated with toxic substances, such as arsenic, lead, diesel fuel, and polycyclic aromatic hydrocarbons (PAHs), to name but a few. Contaminated floodwaters that overflow into residential areas may cause long-term health effects on humans and animals, and may also result in pollution of groundwater reserves, which is a major source for drinking water (Godsil *et al.*, 2009).

Wildfire smoke, which pollutes the air, is especially harmful to persons of all ages with underlying health conditions such as asthma, emphysema, and cardiovascular diseases. Smoke can aggravate pre-existing heart disease that may result in symptoms such as chest pain. In healthy individuals, wildfire smoke usually causes irritation of the eyes, nose, and throat, and/or breathing discomfort or difficulty. Wildfire smoke contains more than 90% of carbon dioxide and water, but hundreds of other chemicals (e.g., carbon monoxide, mercury, sulfur dioxide, PAHs, and nitrogen oxides) can also be present. Some of the chemicals present in smoke are carcinogenic. Prolonged exposure to wildfire smoke causes a considerable risk for both the physical and mental health of those exposed (Curtis and Mills, 2009).

Another indirect longer term health impact of natural disasters is associated with mental health (Figure 4.4). In those areas affected by extreme events, the related trauma tends to have a rather lengthy impact on the population's well-being, both directly and indirectly. Direct consequences may be observed in the form of lifetime disabilities. Indirect outcomes manifest in society through individual breakdowns that lead to stress-related illness, such as depression, sleep disorders, and substance abuse. In addition, disasters may exacerbate existing stress or contribute to acute stress – a condition that can lead to chronic illness and mortality if not properly addressed (Curtis and Mills, 2009). Stress also exacerbates many chronic diseases such as diabetes, pregnancy, heart conditions, even obesity.

As indicated, disaster survivors cope with stress associated with such events in many different ways. These may be reflected through increased alcohol and drug use, increased spousal abuse, and even adverse pregnancy outcomes such as pre-term and low birth weight (LBW) babies (Buekens *et al.*, 2006; Cordero, 1993; Curtis and Mills, 2009). All these tend to increase, particularly in developing countries, after a disaster as a result of post-traumatic stress disorder and depression in the wake of death, destruction, and illness that usually accompany such disasters.

Psychological health impacts of natural disasters do not uniformly affect all segments of a population (Lindell and Prater, 2003). Children, the elderly, people with pre-existing mental illness, racial and ethnic minorities, and families of those who have died in the disaster are more affected than other segments of population. Emergency workers also often suffer from psychological problems because they work long hours even without breaks and witness horrific sights. Studies (e.g., Lindell and Prater, 2003) report that only a very small proportion of the disaster-affected population suffer from post-disaster psychological problems.

There are other impacts of natural disasters. Such events often force governments of affected countries to change policies, shift institutional and organizational arrangements, and impose new regulations (Mileti, 1999). One notable institutional effect of the 1993 Midwest floods in the United States was the buyout and relocation program for flood-prone communities throughout the country. Although buyout and relocation were not new mitigation measures for flood disasters, never before had these been promulgated on such a large scale. This program is seen as part of a more comprehensive and sound approach to floodplain management. This significant change in federal policy and programs was the result of the tremendous losses caused by the 1993 Midwest floods.

There are many other political and institutional implications of natural disasters, some of which are mentioned in the subsequent chapters, including environmental impacts. Natural disasters often become a root cause for political disintegration. For example, the November 1970 cyclone totally devastated the life and living of the coastal region of East Pakistan, which was at that time one of two provinces of Pakistan. The response of the Central Government of Pakistan was inadequate to meet the vast demand for rehabilitation. The people were deeply critical of this neglect. In the December 1970 general election of Pakistan the Awami League – the largest political party, headed by the great nationalist leader of East Pakistan Sheik Mujibur Rahman – won a landslide victory (167 of 169 seats) in East Pakistan with a healthy majority in the 313-seat national assembly to form a national government. Unfortunately, democracy was subverted. The President of Pakistan, General Yahiya Khan, decided not to hand over the power to the majority leader, Sheikh Mujibur Rahman to form a national government; instead he cracked down on the political unrest. After a civil war with West Pakistan that lasted for nine months East Pakistan became the independent country of Bangladesh in December 1971.

Thus, natural disasters, including terrorist attacks, can lead to political turmoil at times and have changed the direction of policy, such as the creation of the Department of Homeland Security in the United States after 9/11. Disasters have also a bearing on taxes, insurance rates, and many other aspects of life. Needless to say, disagreement may often arise regarding disaster recovery policies, which often lead to division and conflict between individuals and groups impacted by a disaster. This issue is discussed in Chapter 6.

4.2.2 Beyond negative impacts

Natural disasters also produce positive impacts, which are not always acknowledged in disaster literature. This literature emphasizes negative impacts of extreme events for several reasons (Few, 2003). Smith and Ward (1998) maintain that positive impacts of natural disasters are ignored because they are less understood and are more difficult to assess than losses. Moreover, the mass media play an important role in creating a negative perception of natural disasters. Images of destruction caused by disasters cover the pages of newspapers and television screens. People rarely hear about the success stories of communities and individuals that make their situations better by constructing a more sound and disaster-resistant infrastructure, developing new mitigation measures to minimize destruction, and creating new coping strategies that enhance the ability to recover after a disaster occurs.

Although several authors (e.g., Few, 2003) take pains to point out that natural disasters are by no means phenomena of solely negative consequences, little is known regarding the benefits and opportunities associated with natural disasters, particularly for disasters other than floods. The objective of this subsection is to highlight the positive impacts of natural disasters. Mileti (1999) maintains that disaster-affected areas often benefit from outside assistance, which pours into such areas and regions (e.g., insurance payouts and federal aid). That is, new capital is brought into the impacted community for reconstruction and rehabilitation. Mileti (1999) further points to a sizable regional economic expansion the year after an event (as with the Northridge earthquakes, Hurricane Andrew, and Hurricane Hugo). Although disasters have negative impacts on gross national product (GNP), this economic indicator does count all reconstruction costs as gains (Mileti, 1999).

Reconstruction, in fact, decreases, to a degree, future hazard risk by destroying weak and old buildings, strengthening new buildings, and preventing rebuilding in hazard-prone areas. Many structures are rebuilt, or replaced after disasters, providing an opportunity to modernize communities affected by disasters as well as enhancing their aesthetic beauty. For example, the downtown area of Charleston, North Carolina was severely impacted by Hurricane Hugo in 1989. This natural disaster spurred improvements in its historic downtown district. The hurricane completely destroyed a lot of older buildings in the downtown area which needed to rebuild and this event provided that opportunity (Loose and Sottili, 2005). Similarly, P'Rayan (2005) reported that the 2004 Indian Ocean tsunami was a blessing in disguise for the administrators of the Indian city of Chennai because it was looking for an opportunity to move a long-lived slum area called Thideer Nagar to a different location.

Parker *et al.* (1997) maintain that the reconstruction phase of disasters presents a host of new opportunities for governments, communities, and individuals to avoid re-creating the conditions that led to disaster in the first place. They stress that once a settlement is established in a hazard-prone area, this "window-of-opportunity" rarely presents itself until a disaster of considerable magnitude occurs, permitting a fresh start. Similarly, other hazard researchers (e.g., Alexander, 2000; Cuny, 1983) believe that the occurrence of a disaster represents a good opportunity to reassess development strategies in order to build disaster mitigation into them.

As noted, almost all natural disasters have some positive impacts. However, with the exception of floods, these impacts have received little attention among hazard researchers. Unlike in the Western world, a "flood" is not always uniformly viewed negatively nor considered a natural disaster in the non-Western world (Paul, 1997). The people of Bangladesh, for example, perceive flooding as both a resource and a hazard (e.g., Rasid and Paul, 1987). A normal flood (*barsha*) resulting from usual monsoon rainfall is considered a resource. It is beneficial in the sense that it makes the land productive by providing necessary moisture and fresh fertile silt to the soil (Brammer, 1990a). Moreover, fish caught during the flood season constitute the main source of protein for many. In fact, millions of Bangladeshis depend on fisheries in the floodplains for their livelihood and the life-cycle of some key fish species depends on the ability to migrate between the rivers and the seasonal floodplains (Paul, 1997). Thus the routine and expected annual floods are not perceived in Bangladesh as disasters.

An abnormal flood (*bonna*) occurs in Bangladesh once every few years and results from excessive rainfall. This is regarded as an undesirable and damaging phenomenon (Paul, 1985). It causes widespread damage to standing crops and properties and costs human and animal lives. However, by analyzing crop production statistics in Bangladesh, Brammer (1990b) concludes that despite considerable crop damage, there are compensatory increases in rice production in areas not affected by the floods, while affected areas experience significantly increased production in the dry season following an abnormal flood. He attributes this increase in rice production to the availability of extra moisture in the year of high floods as well as deposits of fertile silt on floodplains.

This does not only apply in Bangladesh; the economies of many flood-prone countries are highly dependent on the natural recurrence of flooding. Floods of any kind rejuvenate floodplain vegetation and are important drivers of many ecological processes there. Although floods are often destructive, redevelopment on the flat land of the floodplain is much easier than in hilly or mountainous regions (Smith and Ward, 1998). In regions of the world where public water supplies are not abundantly available, floods provide water for domestic use and irrigation. Standing water in ponds and lakes may help to recharge aquifers that are later tapped by wells for water supply (Few, 2003).

On the large plains of semi-arid West Africa, seasonal flooding determines the amount of land cultivated in a given year. Since this region experiences a dry period, greater seasonal flooding provides increased supplies of grazing for cattle, sheep, and goats, enhances the amount of fish available, and ensures a plentiful rice crop. Floodwaters are extremely important to the bamboo and log business of the Philippines; annual floods are used to float these two items to downstream markets that would otherwise be difficult to reach. As noted, floods trigger environmental processes which benefit native plants, fish and animals, often with consequent human economic benefits. Wetland vegetation provides important resting, feeding, and nesting areas for many waterfowl species – usually a direct result of rivers being flyways for migrating birds (Smith and Ward, 1998).

Similar to floods in the developing world, hurricanes in the United States have provided an economic boost in many communities/cities along the Gulf and Atlantic coasts. According to a report published in the *USA Today* (Hagenbaugh, 2004), economists expected that the 2004 Hurricane Charley would boost the economy of Florida in the next few years as the rebuilding effort added dollars and workers to an area that was already bustling. They estimated that the hurricane added more than 1 percentage point to economic growth in the state in the fourth quarter, bringing total growth to 7%. They further believed that this gain would continue into 2005. Despite a long record of hurricane damage, Pilkey and Coburn (2004) reported that beachfront-property owners of the Florida and Alabama coast built new, costly buildings that replaced damaged ones. These owners consistently rebuild after each hurricane because federal, state, and local governments always provide assistance in the event of another hurricane.

Immediately after disasters, particularly tornadoes, hurricanes, and earthquakes, the influx of emergency personnel, first responders, claim adjusters, and roofers provide an economic boom in the affected communities. These newcomers spend untold sums at hotels, restaurants, and stores, bringing unexpected revenues to the impacted community/town. In addition, purchase of substantial amounts of building and repair materials increases dramatically after disasters. This often causes increase in sale tax revenues of the impacted communities. Some sectors of economy, such as the construction industry, generally experience economic growth after disaster events (Ewing *et al.*, 2003; French *et al.*, 2010). Drawing upon empirical data from 89 countries, Skidmore and Toya (2002) claim that climatic disasters are positively correlated with economic growth, whereas geologic disasters are negatively associated with economic growth.

In a different way, the 2004 tsunami worked as a catalyst for addressing many of the pre-existing issues of the affected people of Southern India. Both state governments and NGOs not only addressed the needs that were created by the tsunami but also addressed pre-existing needs of those affected. A few examples of these pre-existing and pre-disaster issues were: insufficient housing, lack of financial stability, health-related issues, absence of land rights, ineffective political structures of the villages, dismal women's rights, and wide gender gap issues.

Another positive impact of the tsunami was that the agriculture bounced back better than expected in impacted areas of Indonesia. Some 40 000 families returned to the "land" after rains flushed out salinity that they feared would reduce soil fertility. Yet, some of these families are boasting record yields due to the nutritional value of silt deposited by flooding associated with the tsunami (Noticias.info, 2006).

Many countries of the world have made some progress in saving lives as well as reducing injuries and property damage from some natural disasters. However, measuring the physical damage and losses in monetary term is difficult primarily for two reasons. First, because of the nature of damage/loss, it is often difficult to assign dollar values, particularly for indirect and intangible damage and losses. In addition, loss data are not readily available, particularly in developing countries. Whatever data are available, much of it is incomplete, fragmentary in nature, and not of high quality, which severely restricts its use. No systematic damage and loss data are available by location or by specific hazard. However, quality of data has been improving over time and many organizations at subnational, national, and international levels have been involved in collecting reliable data on disaster damage and losses.

References

Adams, B.J. and Huyck, C.K. (2005) The emerging role of remote sensing technology in emergency management. In *Infrastructure Risk Management Processes: Natural, Accidental, and Deliberate Hazards*. Reston, VA: American Society of Civil Engineers, pp. 95–117.

Adams, B.J. et al. (2009) Post-tsunami urban damage assessment in Thailand, using optical satellite imagery and the VIEWSTM field reconnaissance system. *Geotechnical, Geological, and Earthquake Engineering* 7: 523–539.

Ahern, M. et al. (2005) Global health impacts of floods: epidemiologic evidence. *Epidemiological Review* 27 (1): 36–46.

Ahsan, R.M. and Khatun, H. (2004) *Disaster and the Silent Gender: Contemporary Studies in Geography*. Dhaka: The Bangladesh Geographical Society.

Albala-Bertrand, J. (1993) *Political Economy of Large Natural Disasters*. New York: Oxford University Press.

Alexander, D. (1998) Does a link really exist? *Natural Hazards Observer* May.

Alexander, D. (2000) *Confronting Catastrophe*. New York: Oxford University Press.

Armenian, H.K. et al. (1997) Deaths and injuries due to the earthquake in Armenia: a cohort approach. *International Journal of Epidemiology* 26 (4): 806–813.

Blake, G. (2008) The gathering storm. *OnEarth* 30 (2): 22–37.

Borden, K.A. and Cutter, S.L. (2008) Spatial patterns of natural hazards mortality in the United States. *International Journal of Health Geographics* 7 (64): 1–13.

Brammer, H. (1990a) Floods in Bangladesh: I. Geographical background to the 1987 and 1988 floods. *The Geographical Journal* 156: 12–22.

Brammer, H. (1990b) Floods in Bangladesh: II. Flood mitigation and environmental aspects. *The Geographical Journal* 156: 158–165.

Brenner, S.A. and Noji, E.K. (1995) Tornadoes injuries related to housing in the Plainfield Tornado. *International Journal of Epidemiology* 24 (3): 133–151.

Brooks, H.E. and Doswell, C.A. III. (2002) Deaths in the 3 May 1999 Oklahoma City tornado from a historical perspective. *Weather Forecast* 17: 354–361.

Brown, S. et al. (2002) Tornado-related deaths and injuries in Oklahoma due to the 3 May tornadoes. *Weather Forecast* 17: 343–353.

Buekens, P. et al. (2006) Hurricanes and pregnancy. *Birth Issues and Prenatal Care* 33 (2): 91–93.

CBS News. (2011) Haiti quake death toll on anniversary. January 12.

CEAH (Center for Epidemiology and Animal Health) (2002) *Animal Health Hazards of Concern During Natural Disaster*. Washington, D.C.: CEAH.

Chambers, D.N. (1975) Procedures for determining the design flood in engineering works. *Proceedings of the Institute of Civil Engineers* 58: 723–726.

Coates, L. (1999) Flood fatalities in Australia, 1788–1996. *Australian Geographer* 30 (3): 391–408.

Cohen, C. and Werker, E. (2004) Towards an understanding of the root causes of forced migration: the political economy of "natural" disasters. The Inter-University Committee on International Migration. Working Paper No. 25.

Coppola, D.P. (2007) *Introduction to International Disaster Management*. Boston: Elsevier.

Cordero, J.F. (1993) The epidemiology of disasters and adverse reproductive outcomes: lessons learned. *Environmental Health Perspectives* 100 (S 2): S131–S136.

Cotton, S. and Ackerman, R. (2007) *Caring for Livestock during Disaster*. Fort Collins: Colorado State University Cooperative Extension.

Cuny, F.C. (1983) *Disasters and Development*. New York: Oxford University Press.

Curtis, A. and Mills, W. (2009) *GIS, Human Geography, and Disasters*. San Diego, CA: University Readers.

Drabek, T.S. (1986) *Human System Responses to Disaster: An Inventory of Sociological Findings*. New York: Springer.

Ewing, B.T. *et al.* (2003) A comparison of employment growth and stability before and after the Fort Worth tornado. *Environmental Hazards* 5 (3–4): 83–91.

FEMA (Federal Emergency Management Agency) (1994) *Seismic Rehabilitation of Federal Buildings: A Benefit/Cost Model*. Washington, D.C.: FEMA.

Few. R. (2003) Flooding vulnerability and coping strategies: local responses to a global threat. *Progress in Development Studies* 3 (1): 43–58.

Fisher, S. (2005) Gender based violence in Sri Lanka in the after-math of the 2004 tsunami crisis: the role of international organizations and international NGOs in prevention and response to gender based violence. A Master Thesis. University of Leeds, Leeds.

Fothergill, A. (1998) The neglect of gender in disaster work: an overview of the literature. In *The Gendered Terrain of Disaster: Through Women's Eyes* (eds E. Enarson and B.H. Morrow). Westport, CN: Praeger Publishers: pp. 22–33.

Fritz, H.M. *et al.* (2009) Cyclone Nargis storm surge in Myanmar. *Nature Geoscience* 2 (7): 448–449.

French, S.P. *et al.* (2010) Estimating the social and economic consequences of natural hazards: fiscal impact example. *Natural Hazards Review* 11: 49–57.

Gautham, S. (2006) Teach the girl to swim tsunami, survival and the gender dimension. http://www.countercurrents.org/gen-gautham/00806.htm (accessed February 2006).

Godsil, R. *et al.* (2009) Contaminants in the air and soil in New Orleans after the flood. In *Race, Place, and Environmental Justice after Hurricane Katrina* (eds R.D. Bullard and B. Wright). Boulder, CO: Westview Press, pp. 115–138.

Government of Bangladesh (2008) *Cyclone Sidr in Bangladesh: Damage, Loss and Needs Assessment for Disaster Recovery and Reconstruction*. Dhaka: Government of Bangladesh.

Government of Yemen (2009) *Damage, Losses and Need Assessment: October 2008 Tropical Storm and Floods, Hadramout and Al-Mahara, Republic of Yemen*. Sana: Government of Yemen.

Hagenbaugh, B. (2004) Storm of rebuilding follows Hurricane Charley. *USA Today* 23, November.

Haque, C.E. (1997) Atmospheric hazards preparedness in Bangladesh: a study of warning, adjustments and recovery from the April 1991 cyclone. *Natural Hazards* 16 (2–3): 181–202.

HPA (Health Protection Agency) (2008) Health Advice: General Information Following Floods. http://www.hpa.org.uk/web/HPAwebFile/HPAweb_C/119447339369 (accessed July 28, 2008).

Hill, A.A. and Cutter, S.L. (2001) Methods for determining disaster proneness. In *American Hazardscapes: The Regionalization of Hazards and Disasters* (ed. S.L. Cutter). Washington, D.C.: Joseph Henry Press, pp. 13–36.

Hofer, T. and Messerli, B. (2006) *Floods in Bangladesh: History, Dynamics and Rethinking the Role of the Himalayas*. Tokyo: United Nations University Press.

IEM (2010) *A Study of the Economic Impact of the Deepwater Horizon Oil Spill*. New Orleans: IEM.

Ikeda, K. (1995) Gender differences in human loss and vulnerability in natural disasters: a case study from Bangladesh. *Indian Journal of Gender Studies* 2 (2): 171–193.

IFRC and RCS (International Federation of Red Cross Society and Red Crescent Society) (2009) *World Disaster Report: Focus on Early Warning, Early Action*. Geneva: IFRC and RCS.

IFRC and RCS (2010) *World Disaster Report: Focus on Urban Risk*. Geneva: IFRC and RCS.

International Organization for Migration (2007) *Migration, Development and Natural Disasters: Insights from the Indian Ocean Tsunami*. Geneva: International Organization for Migration.

Jonkman, S.N. and Kelman, I. (2005) An analysis of the causes and circumstances of flood disasters deaths. *Disasters* 29 (1): 75–97.

Kahn, M.E. (2003) The death toll from natural disasters: the role of income, geography, and institutions. Mimeo, Tufts University.

Kahn, M.E. (2005) The death toll from natural disasters: the role of income, geography, and institutions. *The Review of Economics and Statistics* 87 (2): 271–284.

Kniesner, T.J. et al. (2010) Policy relevant heterogeneity in the value of statistical life: new evidence from panel data quantile regressions. *Journal of Risk and Uncertainty* 40: 15–31.

Krug, E.G., et al. (1998) Suicide after natural disasters. *New England Journal of Medicine* 338: 373–378.

Kuni, O. et al. (2002) The impact on health and risk factors of the diarrhoea epidemics in the 1998 Bangladesh floods. *Public Health* 116: 68–74.

Lindell, M.K. and Prater, C.S. (2003) Assessing community impacts of natural disasters. *Natural Hazards Review* 4 (4): 176–185.

Loose, C. and Sottili, C. (2005) Cancellation policies, refunds and more. *Washington Post*, 10 September.

MacDonald, R. (2005) How women were affected by the tsunami: a perspective from Oxfam. *PLoS Medicine* 2 (6): 474–475.

Massimo, S. et al. (2006) Vulnerability to heat-related mortality: a multicity, population-based, case-crossover analysis. *Epidemiology* 17 (3): 315–323.

Mileti, D.S. (1999) *Disasters by Design: A Reassessment of Natural Hazards in the United States*. Washington, D.C.: Joseph Henry Press.

Miller, G. (2005) Poor countries, added perils for women. *Science* 308: 1576–1577.

Mirza, M. M. Q. et al. (2001) Are floods getting worse in the Ganges, Brahmaputra and Meghna Basins? *Environmental Hazards* 3: 37–48.

Neumayer, E. and Plumper, T. (2007) The gendered nature of disasters: the impact of catastrophic events on the gender gap in life expectancy, 1981–2002. *Annals of the Association of American Geographers* 97 (3): 551–566.

Nishikiori, N. *et al.* (2006) Who died as a result of the tsunami? – risk factors of mortality among internally displaced persons in Sri-Lanka: a retrospective cohort analysis. *BMC Public Health* **20** March.

Noji, E.K. (1997) Earthquake. In *The Public Health Consequences of Disasters* (ed. E.K. Noji) Oxford: Oxford University Press.

Noji, E.K. (2005) Public health issues in disasters. *Critical Care Medicine* **33** (1): S29–S33.

Norris, F.H. *et al.* (2010) Prevalence and consequences of disaster-related illness and injury from Hurricane Ike. *Rehabilitation Psychology* **55** (3): 221–230.

Noticias.info. (2006) Aceh's Recovery Stumbled But Now Gaining Momentum: Report. http://www.noticias.info/asp/aspComunicados.asp?nid=128713@src=0" (accessed February 21, 2006).

Ozerdem, A. (2006) The mountain tsunami: afterthoughts on the Kashmir earthquake. *The Third World Quarterly* **27** (3): 397–419.

Pan, Q. *et al.* (2008) Economic impacts of terrorist attacks and natural disasters: case studies of Los Angeles and Houston. *Geospatial Technologies and Homeland Security* **94**: 35–64.

Parker, D. *et al.* (1997) Reducing vulnerability following flood disasters: issues and practices. In *Reconstruction after Disaster: Issues and Practices* (ed. A. Awotona). Aldershot: Ashgate, pp. 23–44.

Paul, B.K. (1985) Perception of and agricultural adjustment to floods in Jamuna Floodplain, Bangladesh. *Human Ecology* **12** (1): 1–19.

Paul, B.K. (1997) Flood research in Bangladesh in retrospect and prospect: a review. *Geoforum* **28** (2): 121–131.

Paul, B.K. (1998) Coping with the 1996 tornado in Tangail, Bangladesh: an analysis of field data. *The Professional Geographer* **50**: 287–301.

Paul, B.K. (2003) Relief assistance to 1998 flood victims: a comparison of performance between the government and NGOs. *The Geographical Journal* **169**: 75–89.

Paul, B.K. (2005) Evidence against disaster-induced migration: The 2004 tornado in North-Central Bangladesh. *Disaster* **29** (4): 370–385.

Paul, B.K. (2007) 2004 Tsunami relief efforts: an overview. *Asian Profile* **35** (5): 467–478.

Paul, B.K. (2009) Why relatively fewer people died? The case of Bangladesh's Cyclone Sidr. *Natural Hazards* **50**: 289–304.

Paul, B.K. (2010) Human injuries caused by Bangladesh's Cyclone Sidr: an empirical study. *Natural Hazards* **54**: 483–495.

Paul, B.K. *et al.* (2007) Emergency responses for high plains cattle affected by the December 28–31, 2006, Blizzard, Quick Response Research Report 191. The Natural Hazards Center, University of Colorado: Boulder, CO.

Paul, B.K. *et al.* (2010) Cyclone evacuation in Bangladesh: tropical cyclones Gorky (1991) vs. Sidr (2007). *Environmental Hazards* **9**: 89–101.

Paul, B.K. *et al.* (2011) Post-Cyclone Sidr illness patterns in coastal Bangladesh: an empirical study. *Natural Hazards* **56** (3): 841–852.

Picou, J.S. and Marshall, B.K. (2007) Social impacts of Hurricane Katrina on displaced K-12 students and educational institutions in coastal Alabama counties: some preliminary observations. *Sociological Spectrum* **27**: 767–780.

Pilkey, O. and Coburn, A. (2004) Hurricanes' lesson: don't build on the beach. *USA Today* **29** September.

Pradhan, E.K. *et al.* (2007) Risk of flood-related mortality in Nepal. *Disasters* **31** (1): 57–70.

P'Rayan, A. (2005) Intact of tsunami relief and rehabilitation work in India. Worldpress.org. http://www.worldpress.org/Asia/2116.cfm (accessed July 16, 2005)

Raleigh, C. et al. (2008) *Assessing the Impact of Climate Change on Migration and Conflict*. Washington, D.C.: The World Bank.

Ramirez, M. and Peek-Asa, C. (2005) Epidemiology of traumatic injuries from earthquakes. *Epidemiologic Reviews* 27 (1): 47–55.

Rasid, H. and Paul, B.K. (1987) Flood problems in Bangladesh: is there an indigenous solution? *Environmental Management* 11 (2): 155–173.

Rosenberg, M. (2010) Hurricane. http://geography.about.com/cs/hurricanes/a hurricane.htm (accessed January 2, 2011).

Salkowe, R.S. and Chakraborty, J. (2009) Federal disaster relief in the U.S.: the role of political partisanship and preference in presidential disaster declarations and turndowns. *Journal of Homeland Security and Emergency Management* 6 (1): 1–21.

Schmidlin, T.W. (2009) Human fatalities from wind-related tree failures in the United States, 1995–2007. *Natural Hazards* 50: 13–25.

Schmidlin, T.W. and King, P.S. (1995) Risk factors for death in the 27 March 1994 Georgia and Alabama tornadoes. *Disasters* 19: 170–177.

Sen, A. (1981) *Poverty and Famines*. Oxford: Clarendon Press.

Siddique AK et al. (1991) 1998 Floods in Bangladesh: pattern of illness and causes of death. *Journal of Diarrhoeal Disease Research* 9 (4): 310–314.

Siegel, J.M. et al. (1999) Victimization after a natural disaster: social disorganization or community cohesion? *International Journal of Mass Emergencies Disasters* 17: 265–294.

Shultz, M. et al. (2005) Epidemiology of tropical cyclones: the dynamics of disaster, disease, and development. *Epidemiological Review* 27 (1): 21–35.

Skidmore, M. and Toya, H. (2002) Do natural disasters promote long-run growth? *Economic Inquiry* 40)4): 664–687.

Smith, K. (1992) *Environmental Hazards: Assessing Risk and Reducing Disaster*. London: Routledge.

Smith, K. and Ward, R. (1998) *Floods: Physical Process and Human Impacts*. New York: John Wiley & Sons Inc.

Sommer, A. and Mosley, W.H. (1972) East Bengal Cyclone of November 1970: epidemiological approach to disaster assessment. *Lancet* 299: 1029–1036.

Sutter, D. and Simmons, K.M. (2010) Tornado fatalities and mobile homes in the United States. *Natural Hazards* 53 (1): 125–137.

Tapsell, S. (2009) *Developing a Conceptual Model of Flood Impacts upon Human Health. Report T10-09-02 of Flood Site Integrated Project*. Enfield: Flood Research Center.

Thacker, M.T.F. et al. (2008) Overview of deaths associated with natural events, United States, 1979–2004. *Disasters* 32 (2): 303–315.

Thomas, D.S.K. and Mitchell, J.T. (2001) Which are the most hazardous states? In *American Hazardscapes: The Regionalization of Hazards and Disasters* (ed. S.L. Cutter). Washington, D.C.: Joseph Henry Press, pp. 116–155.

Tobin, G.A. and Montz, B.E. (1997) *Natural Hazards: Explanation and Integration*. New York: The Guilford Press.

Toya, H. and Skidmore, M. (2006) Economic development and the impacts of natural disasters. *Economics Letters* 94 (1): 20–25.

UN (United Nations) (2006) *Tsunami Recovery: Taking Stock after 12 Months*. New York: UN.

UN-ECLAC (United Nations Economic Commission for Latin America and the Caribbean) (2003) *Handbook for Estimating the Socio-Economic and Environmental Impact of Disasters*, 2nd edn. New York: UN-ECLAC.

Viscusi, W.K. *et al.* (2010) *Economics of Regulation and Antitrust*. Cambridge, MA: The MIT Press:

Washington Post (2005) The Indian Ocean Tsunami Deaths. Internet site updated on December 22, 2005. http://www.washington.post.com/wp-srv/world/daily/graphics/tsunami_122804.html (accessed January 7, 2006).

Zurick, D. (2011) Post-tsunami recovery in South Thailand, with special reference to the tourism industry. In *The Indian Ocean Tsunami: The Global Response to a Natural Disaster* (eds P.P. Karan and S.P. Subbiah). Lexington, KY: The University Press of Kentucky, pp. 163–182.

5
Disaster Cycles: Mitigation and Preparedness

The current hazards adjustment paradigm and accompanying model of human choice classify hazard adjustments into four temporal phases or stages of emergency management: mitigation, preparedness, response, and recovery (Mileti, 1999; Prater and Lindell, 2000; Tierney et al., 2001). These four stages are collectively known as the disaster cycle or disaster management cycle. This cycle illustrates the ongoing process by which disaster management agencies – with active support from governments, businesses, and other entities – plan to reduce losses from hazards, assure prompt assistance to disaster victims, and take steps to recover from the impacts of disasters as quickly as possible. Appropriate actions and measures at all stages in the cycle lead to greater preparedness, better warnings, as well as reduction in individual and community vulnerability to hazards and disasters.

The four stages of the disaster management cycle are not exclusively distinct from each other, but these are useful categories that have helped organize the planning, activities, research, and policy for comprehensive hazard/disaster management (Mileti, 1999). Figure 5.1 shows that first two phases (mitigation and preparedness) transpire before an extreme natural event occurs, while the last two (response and recovery) take place after an event. Although there are several versions of the diagram illustrating the cyclic nature of activities undertaken in each phase, all activities seek to reduce potential fatalities and injuries, property losses, assure prompt assistant to victims, or achieve rapid and effective disaster recovery.

This chapter is devoted to the mitigation and preparedness phases of the disaster management cycle and the next chapter will cover the remaining two phases of this cycle. More specifically, this chapter will provide an overview of hazard mitigation and preparedness. It is important to note that the four disaster

---

Environmental Hazards and Disasters: Contexts, Perspectives and Management, First Edition. Bimal Kanti Paul.
© 2011 John Wiley & Sons, Ltd. Published 2011 by John Wiley & Sons, Ltd.

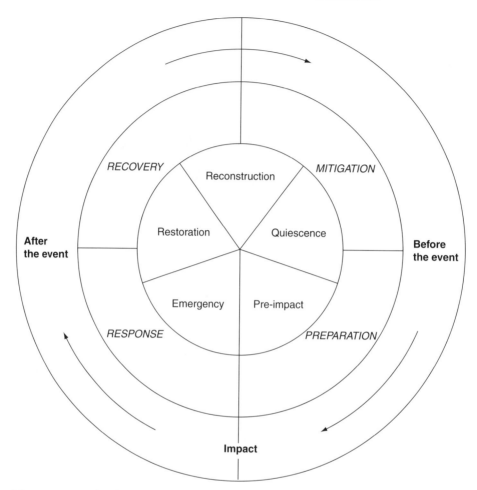

Figure 5.1 The disaster (management) cycle. *Source:* Alexander (2000), p. 3.

management phases do not always, or even generally, occur in isolation nor in this precise order. Often phases overlap and the length of each phase greatly depends on the severity and type of disaster as well as on the resources that can be brought to bear to cope with impacts.

5.1 Mitigation

As noted in Chapter 2, the United Nations General Assembly designated the 1990s as the International Decade for Natural Disaster Reduction (IDNDR). The main objective of the IDNDR was to reduce the loss of life, property destruction, and social and economic disruption caused by natural disasters through the use of technology and dissemination of scientific knowledge (Tobin and Montz, 2004).

Its goal was to improve the capacity of each country of the world to prevent or diminish adverse effects from natural disasters and to establish guidelines for the application of existing science and technology to reduce the impact of natural disasters. The IDNDR recognized mitigation as an essential component of disaster risk reduction efforts for all countries of the world. As the decade progressed, more attention was devoted to mitigation strategies. Thus disaster mitigation means risk (expected loss) mitigation.

Mitigation refers to the policies and activities that attempt to reduce and/or eliminate individual and/or community vulnerability to damage from future disasters. Mitigation measures seek to "treat" the hazard such that it impacts individuals and communities to a lesser degree. Thus natural hazards have the potential to become disasters in the absence of mitigation measures. For this reason, the Federal Emergency Management Authority (FEMA) in the United States considers mitigation as the "cornerstone of disaster management" and it has defined mitigation as: "any cost-effective action taken to eliminate or reduce the long-term risk of life and property from natural and technological hazards" (FEMA, n.d., p. 1). The phrase "cost-effective" is added to this definition to stress that a mitigation measure should save money over the long term. Since the 1970s there has been a growing interest in disaster mitigation programs from both public and private agencies. There is a large body of literature which urges governments to reallocate at least part of their budget from relief to mitigation.

As noted, mitigation measures and actions generally are in place before a disaster occurs and they include a range of options and/or adjustments an individual or community may take to reduce long-term risk to life and property from a disaster event (Maskrey 1989; NCDEM, 1998). However, such measures may be implemented at any time – before a disaster, during a disaster, or after a disaster, as well as during recovery or reconstruction phase. Destruction wrought by future disasters not only depends on the type and magnitude of an extreme event, but also on the types of mitigation measures introduced. These measures tend to be costly, time-consuming, large-scale, high-technology solutions, and managed and financed by large centralized agencies.

Disasters are widely considered opportunities in disguise and mitigation measures can be seen as a fresh start for communities devastated by extreme natural events (Coppola, 2007). Mitigation requires partnership and since a great deal of impetus must come from above, it uses "top-down" approach. Mitigation measures often provide a window of opportunity for local government, in which the most recent disaster sensitizes public opinion and facilitates political agreement through demands for improved future public safety (Alexander, 2000). Local governments in disaster-impacted communities frequently use this opportunity to enforce regulations and to recommend adopting preventive measures. Windows of opportunity are discussed in the next chapter.

5.1.1 Mitigation goals

According to Coppola (2007), there are five major mitigation goals: risk likelihood reduction, risk consequences reduction, risk avoidance, risk acceptance, and risk transfer, sharing, or spreading. These goals are closely associated with the phrase "alternative adjustments to natural hazards" or "theoretical range of adjustments to geophysical events" used in hazard geography (Burton et al., 1968; Smith, 1992). Adjustments are defined as those human activities or measures intended to reduce or minimize negative impacts of an extreme event (White, 1974). These measures may be grouped in many different ways. Burton et al. (1968) for examples, classified these measures as: (i) affect in the cause, (ii) modifying the hazard, (iii) modifying loss potential or burden, (iv) an adjustment to losses by spreading out the losses, (v) planning for losses, and (vi) bearing the losses. The goals of mitigations are briefly discussed below.

5.1.1.1 Risk likelihood reduction

It is possible to reduce the likelihood of occurrence for several hazards. For example, flood frequency can be reduced through constructing dams, dikes, levees, flood walls, and bayous. These measures keep water away from the people up to the design stage of the structure. For some hazards, such as hurricanes and earthquakes, such interventions are not technically possible, at least at present. Mitigation measures that seek to reduce risk likelihood tend to be both structural and nonstructural in nature. The former measures use technological solutions, like dams and levees, while nonstructural measures include legislation, land use planning (e.g., the designation of nonessential land like parks to be used as flood zones), and insurance.

5.1.1.2 Risk consequences reduction

This goal acknowledges the fact that most disasters are unavoidable, but their consequences can be reduced through adopting mitigation measures such as the construction of safe rooms or public cyclone shelters. Since 1998, FEMA has recommended building safe rooms in tornado-prone areas in the United States. These are rooms that are anchored and armored to provide safe shelter during even the strongest of storms. They are made with thick concrete walls and a ceiling reinforced with steel bars. A safe room can be installed in a basement, in the center of the ground level of homes without a basement, or under a garage, the garage floor serving as the ceiling. One can also dig out a space beneath a reinforced entryway, using the large concrete slabs as protection from above (Murphy and Sherry, 2003).

In the United States, safe rooms are in great demand in the tornado-prone state of Oklahoma, where basements are uncommon because of the ubiquitous hard

Figure 5.2 A typical public cyclone center in coastal Bangladesh. *Source:* Author.

clay soils (Murphy and Sherry, 2003). However, even a basement is no guarantee of protection from tornadoes for several reasons. Many basements, for example, are not fully underground, which makes them vulnerable to a degree. In addition, most basements have a wood floor overhead that could collapse when subjected to tornado-force winds.

Public shelters are necessary in areas where disasters such as tornadoes or hurricanes/cyclones are likely. In the United States, it is desirable, but not often present to have community storm shelters near mobile/manufactured home parks. Mobile homes are highly vulnerable in tornado-prone areas. From 1985 to 2007, 43% of all tornado fatalities occurred in mobile/manufactured homes, yet these types of homes made up less than 8% of housing in the tornado-prone areas studied (Sutter and Poitras, 2010).

In 1972, the Bangladesh government began constructing public cyclone shelters in coastal areas to help save lives of coastal residents from cyclones and associated storm surges. These shelters are multi-storied reinforced concrete buildings, raised above ground level (to resist storm surges) and can accommodate 1500–2500 people (Paul *et al.*, 2002). Figure 5.2 illustrates a typical shelter built in the 1970s in coastal Bangladesh. Most of these cyclone shelters are used for multiple purposes during normal and disaster periods. In nondisaster periods, some of these buildings are used as schools, community centers, government offices, and elected local government (i.e., union council) offices. Many government development activities are also carried out in these shelters and are connected with surrounding neighborhoods with flood-resistant access routes.

Coastal embankments are another mitigation measure implemented in Bangladesh to reduce or prevent damage from cyclones and associated storm surges. The height of these embankments ranges from 17 to 20 feet (5–6 m) and they have marine and land-side slopes of 1:7 and 1:3, respectively. The Bangladesh government has also initiated a large afforestation project to provide protection for coastal residents and the beach environment against tidal and storm surges (Government of Bangladesh, 2008). Thus, some hazards have more than one option for disaster consequences reduction. Like the first goal, these measures tend to be both structural and nonstructural. The former includes both new and reinforced structures, which are able to withstand the negative impacts of disasters. These structures make it possible to live in cyclone-prone high-risk areas.

Nonstructural measures include hazard forecasting and warning systems, and evacuation of potential victims from the projected path of an impending disaster. The risk consequences reduction goal also includes public awareness programs that provide residents with information on what to expect, what to do, and when to take action during an advancing disaster. Another way to reduce the consequences associated with a disaster is by identifying hazard zones and restricting or regulating their development (Coppola, 2007; Tobin and Montz, 1997).

5.1.1.3 Risk avoidance

The third mitigation goal is appropriate for disasters whose risks are great even after a partial reduction in either their likelihood or consequences. For these disasters total avoidance is necessary. This usually involves permanent evacuation of people and/or relocation of structures from a high-risk zone to a low- or no-risk zone. Out-migration and buyout programs are examples of risk avoidance mitigation measures. The latter program seeks to physically remove people and structures within a floodplain and then restrict all future construction in that reclaimed land. Risk avoidance is also possible through relocation from volcanoes, avalanche chutes, and landslide zones.

5.1.1.4 Risk acceptance

This goal of mitigation is similar to adjustments which have been termed "bear the losses" – where people are willing to bear the full burden of damage or share it in some fashion with others (Burton *et al.*, 1968). No or extremely few places on the earth's surface are free from natural hazards. A place may not experience all disaster types, but most locales are subject to a select number of hazards. Residents of such places expect these hazards and they generally accept a certain risk associated with the disaster. For example, residents of floodplains expect river flooding and they are willing to bear some damage associated with flooding. It is to be noted that river flooding also has some benefits, such as increasing soil fertility due to the deposition of nutrient-rich silt.

People who accept losses usually do not attempt to alter the cause(s) of disasters. They bear associated losses for several reasons. Often they perform a cost–benefit analysis and find out that an intended reduction measure for a particular hazard is not cost effective. In such a case, they will not pursue that particular measure, but may pursue a similar measure for other hazards for which that particular risk reduction measure will have greater value relative to losses usually incurred. Second, often some reduction mitigation measures will result in one or more undesirable consequences. Such a consequence may be considered more damaging or undesirable than impacts resulting from the hazard itself. Finally, for cultural or other reasons, residents of a particular place and location may prefer to stay there despite certain risks (Coppola, 2007).

5.1.1.5 Risk transfer, sharing, or spreading

The last and most debated mitigation goal is risk transfer, sharing, or spreading of financial losses incurred by disasters. Depending on the availability of disaster insurance, disaster-impacted communities and individuals can spread or transfer disaster loss or risk through insurance coverage. Insurance reduces the financial consequences of a disaster by transferring at least part of any monetary loss of property. It is important to mention that disaster insurance is still not available in many developing countries. Wherever it is available, coverage by insurance may be wholly voluntary or wholly mandatory, and large businesses often self-insure by setting up reserves and diversifying. However, in the United States, federal policy is aimed at encouraging local governments and individuals to insure public and private property against losses due to extreme natural events. The National Flood Insurance Program (NFIP) requires residents of flood-prone communities to carry flood insurance.

Another way to transfer, share, or spread disaster loss is through the provision of relief and rehabilitation assistance, which is more common in developing countries than in developed countries. This assistance is generally offered to disaster victims by both public and private agencies. However, the availability of emergency relief may work against adoption of insurance by prospective disaster victims. The essential features of disaster relief are presented in detail in Chapter 7. Other ways to offset disaster losses include help from friends and relatives in various forms, such as cash, in kind, and labor. This is closely associated with the concept of social capital theory, which is defined by Pierre Bourdieu as the resources that can be derived through one's social network. It refers to the collective value of all "social networks" and the inclinations that arise from these networks to assist each other (Bourdieu, 1985).

5.1.2 Mitigation measure classification

As indicated, there are hundreds of mitigation measures which range from physical measures such as flood levees or safer building designs to nonphysical measures including legislation, training, and even public awareness campaign. Each hazard has a unique set of mitigation options from which disaster managers may choose to engage/develop. Selection of mitigation options depend on costs, available funds, as well as the anticipated social and physical consequences of taking such actions. Mitigation measures can be classified in many different ways. For example, Mileti (1999) classified mitigation measures into five groups: land use planning/regulation, building codes and standards, insurance, engineering, and warning.

5.1.2.1 Land use planning and regulation

The main purpose of this measure is to reduce hazard exposure by land use management or zoning, and restricting construction of structures in hazard-prone areas. This is a regulatory measure where local government is responsible for preparing and following through land use planning and regulation for development in hazard-prone areas, prohibiting new construction in a floodplain, along a coast, or any other hazards-prone area. Among others, land use planning and regulation activities include enforcement of building codes and environmental regulations; public safety measures such as continual maintenance of roadways, culverts and dams; acquisition or relocation of properties such as purchasing buildings located in a floodplain; and retrofitting of structures and design of new construction such as elevating a home or building.

Hazard-prone communities across the globe have implemented many regulatory and land use zoning mitigation measures. For example, after experiencing an F-4 tornado on April 21, 2001, the city of Hoisington, Kansas, enforced two new zoning ordinances that restricted the construction of homes on 50-foot lots in tornado-impacted portions of the city. This enforcement meant residents who were rebuilding had to purchase adjacent lots in order to satisfy the newly established requirements. As a result, housing density in the tornado-affected area became significantly lower than pre-tornado density and is also now lower than in non-impacted portion of the city (Brock and Paul, 2003). This ordinance has reduced exposure by limiting development and density of human occupancy in Hoisington.

Another new zoning regulation that went into effect in Hoisington was that any house constructed west of Alexandria Street could not have a basement because that area was identified by FEMA as being in a floodplain zone. Since everyone wanting to rebuild desired a basement, no one rebuilt their homes in that area. The city then bought this land and converted it into a park with new playground equipment and tennis courts. This change in land use was welcomed by an overwhelming majority of the residents of Hoisington – a town with a population of about 2000 in 2000 (Brock and Paul, 2003). Similarly, Cedar Falls, Iowa, a city of 36 145 people in 2000, bought 128 homes that were severely affected by a flood in 2008. Another 81 homes are in the process of being acquired by the city. All buyout properties will be used as a park or designated as a protected area (Rahman, 2010). In another example of reduced exposure through land use change, several small communities were relocated away from flood-prone areas after the widespread flooding experienced in the US Midwest in 1993 (Mileti, 1999).

Most hazard regulatory measures are initiated by the central government, but such measures are often implemented by the local government. However, local governments are often reluctant to implement strict land use policies and regulatory measures due to a lack of local political will to manage land use, and/or deficiencies in management capacity. Because the spatial extent of different natural hazards is usually well defined, land use planning and regulatory measures are particularly appropriate for specific geographic designations, such as floodplain and coastal zone management.

5.1.2.2 Building codes and standards

Building codes are a collection of laws, regulations, ordinances, or other statutory requirements adopted by a governmental legislative authority regarding the physical structure and/or construction of buildings. This group of mitigation measures is designed to ensure that structures resist the physical impacts created by disasters and their purpose is to establish minimum acceptable requirements necessary for preserving the public health, safety, and welfare as well as the protection of property in the built environment. The primary application of building codes is to regulate new or proposed construction (Mileti, 1999). Although such codes may have little application to existing buildings (unless they are undergoing alteration), the effectiveness of zoning and other regulatory measures can be enhanced through the use of building codes.

There are hundreds of standards in the United States addressing virtually every construction application – from design practices and test methods to material specification. Local and state building codes often vary considerably. Local governments have traditionally enacted comprehensive building codes that regulate all construction. Building codes also differ from one country to another because building materials and principles differ across countries, not to mention that some disasters are area/region-specific. For example, there is no need to construct public cyclone shelters in areas other than coastal zones, which are frequently affected by tropical cyclones and storm surges. When properly applied and enforced, building codes offer a great deal of protection from a wide range of disasters. Implementation of building codes, however, involves additional costs for individuals and/or communities to bear.

Building codes are not restricted only to structural design, but also apply to construction methods and materials. In tornado-prone areas of the United States, FEMA recommends using steel brackets to connect foundations to floors and, in two-story homes one floor to the next, using brackets to tie the roof interior to side walls, building exterior walls of poured concrete (or use steel framing), and retrofitting masonry chimneys with vertical reinforced steel if they extend more than six feet (2 m) above the roof (Murphy and Sherry, 2003).

Many countries of the world have introduced building codes and have updated them fairly regularly with improved knowledge of building design and construction. For example, the Bangladesh government formed a committee of experts to prepare a national seismic zoning map, and outline building codes for earthquake-resistant structure design in 1977. Fifteen years later in 1992, the government appointed a team of consultants to prepare a national building code for Bangladesh. This team developed the Bangladesh National Building Code (BNBC) in 1993 and revised the 1977 seismic zoning map to better serve the design and engineering community (Ali and Choudhury, 2000). The 1993 codes have since been revised. However, like many other developing countries, building codes in Bangladesh have not been completely or even largely successful, primarily because of problems with compliance and enforcement.

Apart from costs, there are other problems associated with building codes, particularly in areas subject to multiple hazards. For example, in parts of California, a wood-frame structure with a cedar shake roof provides relatively good resistance to earthquake shaking, but such a structure would prove to vulnerable to that state's frequent wildfires. Thus the effective building codes must take into account the local context and, in particular, the hazardousness of a place.

5.1.2.3 Insurance

Most property owners in developed countries rely on private insurance to protect themselves against financial losses from some, but usually not all, natural disasters. Although insurance does not reduce actual disaster consequences or reduce hazard likelihood, it allows associated economic losses to be shared across a wide population. It should be noted that insurance often offers no incentives to undertake mitigation.

5.1.2.4 Engineering

This type of mitigation is frequently called structural mitigation, and involves physical measures taken to reduce disaster risk by erecting structures, such as cross-dams, and building a tornado safe room. Engineering measures also attempt to strengthen buildings to better withstand future disasters, such as tornadoes and earthquakes. This can be accomplished in two ways: the structure may be demolished and rebuilt to accommodate new hazard information or it may be modified so that it resists the anticipated external force (retrofitting). Thus, structural measures involve measures that relate to the existing natural and built environments and/or which address future developments.

Structural measures involve some form of construction, engineering, or other mechanical changes or improvements aimed primarily at reducing damage wrought by disasters. These measures are generally expensive and include a full range of regulation, compliance, enforcement, inspection, maintenance, and renewal issues (Coppola, 2007). Some structural mitigation measures, such as the construction of dams to control flooding, may have adverse effects on ecosystems.

In contrast to structural measures, nonstructural measures tend to be less costly and relatively easy for communities and individuals with limited financial or technological resources to implement. These measures are typically policy-related and include application of zoning restrictions, acquisition of land in the floodplain, and promoting citizen hazard risk awareness. Design and construction guidelines also fall into this category. Nonstructural mitigation measures attempt to distribute the population and/or the constructed environment in such a way that exposure to disaster losses is limited (Mileti, 1999). These are measures also aimed at reducing the likelihood of hazard events. Nonstructural mitigation measures are often considered mechanisms where humans adapt to nature.

5.1.2.5 Warning

A warning system is necessary to detect an impending disaster, communicate that information to people at risk, and encourage at risk population to take appropriate action. Disaster warning systems can make the difference between life and death. Lindell *et al.* (2005) maintain that the remarkable reduction in US hurricane casualties since the 1900 Galveston hurricane was due to improved forecasting and warning systems coupled with timely and effective evacuations of coastal residents from imminent threats. However, in many countries, no comprehensive national warning system is available that covers all hazards and all places within a given country. Hazard warning is considered here as a component of preparedness and therefore this topic will be discussed under the preparedness phase of the disaster management cycle.

There are other ways to classify mitigation measures. As indicated, local government entities together with active support from emergency agencies at the state and/or federal level generally introduce new hazard mitigation measures to better prepare for future disasters. Compliance with these measures may be either mandatory or voluntary. Imposition of new building codes and construction practices, new development restrictions, and changes in land use regulations are examples of natural hazard mitigation mandates. In addition, local government and public agencies often recommend adopting preventive measures aimed at minimizing loss of life, injury, and/or destruction of property.

Mitigation measures can also be classified as active and passive. For active measures, authorities promote desired actions by offering economic or other incentives. Economic incentives can take the form of subsidies, low-interest loans for retrofitting, or tax credits. For example, the construction of a 14-foot by 14-foot (4.3 m × 4.3 m) tornado safe room required between US$11 500 and US$13 500 (FEMA, 2008). Homeowners usually adopt only the least expensive mitigation options unless financial incentives are provided by public authorities (Mileti, 1999). Considering this, FEMA and some states have provided partial funding to tornado victims in order to construct safe rooms. In Oklahoma after a devastating series of tornadoes in 1999 more than 6000 people took advantage of US$2000 FEMA rebates and state incentives and built safe rooms (Pattan, 2003).

As indicated, incentive efforts partially compensate disaster victims for the losses incurred, encourage adoption of mitigation measures by lowering the cost of implementation, and reduce individual and community vulnerability from future disasters (Mileti, 1999). For passive mitigation measures, however, authorities prevent undesirable actions (e.g., construction of houses in 500-year floodplain zone) by using legal control and penalties.

5.1.3 Critique of mitigation

There are several criticisms of mitigation. Mitigation projects tend to be large scale and high technology oriented, and thus very costly. Finding the fund to implement projects is often very challenging. A top-down approach is generally used to design and implement such projects. Mitigation projects are generally developed and managed by large centralized agencies without seeking input from disaster survivors. Because of their top-down approach, mitigation projects generally favor the rich and powerful at the expense of the most vulnerable.

The emphasis of most mitigation programs has been on physical measures to address the immediate threat of a natural hazard and not on social changes to address the problem. Such programs only deal with reducing the risks of specific hazards and not on reducing the vulnerability of marginalized groups. So far, mitigation strategies have been designed to reduce the probability and impact of hazards, along with implementing anti-exposure measures, such as restricting development in floodplain zones. Anti-vulnerability mitigation measures (e.g., land reform and poverty alleviation) have received little, if any attention among hazard researchers as a means to reduce hazard risk. Reduction in vulnerability is necessary to build future resilience against hazards and disasters.

Also, mitigation programs often do not take into account all the actual needs and demands of those affected by disasters. One of the greatest obstacles to mitigation is that those most impacted by a disaster often resist any alteration to their place of residence by land use changes and/or imposing new regulations for construction, including amending building codes. Citizens may have a vision of their community, their property, or their home that they are reluctant to alter. However, there are important reasons to mitigate against hazards. The most important is that it makes communities safer places to live. Increased safety brings a host of benefits that become apparent after a disaster, for example the cost of reconstruction is lower and the time it takes is shorter; businesses close for less time; public facilities require fewer repairs; and the community is running again sooner. In the long term, the community will be subject to fewer disasters.

It appears from the above that in addition to dealing with reducing the effects of disasters, mitigation must also address the underlying causes of vulnerability of people living in hazard-prone areas. Along with physical measures, such as reinforcing buildings or raised dykes, mitigation practices must be well integrated with developmental programs which should emphasize the structural causes of disasters. This requires addressing the real causes of poverty and underdevelopment through redistribution of wealth, reformation of existing land-tenure systems, and other measures. The IDNDR has encouraged incorporating these into disaster risk reduction efforts. Mitigation needs to be used as a vehicle for affecting the underlying causes of disaster vulnerability. Furthermore, mitigation strategies must also be aimed at multiple hazards, because so many disasters are the consequence of multiple events. As noted in Chapter 2, geographic information systems (GIS) and other technological advances have the capability to accomplish this important task. Tobin and Montz (2004) claim that unless multiple hazards are incorporated into hazard mitigation activities, efforts to reduce human vulnerability may well be wasted. Mitigation planning has benefited from the availability of disaster information on the Internet (Gruntfest and Weber, 1998).

5.2 Preparedness

Preparedness refers to the degree of alertness and readiness of an individual, a household, or a community against an impending disaster. Among many activities, preparedness includes such activities as formulating, testing and exercising disaster plans, providing hazard warnings, communicating with the public and others regarding disaster vulnerability and what to do to reduce it, evacuating people from harms way, conducting emergency response drills, and providing disaster training for emergency responders and the general public (Tierney et al., 2001). The acquisition of equipment to support emergency action is also an important activity of hazard preparedness. Such activities ensure that the resources necessary for carrying out an effective response in shortest possible time are in place prior to the onset of a disaster or that they can be obtained promptly when needed. Preparedness minimizes the negative consequences of extreme events. It does not only facilitate effective response, but also enhances the ability of social units to effectively respond when a disaster does occur.

Hazard preparedness is often divided into two categories: physical preparedness and social preparedness (Tierney et al., 2001). The former centers on taking actions to ensure that facilities and structures can withstand disaster impacts and the buildings and their contents do not become a serious threat to life and property should a disaster strikes. Social preparedness includes actions (e.g., understanding what state and federal programs are available at the time of disaster, planning for situations involving hazard warning and evacuation, establishing emergency record-keeping systems, and developing disaster plans) to ensure that community organizations are able to adequately respond to the needs of victims in the event of a disaster. Both physical and social preparedness are seen as part of a five-phase preparedness cycle that consists of: (i) raising awareness, (ii) conducting hazards and vulnerability assessments, (iii) improving knowledge about hazard and how to cope with them, (iv) planning, and (v) practice (Tierney et al., 2001).

The preparedness process usually begins with a vulnerability analysis that attempts to identify what hazards could occur in a particular place and a risk analysis to determine the likely problems that an extreme event could impose. Numerous public and private emergency response agencies as well as individuals conduct disaster preparedness activities. Each agency and personnel has unique functions to perform and unique responsibilities to accomplish before disasters strike. Others along the path of an impending hazard need to be compliant with the directions provided by emergency management personnel.

5.2.1 Activities

As indicated, preparedness includes a range of activities and/or actions that are undertaken (at different levels) immediately and/or any time before a disaster which saves lives and minimize property damage. A number of preparedness activities are discussed below. This discussion is followed by an analysis of preparedness activities at four distinct levels: households, organizations, communities, and state and national preparedness. Factors that influence preparedness at each of these levels is also discussed.

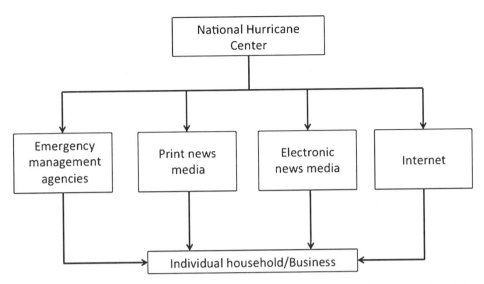

Figure 5.3 Dissemination of impending danger by the NHC to general public. *Source:* Modified after Lindell *et al.* (2005), p. 4.

5.2.1.1 Warning

Hazard warnings include a range of activities such as effective monitoring and accurate forecasting of a forthcoming event, developing warning messages, and disseminating the messages to those at risk. Before issuing a hazard warning, responsible authorities must identify what hazard is about to occur, is currently happening, or has already taken place. Detection and emergency assessment procedures differ by hazard types. For example, the National Hurricane Center (NHC) in the United States monitors the formation of hurricanes and their progression by collecting data from satellite images, radar images, and telemetry data from "hurricane hunter" aircraft. These data are processed by storm forecasting models and evaluated by NHC staff, who then assess the need to declare hurricane watches and warnings in addition to projecting hurricane intensities and landfall sites. The NHC notifies individual households or businesses through state and local emergency management agencies, the news media, and the Internet of any impending danger and provides continuing information updates about its assessments to these parties (Figure 5.3).

For providing tsunami warning, buoys monitor the heat and level of the ocean to track surface temperatures and the presence of tsunamis. Some hazards can be even detected through more than one means. Doppler weather radars, for example, are used to detect tornadoes. In conjunction with these radars, storm spotters also detect and report data about the same tornadoes. However, some hazards, such as earthquakes, are difficult to detect or predict. Although seismographs identify where earthquakes have occurred as well as their magnitude, they cannot predict when such events will occur. It is, however, essential to constantly monitor hazards and some agencies have the responsibility of performing this difficult mission. For example, the National Weather Service (NWS) in the United States is responsible for monitoring weather-related hazards, such as severe thunderstorms, tornadoes, floods, blizzards, and avalanches. The United States Geological Survey (USGS) monitors and issues warnings for earthquakes and volcanoes.

Once a hazard has been identified, responsible authorities perform a number of functions simultaneously. These functions include seeking additional information regarding the hazard, opening of emergency operation centers (EOC), initiating the response, and informing and issuing hazard warnings to potential victims of the hazard. Hazard warnings are advanced notifications of an impending hazard that allow people to take measures to protect themselves and their property. Warnings must be designed to reach all the at-risk population, no matter where they are or what time it is, and these warnings should instruct this population on what to do before, during, and after the disaster. Most warnings are issued before a hazard has occurred. However, warnings may also be disseminated after a disaster in order to help disaster survivors ascertain the best way to respond (McEntire, 2007).

According to Lindell and Perry (1992) the warning process consists of four steps:

1. Risk identification. Based on information collected from various sources (e.g., weather sensors, water flow sensors, seismicity and ground deformation sensors, and air and water monitoring devices), the responsible authority needs to determine whether a threat really exists.

2. Risk assessment. Based on all available information, the responsible authority then decides whether protection is needed. If the decision is made in favor of protection, there is a need to identify the at-risk population which will receive the warning message. This message should be targeted to those at risk, while simultaneously reducing the likelihood for those not at risk of experiencing the impending disaster.

3. Risk reduction. In this third step the responsible authority should determine if protection is possible, and if so, how to best achieve it.

4. Protective response. In this last step the responsible authority must determine which protective actions the target population should be instructed to take. This authority, with the help from local emergency responders (e.g., police and fire officials, and volunteers), should assist the public in effectively taking the recommended protective actions.

Many different terms are used to describe the severity of a hazard warning. To help people prepare for hazards, the NWS in the United States uses the terms "watch" and "warning." A watch means that conditions are favorable for development of the indicated hazard. Watches are issued so that at-risk population may begin taking precautionary measures as far in advance as possible. A warning, on the other hand, indicates that the specific hazard is imminent, is taking place, or has occurred. After receiving hazard warnings, the public should take immediate action to protect themselves and their property. Warnings and/or watches are not possible for several hazards, such as earthquakes or terrorist attack. However, Emrich and Hughey (2011) reported that Japan's early warning system for earthquakes can detect the slightest tremors up to 90 seconds before real impact. In general, severe weather systems can be monitored more effectively and generally allow for advance warning.

In addition to the above two terms, three other terms are also used to describe hazard warnings (Coppola, 2007). Like warnings, "advisories" are given for a hazard that is currently occurring or is about to occur. Advisories are issued for events that are less severe than warnings in terms of the expected consequences to life and property. An "outlook" is a prediction of a hazard event in the near future and thus does not require taking any immediate precautionary action. Finally, "statements" are used to provide detailed follow-up information to warnings, advisories, or watches (Coppola, 2007).

Lead time refers to the amount of time (in minutes, hours, or days) between issuing a warning and actual occurrence of a disaster, or the amount of time citizens will have to act after receiving a hazard warning. Lead times differ by hazard types as well as by technological capabilities. For example, lead time is shorter for tornadoes compared to hurricanes. Lead time is important for taking precautionary measures against impending disasters. For quick-onset disasters, such as tornadoes, at-risk populations have generally only a few minutes to take safer shelter because a tornado warning is issued only minutes before an actual touchdown. In contrast, for slow-onset disasters, such as hurricanes/cyclones, floods, and droughts, people at risk may have hours and even days to prepare for such events. Hurricane/cyclones warnings allow coastal residents to buy additional groceries and supplies before landfall, and/or provide adequate time to leave for safer inland areas. Because of improvements in technology, the lead time has been increasing over time for all disasters. However, early warning systems have been developed to varying capacities for different hazards, such as droughts, tornadoes, hurricanes/cyclones, landslides, floods, chemical releases, wildfires, and terrorist threats.

It is important to mention that any warning message given should be clear and contain accurate information. It should provide detailed information regarding the impending disaster, including when the disaster will occur, how long the disaster will last, what the expected impact and severity of the disaster will be, potential power outages, closed streets/areas, and projected damage to homes and property. The amount of information to be provided in a warning may differ by hazard type. For example, hurricane/cyclone warnings should contain information such as current position of the hurricane/cyclone in terms of latitude and longitude coordinates, intensity, estimated central pressure, direction and wind speed, maximum sustained wind speed and radius of maximum sustained wind, areas likely to be affected, time of landfall, place of landfall, and expected storm surge height. The language to be used in warning message must be simple so that illiterate persons can understand the message. If possible, the local dialect should be used to warn prospective disaster victims.

In developed countries early warnings may come from many different official and/or unofficial sources. In the United States, weather-related hazard warnings are provided by the NWS and the National Oceanic and Atmospheric Administration (NOAA). In addition to these official sources, state and local government agencies and other official sources can also issue warnings and distribute them. Warning messages, however, must be repeated, consistent, and disseminated from more than one source, preferably the credible one. If a warning is issued once, or if the message contradicts other warnings, its intended impact on the behavior of the at-risk population will be reduced.

Types of warning systems

Both technological and nontechnological methods are used to broadcast hazard warnings. The responsible authority generally uses several warning systems for the same hazard in order to inform the impending danger to all the at-risk population. For example, outdoor sirens, weather radios, and electronic media are used for tornado warnings. Each type of warning system has strengths and weaknesses. It is advisable to use as many means as possible to issue a warning as long as the message is consistent. This is necessary to increase the chance that most of the at-risk population will receive the warning message.

Emergency alert systems (EAS) are used in the United States for a variety of hazards and may be issued by way of television and radio. They use digital technology to broadcast messages. This digital data contains information on the type of message, valid time, and states/counties for which messages are valid. An EAS message quickly warns a large number of people. A significant advantage of this is that it is a nationwide system for emergency broadcasts. A disadvantage is the training required for operating EAS equipment. In addition, these messages are broadcast through radio and television so only the people who are listening to the radio and watching television at the time of broadcast will receive the warning. Also, power outages can render television and radio useless (Csiki, 2003).

As noted, mainstream media (radio and all types of television) are the primary methods used to communicate hazard warnings to the public in the United States. The government distributes warnings to local stations via EAS. The NWS distributes warnings to the Weather Channel and the Internet directly. Radio and television stations are not required to interrupt programming to broadcast warnings. However, television channels can often scroll the warning text at the bottom of the screen and radio stations may briefly mention the warning.

Advantages and disadvantages are similar to EAS. In addition, television is important as images – moving or still – reinforce a message of danger for people, providing a distinct advantage. Conversely, a disadvantage is that people have to be tuned in, which usually occurs only a couple of times per day (Csiki, 2003).

In addition to commercial radio, NOAA Weather Radio (NWR) is used in the United States to disseminate weather-related hazard information directly to the public. NWR is a network of government (NWS) radio stations that is the primary method of distributing warning information directly to the public. State and local governments can also use NWR to broadcast other warnings. An advantage of NWR is that warnings are provided directly to the public verbatim. Disadvantages are that coverage is not yet universal, messages cannot be delivered to specific individuals within a county, and the radio-based signal is subject to periodic tropospheric ducting and skip episodes, rendering it useless for warning capability during those events. Moreover, most people do not have weather radios.

Sirens are another warning system used to warn people for tornadoes, floods, and nuclear attack. Sirens are either operated on commercial power or rechargeable battery backup. The latter sirens can be used even when power is lost. In addition to broadcasting an alert, some sirens can also broadcast a voice message immediately following the alert. The broadcast format varies between localities. Most locations only sound the siren when a warning is issued for the community, while in some locations the siren is sounded when the hazard is in an adjacent county. However, because sirens are used for multiple hazards, and as there are different sounds for each hazard, people may become confused upon siren activation. Moreover, sirens often cannot be heard indoors, unless people are physically near the siren (Csiki, 2003), and people living next to the siren "bear the brunt" of siren noise. Sirens also require a significant amount of maintenance and many communities cannot afford them.

Reverse 911 is the automated dialing and delivery of hazard warnings from a single site to homes and businesses in a designated jurisdiction. Advantages are that reverse 911 can deliver a message to the people who must receive it and no one else. Disadvantages are that one must have an operable phone to receive the warning, people can mistake the warning for a telemarketing or other advertising/common call, and it does not work if telephones are rendered inoperative. Also, Reverse 911 systems are not available in all communities within the United States.

Currently, many other technology-oriented warning systems, such as cell phones, Internet, pagers, palm pilots, intercoms, teletype writers, wireless telephone devices, and strobe lights are available. Some of these are used to warn people in large buildings and to make the deaf aware of impending hazards. There are also many nontechnological hazard warning devices, but they are not widely used. Police and fire department personnel may knock on doors and directly deliver a hazard warning message to residents. Door-to-door warnings permit answering of questions, but require significant manpower and are very time consuming. Instead of door-to-door warnings, firefighters and police officers can generally use bullhorns to broadcast a warning or evacuation order.

It is worth noting that a single warning system cannot be effective for all hazards nor effectively warn all people in a community. Therefore a community should depend on more than one warning system for any given hazard. This is so because each system operates alone and can rarely, if ever, reach the full desired audience. Public perception and compliance with warnings is another reason to have more than one warning system. Sorensen (2000) maintains that when a hazard warning is first issued, the public usually does not take immediate action. Instead, they attempt to find other sources of information before taking appropriate action.

Coppola (2007) maintains that an effective warning system involves three distinct processes that all work together to form a network of communication for those in the area targeted for a hazard warning. These processes are: planning, public education, and testing and evaluation. The first process involves making decision regarding how and when the public will be warned, what should be the warning message, and what authority and equipment is needed to issue the warnings. An important component of the planning process is to make sure that everyone living within the threat area will receive the warning. This includes people who are at home, in school, at work, or other places. In addition, people who are disabled or traveling (e.g., in cars and on trains) as well as those who speak a different language or are illiterate should get the hazard warning.

In order to increase the compliance with hazard warnings, the public must be educated through disaster education programs or by other means. No matter how well devised a disaster plan, if people cannot differentiate between a "watch" and "warning," or do not fully understand what to do after receiving a warning, the planning will have little or no value. The public should be made familiar with what a hazard warning will sound like and what they are expected to do when a warning is issued; such familiarity will result in more cooperation and compliance with future hazard warnings.

The final component of a hazard warning system is testing and evaluation. The warning system must be continuously tested, evaluated, amended, re-tested, and re-evaluated in order to maintain an effective system as well as to ensure that the public is not exposed to the warning protocol for the first time during a disaster. Testing is a means of allowing potential disaster victims to experience the warning system in a low-stress environment and to hear the actual wordings or sound the warning system issues. Hearing and recognizing the warning and knowing its message beforehand will reduce confusion and panic when an actual warning for a hazard is issued.

Testing also allows people to participate in a drill that simulates the disasters and allows public officials to evaluate their plans. The process of evaluating a warning system is perhaps the most critical aspect. Such evaluations provide opportunities for the system to be observed in action to ensure that every component performs satisfactorily. If not, it provides opportunity to make necessary adjustments. It is to be noted that the development of GIS has facilitated disaster forecast and warning in a timely manner (Tobin and Montz, 2004).

5.2.1.2 Evacuation

Evacuation is the primary protective action utilized in disasters such as floods, tsunamis, large wildfires, hurricanes/cyclones, and volcanic eruptions. It refers to the movement of people away from a potential or actual hazard impact area for the primary purpose of ensuring safety. Evacuation can take place before, during, or after a disaster occurs. For example, a study conducted following Hurricane Katrina in seven southeast Louisiana parishes found that 74% of respondents did evacuate before the storm's landfall and almost 87% of those evacuees traveled a distance of 75 or more miles (112 km) away. Another 14% of all respondents evacuated after the storm (Harrington et al., 2006). However, the purpose of an evacuation is to reduce the loss of life and the chance for injury. Although evacuation is an important preparedness measure for many hazards, it is widely practiced for an approaching hurricane/cyclone and thus coastal residents are relatively frequently subject to evacuation from the projected path of these natural events.

It is very common to evacuate potential victims away from their homes into personal or established community shelters located within or outside the high-risk area in advance of a hazard's arrival. Thus, in most instances, evacuation orders are issued along with hazard warning messages. There are two types of evacuation orders: voluntary and mandatory evacuation. Under the former evacuation mandate, each individual/household decides whether or not to evacuate independent of others. Mandatory evacuation is a somewhat misleading term because not all people comply with this evacuation order. It creates a sense of urgency, yet fails to be enforced at a legislative level. However, studies (e.g., Dilmener, 2007) have reported that individuals who receive a voluntary evacuation order are twice as likely to evacuate than those who did not receive an order at all; while those who received a mandatory evacuation order are almost five times more likely to evacuate than those receiving no evacuation notice.

Evacuation is rarely an individual process. Even in single-person households, the first response to the initial evacuation warning is to seek further information on the validity of the threat from authorities, and/or consulting with a neighbor, friend, family member, relative, or co-worker. Evacuations usually take place in a group context. In business settings, co-workers typically evacuate in groups (Aguirre et al., 1998). The time period required for evacuating people is highly variable in that an evacuation may last for nearly any amount of time, may occur more than once, or even sequentially should there be secondary hazards or a reoccurrence or escalation of the original hazard. Furthermore, evacuations can range in geographic scope from a single residential subdivision threatened by a landslide, to multiple states/countries threatened by a hurricane/cyclone.

Similarly, the number of people involved in an evacuation has ranged from only a few persons to over a million people. For example, over 3 million people were evacuated in both Hurricanes Floyd and Rita (Sorensen and Vogt, 2006).

Once a warning is given, people who are willing to evacuate from the threatened area take time to prepare to implement the protective action. This time is referred to as mobilization time or preparation time. Implementation (or departure) time is defined as when the physical evacuation is undertaken. Mobilization times are highly variable and seem to depend on the perceived time to impact and the level of urgency to respond (Lindell and Perry, 1992). For rapid-onset events, departure times are within fairly narrow range. In slower events, departure times are more spread out, but can vary by location. With Hurricane Floyd, for example, there was a relatively steep departure curve in the Charleston, South Carolina area. Further north in Myrtle Beach, however, departure rate was more spread out (Sorensen and Vogt, 2006). Clearance time is the sum of the times between when a warning is received, mobilization time, and trip travel time, which is the time from evacuation trip departure until the destination is reached. Clearance times are generally shorter for rapid-onset events and much longer for other event such as hurricanes, particularly in populated areas. For Hurricane Floyd, it took evacuees an average of about 9 hours to reach their final destinations (Sorensen and Vogt, 2006).

Types of evacuation

Hazard evacuations can be classified in many different ways. One way is to dichotomize between horizontal and vertical evacuation. In the former, people are moved away from a potential threatened area. The vast majority of evacuations are horizontal in nature, particularly in the United States where people often move hundreds of miles inland when a hurricane approaches. In contrast, some evacuations are vertical in nature. This means, people do not leave the threatened area, but move to safer places within the area. As noted, public cyclone shelters in coastal Bangladesh are constructed within cyclone high-risk areas. These shelters are located no more than 5–6 miles (8–10 km) from coastal villages. In addition to moving to public cyclone centers, coastal residents of Bangladesh also take shelter against an advancing cyclone either in nearby public buildings (e.g., mosques, colleges, schools, and government offices) or in neighbors' houses that are perceived as being structurally stronger than their own (Paul and Dutt, 2010).

As in Bangladesh, many people in typhoon-prone areas of the Philippines do not evacuate the risk area, but stay in their typhoon-resistant homes, called core shelters. This is a low-cost house, which people can build themselves with local materials. It is known as a core shelter because it provides a basic typhoon secure unit or core into which families can retreat in the event of a typhoon, together with any stock and other possessions that they wish to protect (Diacon, 1997). Public cyclone shelters in Bangladesh and core shelters in the Philippines are, therefore, fundamentally different from shelters used in the United States for advancing hurricanes. Although the horizontal evacuation is widely practiced in the United States, schools or other public buildings located within the threat area are often temporarily designated as places of safety to shelter people from impending hurricanes. For example, the Superdome and the New Orleans Convention Center located near downtown New Orleans were used to shelter thousands of Hurricane Katrina victims in 2005.

There are advantages and shortcomings of both horizontal and vertical evacuations. In horizontal evacuation, individuals or groups travel to a place that is perceived as completely removed from the immediate danger. Such places often involve staying with their friends and relatives or in a hotel/motel. Cars and buses are usually used for horizontal evacuations. Unfortunately, not all people have a car. For example, nearly 33% of New Orleans residents did not own a car at the time when Hurricane Katrina made landfall along the Louisiana coast in 2005. Most of these people together with others who had no money for gas, food, or hotel rooms defied the evacuation orders and stayed home (Harrington et al., 2006). Lu et al. (2004) estimated that household costs for gas, food, and lodging costs averaged US$267.57 per household in Hurricane Lili which was the deadliest and costliest hurricane of the 2002 Atlantic hurricane season for the United States.

Horizontal evacuation is not only expensive, this type of evacuation also requires large numbers of law enforcement personnel to control traffic, prior planning, and coordination. The sheer number of evacuees may create serious traffic congestion problems even when all lanes of highways are opened in only one direction to facilitate the evacuation. For example, some Hurricane Rita (2005) evacuees had to stay on the overburdened highway for up to 12 hours (McEntire, 2007). Evacuating large numbers of people in buses greatly reduces traffic jams because each evacuee is not driving his or her own car. Buses are generally used to move small groups of people, but using this means of evacuation requires coordination between buses and drivers, either of which may not be available on short notice. Moreover, heavy congestion and traffic jams can cause increased anxiety which increases the likelihood of vehicle accidents. Neither cars nor buses can be used effectively when roads and bridges are out.

Boats are rarely used for evacuation purposes. However, boats along with helicopters and small planes may be used for islands and flood-prone areas. The availability of boats is often a problem for this method of evacuation. Evacuating by helicopter or plane may be the only option to move people or provide emergency assistance for those who have been cut off due to damaged roads and/or bridges. During the 2010 devastating flood in Pakistan, helicopters were often used to evacuate people from their flooded homes. As with buses and boats, it is difficult to find an adequate number of helicopters, planes, and pilots on short notice. Unlike cars, buses, boats, and helicopters/planes, walking is the main method of evacuation for vertical evacuation when evacuation distances are relatively short. For disabled persons, relatives and friends are often needed to help them to evacuate.

However, evacuations may also be dichotomized as short or long term. Depending on the nature of damage, disaster victims may be able to return to their homes within hours or a few days after the danger has passed. In other cases, it may take weeks or months to make their return. Some people evacuated after Hurricane Katrina may never return home. Approximately 72% of Katrina victims had returned to their homes by 2008 – three years after this event (Phillips, 2009).

Although evacuation behavior has been closely associated with officials issuing hazard warnings, people often evacuate before an official warning is issued. This type of evacuation is called "early" or "spontaneous" evacuation. For example, a significant proportion of households threatened by Hurricane Lili decided to evacuate before authorities issued an official evacuation warning (Lu et al., 2004). People may also evacuate from outside the official evacuation zone, which is called "shadow evacuation." This type of evacuation is more common for events associated with hazardous material accidents. Phased evacuation, on the other hand, stages the evacuation in a sequential manner. The timing in which different risk zones are warned to evacuate depends on the nature of the evacuation problem that the phasing is attempting to eliminate. In hurricane-prone areas plans have generally been developed to evacuate coastal areas prior to inland areas. Other strategies can involve a combination of sheltering and evacuation such as sheltering people nearest to the hazard and evacuating people more distant or, the reverse, evacuating people closest to the hazard and sheltering more distant populations (Sorensen and Vogt, 2006).

Evacuation behavior

People do not always comply with evacuation orders. Evacuation rates generally vary for different hazard types, and for different levels of risk (defined geographically). Evacuation rates are very high for most hazardous material accidents, where compliance may be in the high 90% range. Evacuation rates are typically low for slow-onset events such as riverine floods. Evacuation rates vary in hurricanes depending on the strength of the storm and its location. In high-hazard storm surge areas evacuation rates may be as high as 90% for major storms. Evacuation rates are much lower for smaller hurricanes and in lower risk zones.

Evacuation rates also differ between developed and developing countries. After providing early warnings five days before the landfall of Cyclone Sidr in 2007 and issuing emergency evacuation orders almost 27 hours before it struck the southwestern coast of Bangladesh, only nearly 40% of the people had evacuated to shelters in place prior to the landfall (Paul and Dutt, 2010). According to a study conducted by researchers from Harvard University, three-fourths of those in the New Orleans Metropolitan Area and nearly one-third of those in

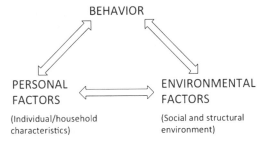

Figure 5.4 The reciprocal determination of behavior, person, and environment. *Source:* Modified after Bandura (1986).

Alabama, Louisiana (exclusive of the New Orleans Metropolitan Area), and Mississippi evacuated prior to the time Katrina hit (HKCAG, 2006). Another study claims that nearly 70% of all respondents did evacuate before Katrina's landfall. Twenty-one per cent of all respondents evacuated after the storm and 10.4% did not evacuate at all (Harrington *et al.*, 2006). This study selected 491 respondents from seven southeastern Louisiana parishes.

Several overlapping theories and perspectives exist in the hazard literature regarding individual responses to emergency warnings and evacuation orders. These theories can provide a general framework for understanding why prospective disaster victims do or do not respond to evacuation orders issued for an approaching extreme event. Social cognitive theory (SCT) provides a basis for understanding the personal and environmental factors that shape individual behavior toward an extreme natural event (Bandura, 1986). Figure 5.4 illustrates human behavior as being reciprocally determined by internal personal and environmental factors in which each person deals with the emergency (Bandura, 1986). Questions associated with SCT's personal factors (e.g., landholding size, home ownership, annual income, race/ethnicity, immigration status, age, level of education, and past warning and disaster experience) should provide an understanding of how people cognitively interpret the approaching extreme event's severity and duration, their perceived susceptibility to harm, the skills required to avert harm, their confidence in performing necessary actions (self-efficacy), their goals at the time of the emergency, and their expectations of surviving the emergency.

SCT's environmental factors, on the other hand, provide an understanding of the knowledge, attitudes, and practices of family, friends, co-workers, and others in a person's social network as he or she deals with the crisis (i.e., social environment) and the extent of information individuals possess about what to do, where to go, and the steps they must take to increase safety and avoid harm. Individuals generally inquire of members in their social network to corroborate warning information, a process which is known in hazard literature as "confirmation behavior" or "normalcy bias." They also frequently seek to reinforce the warning message by observing environmental cues and checking alternative warning sources. Individuals often fail to heed warning and/or evacuation orders if they are issued by what they perceive as an untrustworthy source, or even because members of their social network are not evacuating (Sorensen and Mileti, 1987).

Social and structural environments, taken together, provide a basis for understanding differences in the cognitive representation of the event and its effects among people who evacuate versus those who do not. Finally, SCT presents a balanced and optimistic view of the human condition: people and their behaviors are shaped by their environments, and people also shape their environments through their behavior and expectations. A social cognitive perspective of crisis management holds that individual behavioral changes can be facilitated by modifying people's personal factors and by altering environmental factors to encourage more appropriate behavior with regard to future emergencies (Bandura, 1986).

Considering that any emergency involves at least a potential threat to life and/or property, a prevalent feeling and consideration among most residents is some form of fear. Fear poses special problems for managing an emergency. Terror management theory (TMT) provides a means of investigating people's motives regarding their decision to either stay or leave (Greenberg et al., 1997; Pyszczynski et al., 1999). Feelings of fear may induce prospective disaster victims to engage in different defense mechanisms. People who develop constructive defense mechanisms are likely to respond to fear in a self-protective and socially acceptable manner. However, some individuals may reduce fear by denying their vulnerability to the hazard threat, distorting its immediacy, distracting themselves from it, dismissing the source of hazard information, or using other mechanisms that do not result increased safety from the hazard (Chaplin, 2000; Greenberg et al., 1993; Pyszczynski et al., 2003). Such mechanisms tend to be self-destructive and anti-social. The specific defense mechanisms in which an individual will engage likely depend on cognitive information he or she has about the emergency. Specifically, if an individual judges the threat as real but does not feel that the recommended response can be carried out effectively, he or she is more likely not to comply with evacuation orders.

In addition to SCT and TMT, a number of other approaches to environmental hazards, such as environmental inequality, vulnerability, and political ecological paradigms (Wisner et al., 2004), general risk communication model (GRCM) (Sorensen and Mileti, 1987), and general hazards risk communication model (GHRCM) (Blanchard-Boehm and Cook, 2004), have focused on social, psychological, economic, demographic, and geographic variables influencing public response to hazards, hazard warnings, and evacuation (also see Dow and Cutter, 2000; Drabek, 1996; Lindell and Perry, 1992; Quarantelli, 1994; Sorensen, 2000).

The environmental inequality paradigm, the political ecological paradigm, and the vulnerability paradigm concede that public sources discriminate against marginal groups (e.g., the poor, women, immigrants, and ethnic and religious minorities) in the provision of risk communication (Bolin and Stanford, 1999; Cutter et al., 2000; Liverman, 1990; Wisner et al., 2004). Hazard literature clearly suggests that disabled persons, elderly residents, and members of marginal groups are particularly vulnerable to extreme natural events because it is more difficult for them to receive warnings and to respond appropriately. Many other factors, such as income level, education, and occupation, also contribute to vulnerability of people living in hazard-prone areas.

Both the GRCM and GHRCM models consider individual response to a hazard warning as a social process (Blanchard-Boehm and Cook, 2004; Mileti and Peek, 2000). After receiving a hazard warning, people typically go through a social psychological process to form personal assessments of the risks they face and formulate ideas about what to do before they take protective action. Mileti (1995) maintains that warning systems that are not designed to take the social psychology of public warning responses into account are much less likely to elicit compliance with evacuation orders. As noted, believing or accepting a hazard warning message depends on many factors, including the source of the warning, characteristics of the message, its compatibility with the receiver's existing beliefs, and the characteristics of the receiver. Mileti (1999) claims that people think of protective response to extreme events in terms of four principal characteristics: efficacy, cost, time requirements, and implementation barriers. These characteristics are also applicable in the context of compliance with hazard evacuation orders.

Heuristic cues affect behavior, in addition to affective and cognitive factors. Heuristic cues refer to shortcuts in the decision-making process that can influence behavior (Chaiken, 1987; Malhotra and Kuo, 2008; Petty and Cacioppo, 1986). For instance, a person may fail to heed a warning because it was issued by an untrustworthy source, or because neighbors, friends, and/or relatives were found not to be evacuating. Customs, values, and preferences of the hazard warning recipient are also likely to affect the evacuation behavior. Consequently, a profile of receiver characteristics is required to understand why someone might be likely to heed – or fail to heed – a hazard warning to evacuate (Sorensen and Mileti, 1987). In such a profile, information is developed about the receiver's level of acculturation, immigrant status, socio-economic condition, racial/ethnic background, gender, home ownership, interpersonal and political trust, geographical location, potential barriers to evacuation (e.g., lack of transportation, distance to shelter, and conditions of shelter as perceived by the prospective evacuee), individual responses to the media and media preferences (Witte, 1994). In developed countries like the United States, the presence of household pets also affects the decisions of those at risk to evacuate (Dow and Cutter, 2000; Heath et al., 2001).

Given the theories and perspectives presented above, it is not surprising that the disaster literature stresses that a myriad of personality traits and individual circumstances dictate who has access to hazard warnings and who would likely comply with such warnings. The poor are less likely than the more affluent to have access to multiple media or other technologies, thus inhibiting their receipt of hazard warnings. The elderly have a tendency to dismiss warnings in a cognitive process framed by situational factors, such as compromised mobility and media access (Gruntfest, 1987). Individuals will often ignore warnings if previous warnings proved wrong in the past, which is termed in disaster literature as the "cry wolf effect" or "false alarm effect" (Paul and Dutt, 2010). Mileti and Sorensen (1990) claim that people directly affected by disasters are significantly more responsive to future hazard predictions and warnings than those who have never been affected by a disaster.

Other individual traits that lead to evacuation order failures include: the inability to evacuate due to the lack of transportation or travel funds, or commonly held beliefs that "it won't happen to me/us" or "God will protect us" (Biddle, 2007). Evidence (e.g., Ikeda, 1995; Dow and Cutter, 1998) suggests that prospective disaster victims often defy evacuation orders and stay home because they are afraid of having their homes looted during the evacuation. Concerns, such as being unable to return home after an evacuation due to the imposition of a curfew, work responsibilities, as well as concern for the well-being of pets and household members, have also prompted individuals to remain at home. All the above act as barriers to evacuation.

Sorensen and Mileti (1987) reported that a response to emergency warnings is impacted by a family's preparation for emergencies, having children, consistency and clarity of the warning message, and gender of the respondent. Predictors for noncompliance with evacuation requests are optimism concerning outcome (Dow and Cutter, 2000), being a racial/ethnic minority, crime and fear of property loss, and lack of credible information regarding a disaster (Sorensen and Mileti, 1987). In the context of hurricanes, Dow and Cutter (2000) maintain that the factors that influenced evacuation response were magnitude of the storm, proximity of the storm to respondent, elected official decrees, and past hurricane experience. Bateman and Edwards (2002) added that women are more likely than men to evacuate because women perceive their residence to be at greater risk, and were more likely to have developed a household evacuation plan. They came to this conclusion after analyzing data collected from a cross-sectional survey of 1050 coastal North Carolina households affected by Hurricane Bonnie, which made landfall on August 25, 1998.

Studies (e.g., Haque, 1995; Haque and Blair, 1992) conducted in Bangladesh suggest that households often decide in favor of partial evacuation where some, but not all household members comply with evacuation orders. Concerns over the well-being of relatively more vulnerable family members often leads to partial evacuation. One or more family member stays at home to take care of a more vulnerable member(s). Haque (1995) noted that only the most vulnerable family members (e.g., women, children, the disabled, and the elderly) were evacuated prior to the landfall of the 1991 Bangladesh cyclone. He also reported that at least one adult male member stayed home to protect property from looting. Although the keeping of small animals as pets is almost non-existent in rural

Bangladesh, previous studies (e.g., Haque, 1997; Haque and Blair, 1992; Paul and Dutt, 2010) reported that coastal residents who owned cattle were reluctant to evacuate to a public cyclone shelter, at least not until they had moved their cattle to a safe place.

It is clear from the above that there are many reasons why people refuse to evacuate. These reasons are: non-awareness of the evacuation order, evacuation orders that are not clear, neighbors' and friends' non-evacuation, not taking the evacuation order seriously, fear of looting, fear of re-entry, job responsibilities, family obligations, distance to shelter, past evacuation experience, pet/cattle ownership, educational level, age, gender, and cost involved. An individual's decision to evacuate is a complex process, which is influenced by a combination of factors. Education about the need for evacuation in developing countries, and evacuation assistance to those without means in countries where horizontal evacuation is very common can increase the evacuation rate.

Evacuation is not only expensive, it is also dangerous – people can and do die while evacuating. For these reasons, hazard experts often recommend sheltering in place. In-place shelters should be designated long before the occurrence of an extreme event and people should be aware of the location of all such shelters. Not surprisingly, an evacuation is counterproductive if it is not necessary. In such a case, people may lose faith in evacuation orders and subsequently may not comply with such orders. Computer models can help in this regard. HAZUS identifies vulnerable areas and estimates potential damage. CAMEO/ALOHA provides plume models for hazardous chemical releases, and SLOSH generates storm surge models and potential inundation areas. However, the parameters used in these models are often subjective and thus fail to accurately designate the entire area under threat.

5.2.1.3 Exercise

One of the most effective ways to prepare for a disaster is to conduct practice exercises on a regular basis. The primary goal of such exercises is to improve operational readiness for people who participate in disaster response and recovery efforts. Exercises enhance a community's overall emergency management capabilities and provide useful tools for local programs to train emergency personnel and evaluate their operational readiness. Through exercises it is possible to identify shortcomings of emergency operations plans (EOP), reveal resource gaps, improve coordination and individual performance, and clarify roles and responsibilities of personnel involved in disaster activities. Disaster exercises also foster cooperation among government agencies and private sector resources, increase general awareness of proficiencies and needs, and verify the availability and utility of equipment (McEntire, 2007).

According to FEMA, there are five types of exercises (see McEntire, 2007). An *orientation exercise* is considered to be the foundation for emergency management exercises and will lay the groundwork for a comprehensive exercise program. It is attended by all individuals or personnel with a role or interest in the goals or mission identified in the EOP. An orientation exercise motivates active participation in emergency response operations and lasts about an hour in an office setting. The presenter generally addresses the participants in an informal classroom environment.

A *drill* is an exercise that focuses on a single or relatively limited component of a community's response system. It is intended to improve the ability of staff to react in an effective and organized fashion in the event of a disaster by providing them experience using rescue equipment and coordinating efforts among the various agencies involved. Drills are undertaken to test, evaluate, and improve hazard response and recovery operations. Although the amount of time required to carry out a drill depends on the function or action being tested, drills generally last only 1 or 2 hours.

A *tabletop exercise* is a planned activity in which disaster management officials practice full activation of the emergency response plan in an office setting. It is usually informal and is designed to elicit constructive discussion from the participants on all plan components. Participants are encouraged to discuss plans in depth and identify problems and weaknesses of the plan. Tabletop exercises are conducted by a facilitator and usually last few hours.

A *functional exercise* is designed to enhance individual and organizational skills required in emergency management. It is based on a simulation of a realistic emergency situation that includes a description of the situation (scenario), a master sequence of events list, and communications between simulators. Unlike the previous three types of exercises, a functional exercise is time-dependent, thereby introducing stress to the scenario, and requiring participants to actually act out their emergency management roles and responsibilities. This type of exercise tests a limited number of response and recovery capabilities, and thus does not require full activation of the emergency response plan. It is also limited to a few agencies, and police and public health officials do not generally attend such exercises. These exercises may take place at the emergency office center (EOC), in the field, or at both locations. Functional exercises often involve testing the use of equipment.

A *full-scale exercise* is a very large simulation which encompasses a majority of the emergency management functions. This type of exercise involves the actual mobilization and deployment of appropriate personnel and resources to demonstrate operational abilities. It requires inclusion of most emergency response organizations working in real time, and use all of the required equipment and procedures. It lasts nearly an entire day, and creates a significant degree of stress. The last two types of exercises are expensive, complex, and require months of preparation (Coppola, 2007).

For all five types of exercises, there is a need to schedule the date, location, and personnel. The organizer also needs to specify the purpose, expected actions, messages, and evaluation criteria for the exercises. Failing to adequately address these components will limit the effectiveness of exercises. It is important to note that such exercises are on going; they should be repeated at least once a year, if not more. This is necessary because future exercises should be built on the strengths and weaknesses of previous exercises. Unsurprisingly, funding is the main problem for conducting such activities on a regular basis.

5.2.1.4 Household preparedness

Household level preparedness activities include: developing an emergency plan for the household, discussing preparedness plan (evacuation route) with family members, storing food and water, securing the home and contents, having a first-aid kit at home, making sure that there is a battery-powered radio on hand and a working flashlight, and taking other steps to prepare and anticipate whatever problems a disaster might create. Purchasing hazard insurance and making structural changes to the home are also household-level preparedness activities. Such household preparedness activities are disaster-specific. For example, having a helmet for each family member is a preparedness measure for households who live in apartment buildings in earthquake-prone areas. Similarly, installing window protection that can reduce or eliminate potential impacts caused by high winds is a preparedness measure for residents of hurricane/cyclone-prone coastal areas.

Existing hazard literature clearly suggests that the number of preparedness measures differs by disaster type. The number of measures also differs for a given disaster because of variations in cultural and development contexts. For example, Turner *et al.* (1986) provided a list of 16 household earthquake preparedness measures for California, USA. But Paul and Bhuiyan (2010) selected a battery of 10 earthquake preparedness measures to examine seismic risk perception among residents of Dhaka, Bangladesh. These studies also show that the extent of adoption and willingness to adopt differs greatly by recommended measures. Some respondents acknowledged the risk, nevertheless, had undertaken no preparedness measures at all (Paul and Bhuiyan 2010).

Correlates of social vulnerability are also related to household preparedness. For example, minorities and low income households display lower levels of disaster preparedness (Coppola, 2007; Horney *et al.*, 2008). These associations can be explained in terms of decreased ability to receive and understand complex hazard information among minority groups and low income households. Mileti and Darlington (1997) reported that earthquake preparation was less common in the United States among minorities than among white populations. Similar findings are reported for African American households with respect to hurricane preparation supplies (Norris *et al.*, 1999). These studies indicate that minority households simply lacked the economic resources to better prepare for disasters.

The above findings, however, are not completely consistent among hazards nor across regions. Some researchers have found no racial variation with respect to flood preparation (e.g., Lindell *et al.* 1980). Tierney *et al* (2001) even claim that household preparedness levels are attributable in part to variations in the sources (e.g., media friends and community members) of information people use for assessing the danger a given hazard may pose. This, in turn, is related to the content of that information and to the impression it makes. If potential victims of an impending disaster think that the information is credible, they are then more likely to adopt preparedness measures relative to those that question the validity of the hazard information received.

Disaster preparedness has been found to be significantly associated with the level of hazard awareness. People who had heard, understood, and personalized the risk were much more likely to adopt preparedness measures than those who had no knowledge about an impending danger (Tierney *et al.*, 2001). These researchers also found that the extent of adoption of preparedness measures was related to recent disaster experience, the amount of damage households had experienced, and personal contact with friends, relatives, and others who were trying to prepare for an impending disaster. Disaster experience may alter individual perceptions of hazards, and may also change individual attitudes and behavior concerning hazard preparedness (Uitto, 1998). Memory is similar to experience and it is affected by variables such as age and the actual severity of the experienced disaster. Having school-going children, being married, educational attainment, the number of children at home, advanced age, owning a home, and having lived longer in the community and current homes were all also found to be positively related to levels of household preparedness (Tierney *et al.*, 2001).

Several studies (e.g., Mulilis and Duval, 1995; Paton, 2003) have examined the role cognitive variables play in adoption of preparedness measures. The studies found that those who appraise their personal resources (self-efficacy and response efficacy) as sufficient are more likely to adopt earthquake preparedness measures, rather than using emotion-focused strategies. Lindell and Whitney (2000) reported that respondent's rating of the efficacy of different preparedness measures in protecting persons and property was highly correlated with adoption of these measures. Community attachment was also found to be positively associated with preparedness measures. Lindell and Prater (2000) noted variations in the mean number of earthquake preparedness measures adopted between residents of highly vulnerable southern California and less vulnerable western Washington.

The existing hazard literature clearly shows that most of the household-level preparedness studies are concerned with earthquakes and hurricanes/cyclones. There is a gap in existing knowledge regarding the extent and determinants of preparedness for other natural hazards such as floods, blizzards, volcanic eruptions, and tornadoes. Further research is also needed to better understand what motivates people in high-risk hazard regions to increase and/or sustain preparedness efforts during periods of relative normalcy.

5.2.1.5 Organizational preparedness

Both public and private organizations have disasters preparedness programs. Preparedness activities of these organizations include: developing emergency response plans and warning systems, identifying evacuation routes and shelters, acquiring needed equipment, supplies and materials, and maintaining emergency supplies and communication systems. Other preparedness activities include notifying and mobilizing key personnel, training employees and response personnel on what to do in an emergency situation, conducting exercises and drills, and informing citizens of risks, preparedness options, as well as the content and meaning of warning messages through education programs. There has been a tendency to focus on organizational preparedness only for specific kinds of hazards; this needs to be broadened to include all citizens as well as all hazards they might be impacted by.

Organizations involved in preparedness include local emergency management agencies, and other crisis-oriented organizations (e.g., police and fire departments, and emergency medical service (EMS) agencies), and private business organizations. Both emergency and non-emergency organizations have preparedness responsibilities. There is a tendency of private and non-emergency organizations to be less than enthusiastic about disaster preparedness because available resources are generally among other, more pressing concerns. Moreover, organizations that are experiencing financial difficulty will tend to downplay any preparedness responsibilities (Tierney *et al.*, 2001).

A number of factors, such as type and size of the organization as well as disaster experience, are positively associated with hazard preparedness. There is some evidence suggesting that nationally based organizations with multiple locations have higher levels of preparedness than individual local firms (Drabek, 1996). In terms of the age of organizations, studies found both positive and negative associations with hazard preparedness (Tierney *et al.*, 2001). Gillespie and Streeter (1987) found that preparedness was positively associated with the internal structure of organizations. Specifically organizational capacity and formalization of roles and procedures within the organization were linked to enhanced preparedness. Public and official support also facilitate hazard preparedness within organizations.

5.2.1.6 Community preparedness

Like household and organizational preparedness, community preparedness encompasses a wide range of activity such as formulating plans, providing training for disaster responders and the general public to improve their understanding of what to do in a disaster, and conducting emergency drills and exercises. Because of the infrequent nature of extreme events, hazard preparedness is low in most communities and few resources are allocated to disaster preparedness. According to Perry and Lindell (2003), preparedness results from a process in which a community (i) assesses its susceptibility to the full range of environmental hazards, (ii) identifies material and human resources available to cope with these threats, and (iii) defines the organizational structures by which a coordinated response is to be made.

Communities that experience disasters frequently are more likely to be prepared for such events than communities subject to infrequent disasters. Larger communities tend to be more prepared than the smaller communities because of differences of available resources and manpower between them.

5.2.1.7 State and national preparedness

In a country such as the United States where a federal system of government exists, state government plays a key role in disaster preparedness, both supporting communities impacted by disasters and coordinating with the federal government on a wide range of disaster-related tasks (Tierney et al., 2001). State governments are required to develop their own disaster plans. Irrespective of political system, governments are involved in the diverse range of preparedness actions which are divided by Coppola (2007) into five general categories: planning, exercise, training, equipment, and statutory authority. Excepting exercise, all other categories are briefly discussed below.

Planning

Like households, organizations, and communities, most governments have an emergency plan, called emergency operations plan (EOP). This document is also referred to as contingency plan, continuity of operations plan, emergency response plan, and counter disaster plan. An EOP (or equivalent) is required at every level of government, from local to national. An EOP can also be developed for individual entities, such as hospitals, malls, or schools. Beyond the national level, EOP can also be created for countries of a particular geographic region, continent, or for the entire globe (Coppola, 2007).

Irrespective of level, all EOPs should clearly specify the people and agencies who will be involved in the response to extreme natural events, and distribute the responsibilities and actions among these individuals and agencies. Through an EOP, state and local governments should know in advance what needs to be done, how they will accomplish these tasks, what equipment they will use, and what resources available within and outside the concerned jurisdictions. This planning document should be based upon accurate knowledge of the threat and be flexible to accommodate changes in the threat environment and/or changes with the introduction of new or improved equipment for responding to hazards and disasters. (Perry and Lindell, 2003).

Training

Training is an important component to government preparedness. Training is essential for emergency response officials as well as for the public-at-risk. Officials need adequate training to effectively handle emergency situations without putting themselves in difficult situations. However, disaster training is not universally available – it is generally unavailable in many developing countries. Depending on the type of disaster, training is also needed for the at-risk population. For example, residents living close to nuclear power plants may be given potassium iodide tablets and instructed on their usage (Perry and Lindell, 2003). Training is likely to yield high dividends in terms of the effectiveness of emergency response. It can also become an important source of feedback regarding potential problems with the EOPs.

Equipment

The role that equipment plays in hazard forecasting and warning dissemination, search and rescue, risk communications, and debris management cannot be overestimated. For performing all these and other hazard-related activities, a host of adequate and appropriate equipment is needed: rescue equipment, fire suppression equipment, personal protective equipment (PPE), and other emergency and disaster response support equipment. Rescue equipment saves humans and animals trapped or unable to free themselves from life-threatening situations. Among other things, rescue equipment includes shoring and other support devices to stabilize collapsed buildings or mineshafts, digging, cutting, spreading, and other manipulation devices, vehicles (trucks, boats, helicopters, airplanes), and ropes, strapping and other items to extract victims from hard-to-reach locations, as well as imaging, listening, and locating devices (Coppola, 2007).

Fire suppression equipment is used to control the spread of fires affecting all forms of structures and vehicles. Fire suppression equipment includes vehicles, extinguishers, hose assemblies, chemicals, ladders, cranes, and cutting and spreading tools. PPE is used to protect responders from life-threatening situation while performing their duties. Among other things, PPE includes glasses, bulletproof vests, and oxygen and gas masks. All types of radios, telephones, and internet connections are also needed for communication between responders and emergency operations centers personnel and other emergency offices and individuals. Unfortunately, equipment for disaster response and preparedness requires some significant financial investment and many countries, particularly developing ones, are unable to purchase such equipment.

Statutory authority

As noted, many government officials and agencies are involved in disaster preparedness programs. Statutory authority ensures that disaster response agencies and functions are established, staffed, and receive necessary funding. This facilitates completion of duties by emergency personnel both during disaster and nondisaster periods. Stationary authorities are required to keep abreast of all facets of disaster management as required by new information or expanded needs (Coppola, 2007). New and changing circumstances can bring about changes in the emergency management system. For example, after the terrorist attacks of September 11, 2001, the US Federal government created the Department of Homeland Security (DHS) and FEMA became a part of the DHS.

Mitigation and preparedness activities take place at various levels. Successful implementation of these activities not only depends on available resources, but also on culture and political systems. These factors together determine what particular activities are effective and/or preferred over others. Unfortunately, our knowledge about these determinants is limited due to the lack of adequate research on both the mitigation and preparedness phases of the disaster management cycle. Further studies are, therefore, much needed focusing on these two stages of the cycle.

References

Aguirre, B. *et al.* (1998) A test of emergency norm theory of collective behavior. *Sociological Forum* 13: 301–320.
Alexander, D. (2000) *Confronting Catastrophe*. Oxford: Oxford University Press.
Ali, M.H. and Choudhury, J.R. (2000) Assessment of seismic hazard in Bangladesh. In *Disaster in Bangladesh: Selected Readings* (ed. K. Nizamuddin). Dhaka: Disaster Research Training and Management Center, University of Dhaka, pp. 109–126.
Bandura, A. (1986) *Social Foundations of Thought and Action: A Social Cognitive Approach*. Englewood Cliffs: Prentice Hall.
Bateman, J.M., and Edwards, B. (2002) Gender and evacuation: a closer look at why women are more likely to evacuate for hurricane. *Natural Hazards Review* 3 (3): 107–117.
Biddle, M.D. (2007) Warning reception, response, and risk behavior in the May 3 1999 Oklahoma City long-track violent tornado. Unpublished PhD dissertation, Department of Geography, University of Oklahoma, Norman, OK.
Blanchard-Boehm, D.R. and Cook, M.J. (2004) Risk communication and public education in Edmonton, Alberta, Canada on the 10th anniversary of the "Black Friday" Tornado. *International Research in Geographical and Environmental Education* 13 (1): 38–54.
Bolin, R. and Stanford, L. (1999) Constructing vulnerability in the first world: the Northridge Earthquake in Southern California, 1994. In *The Angry Earth: Disaster in Anthropological Perspective* (eds A. Oliver-Smith and S.M. Hoffman). New York: Routledge, pp. 89–112.
Bourdieu, P. (1985) Social space and symbolic power. *Sociological Theory* 7 (1): 14–25.
Brock, V.T. and Paul, B. K. (2003) Public response to a tornado disaster: the case of Hoisington, Kansas. *Papers of the Applied Geography Conferences* 26: 343–351.
Burton, L. *et al.* (1968) The Human Ecology of Extreme Geophysical Events. Natural Hazard Research Working Paper, No. 1. University of Toronto: Toronto.

Chaiken, S. (1987) The heuristic model of persuasion. In *Social Influence: The Ontario Symposium* (eds M.P. Zanna *et al.*). Hillsdale, NJ: Lawrence Erlbaum Associates, pp. 3–39.

Chaplin, S. (2000) *The Psychology of Time and Death*. Ashland, OH: Sonnet Press.

Coppola, D.P. (2007) *Introduction to International Disaster Management*. Boston: Elsevier.

Csiki, S. (2003) Existing methods and technologies for communication of hazard warnings to the American public. *Papers of the Applied Geography Conferences* **26**: 361–370.

Cutter, S.L. *et al.* (2000) Revealing the vulnerability of people and places: a case study of Georgetown County, South Carolina. *Annals of the Association of American Geographers* **90** (4): 713–737.

Diacon, D. (1997) Typhoon resistant housing for the poorest of the poor in the Philippines. In *Reconstruction after Disaster* (ed. A. Awotona). Aldershot: Ashgate, pp. 130–147.

Dilmener, R.S. (2007) *A Theory of Evacuation as a Coordination Problem*. Durham, NC: Duke University Press.

Dow, K. and Cutter, S.L. (1998) Crying wolf: repeat to hurricane evacuation orders. *Coastal Management* **26**: 238–252.

Dow, K. and Cutter, S.L. (2000) Public orders and personal opinions: household strategies for hurricane risk management. *Environmental Hazards* **2** (2): 143–155.

Drabek, T.E. (1996) The social dimensions of disaster: a FEMA Higher Education Course. US Federal Emergency Management Agency, Emmitsburg, MD.

Emrich, C.T. and Hughey, E.P. (2011) A hazard geographer's perspective on recent disaster events. *AAG Newsletter* **46** (4): 16–17.

FEMA (Federal Emergency Management Agency) (n.d.) Hazard Mitigation Planning Made Easy.

FEMA (2008) *Taking Shelter from the Storm: Building a Safe Room for Your Home or Small Business*. Washington, D.C.: FEMA.

Gillespie, D. and Streeter, C.L. (1987) Conceptualizing and measuring disaster preparedness. *International Journal of Mass Emergencies and Disasters* **5**: 155–176.

Government of Bangladesh (2008) *Cyclone Sidr in Bangladesh: Damage, Loss and Needs Assessment for Disaster Recovery and Reconstruction*. Dhaka: Government of Bangladesh.

Greenberg, L. *et al.* (1993) Effects of self-esteem on vulnerability-denying defensive distortions: further evidence on anxiety-buffering function of self-esteem. *Journal of Experimental Social Psychology* **29** (3): 229–251.

Greenberg, L. *et al.* (1997) Terror management theory of self-esteem and cultural worldview: empirical assessments and conceptual refinements. In *Advances in Experimental Social Psychology* (ed. M.P. Zanna). San Diego, CA: Academic Press, pp. 61–139.

Gruntfest, E. (1987) Warning dissemination and response with short lead-times. In *Hazard Management: British and International Perspective* (ed. J. Handmer). London: Geo Books, pp. 191–202.

Gruntfest, E. and Weber, M. (1998) Internet and emergency management: prospects for the future. *International Journal of Mass Emergencies and Disasters* **16** (1): 55–72.

Haque, C.E. (1995) Climate hazards warning process in Bangladesh: experience of, and lessons from, the 1991 April cyclone. *Environmental Management* **19** (5): 719–734.

Haque, C.E. (1997) Atmospheric hazards preparedness in Bangladesh: experience of, and lessons from, the 1991 April cyclone. *Natural Hazards* **16** (2–3): 181–202.

Haque, C.E. and Blair, D. (1992) Vulnerability to tropical cyclones: evidence from the April 1991 cyclone in coastal Bangladesh. *Disasters* **16** (3): 217–229.

Harrington, L. *et al.* (2006) Southeastern Louisiana evacuation/non-evacuation for Hurricane Katrina. In *Learning from Catastrophe: Quick Response Research in the Wake of Hurricane Katrina* (ed. Hazards Research Center). Boulder, CO: University of Colorado at Boulder, pp. 327–352.

Heath, S.E. *et al.* (2001) Human and pet-related risk factors for household evacuation failure during a natural disaster. *American Journal of Epidemiology* **153** (7): 659–665.

Horney, J. *et al.* (2008) Factors associated with hurricane preparedness: results of a pre-hurricane assessment. *Journal of Disaster Research* **3** (2): 1–7.

HKCAG (Hurricane Katrina Community Advisory Group) (2006) *Overview of Baseline Survey Results: Hurricane Katrina Community Advisory Group*. Cambridge, MA: Harvard University Press.

Ikeda, K. (1995) Gender differences in human loss and vulnerability in natural disasters: a case study from Bangladesh. *Indian Journal of Gender Studies* **2** (2): 171–193.

Lindell, M.K. and Perry, R. (1992) *Behavioral Foundations of Community Emergency Planning*. Washington, D.C.: Hemisphere Publishers.

Lindell, M.K. and Prater, C.S. (2000) Household adoption of seismic hazard djustments: a comparison of residents in two states. *International Journal of Emergencies Disasters* **18**: 317–338.

Lindell, M.K. and Whitney, D.J. (2000) Correlates of household seismic hazard adjustment adoption. *Risk Analysis* **20** (1): 13–25.

Lindell, M.K. *et al.* (1980) Race and disaster warning response. Research Paper. Battelle uman Affairs Research Center.

Lindell, M.K. *et al.* (2005) *Organizational Communication and Decision Making in Hurricane Emergencies*. College Station, TX: Hazard Reduction & Recovery Center, Texas A&M University.

Liverman, D. (1990) Drought in Mexico: climate, agriculture, technology and land tenure in Sonora and Puebla. *Annals of the Association of American Geographers* **80** (1): 49–72.

Lu, J.C. *et al.* (2004) *Household Evacuation Decision Making in Response to Hurricane Lili*. College Station, TX: Texas A&M University Hazard Reduction and Recovery Center.

Malhotra, N. and Kuo, A.G. (2008) Attributing blame: the public's response to Hurricane Katrina. *The Journal of Politics* **70** (1): 120–135.

McEntire, D.A. (2007) *Disaster Response and Recovery*. New York: John Wiley & Sons Inc.

Maskrey, A. (1989) *Disaster Mitigation: A Community Based Approach*. Oxford: Oxfam.

Mileti, D.S. (1995) *Natural Hazards Warning Systems in the United States*. Boulder, CO: University of Colorado.

Mileti, D. (1999) *Disasters by Design: A Reassessment of Natural Hazards in the United States*. Washington, D.C.: Joseph Henry Press.

Mileti, D.S. and Darlington, J.D. (1997) The role of searching behavior in response to earthquake risk information. *Social Problems* **44** (1): 89–103.

Mileti, D.S. and Peek, L. (2000) The social psychology of public response to warnings of a nuclear power plant accident. *Journal of Hazardous Materials* **75**: 181–194.

Mileti, D.S. and Sorensen, S. (1990) *Communication of Emergency Public Warnings: A Social Science Perspective and State-of-the Art Assessment.* Oak Ridge, TN: Oak Ridge National Laboratory.

Mulilis, J.P. and Duval, T.S. (1995) Negative threat appeals and earthquake preparedness: a person-relative-to-event (PrE) model of coping with threat. *Journal of Applied Social Psychology* **39** (1): 1319–1339.

Murphy, K. and Sherry, M. (2003) After devastation come the lessons: safer homebuilding again in FEMA spotlight. *The Kansas City Star*, May 15: A1 and A4.

NCDEM (North Carolina Division of Emergency Management) (1998) *Tools and Techniques: Putting a Hazard Mitigation Plan to Work.* Raleigh, NC: NCDEM.

Norris, F.H. et al. (1999) Stability and change in stress, resources, and psychological distress following natural disasters: findings from Hurricane Andrews. *Anxiety, Stress, and Coping* **12** (4): 1–22.

Paton, D. (2003) Disaster preparedness: a social-cognitive perspective. *Disaster Prevention and Management* **12** (3): 210–216.

Pattan, A. (2003) Grassroots homeland security. *Natural Hazards Observer* **27** (5): 1–3.

Paul, A.K. et al. (2002) Role of shelter center for cyclone hazard mitigation in Cox's Bazar, Bangladesh. *The Chittagong University Journal of Science* **26** (1&2): 113–123.

Paul, B.K. and Bhuiyan, R.H. (2010) Urban earthquake hazard: perceived seismic risk and preparedness in Dhaka City, Bangladesh. *Disasters* **34** (2): 337–359.

Paul, B.K. and Dutt, S. (2010) Hazard warnings and responses to evacuation orders: the case of Bangladesh's Cyclone Sidr. *Geographical Review* **100** (3): 336–355.

Perry, R.W. and Greene, M.R. (1982) The role of ethnicity in the emergency decision-making process. *Sociological Inquiry* **52**: 309–334.

Perry, R.W. and Lindell, M.L. (2003) Preparedness for emergency response: guidelines for the emergency planning process. *Disasters* **27** (4): 336–350.

Petty, R.E. and Cacioppo, J.T. (1986) *Communication and Persuasion: Central and eripheral Routes to Attitude Change.* New York: Springer-Verlag.

Phillips, B.D. (2009) *Disaster Recovery.* Boca Raton: CRC Press.

Prater, C.S. and Lindell, M.K. (2000) Politics of hazard mitigation. *Natural Hazards Review* **1** (2): 73–82.

Pyszczynski, T. et al. (1999) A dual process model of defense against conscious and unconscious death-related thoughts: an extension of terror management theory. *Psychological Review* **106** (4): 835–846.

Pyszczynski, T. et al. (2003) *In the Wake of 9/11: The Psychology of Terror.* Washington, D.C.: American Psychological Association.

Quarantelli, E.L. (1994) Preparedness and disasters: a very complex relationship. Preliminary paper no. 209. University of Delaware, Newark, DE.

Rahman, M.K. (2010) Preparation and response to the flood of 2008 and windstorm of 2009 in Cedar Falls, Iowa. MA thesis, University of Northern Iowa, Cedar Falls, Iowa.

Smith, K. (1992) *Environmental Hazards: Assessing Risk and Reducing Disaster.* London: Routledge.

Sorensen, J.H. (2000) Hazard warning systems: review of 20 years of progress. *Natural Hazards Review* **1** (1): 119–125.

Sorensen, J.H. and Mileti, D.S. (1987) *Public Response to Emergency Warnings.* Reston, VA: USGS.

Sorensen, J. and Vogt, B. (2006) *Interactive Emergency Guidebook.* http://emc.gov/CSEPPweb/evac_files/index.htm (accessed December 22, 2010).

Sutter, D. and Poitras, M. (2010) Do people respond to low probability risks? Evidence from tornado risk and manufactured homes. *Journal of Risk and Uncertainty* **40**: 181–196.

Tierney, K.J. *et al.* (2001) *Facing the Unexpected: Disaster Preparedness and Response in the United States*. Washington, D.C.: Joseph Henry Press.

Tobin, G.A. and Montz, B.E. (1997) *Natural Hazards: Explanation and Integration*. New York: The Guilford Press.

Tobin, G.A. and Montz, B.L. (2004) Natural hazards and technology: vulnerability, risk, and community response in hazardous environments. In *Geography and Technology* (eds S.D. Brunn *et al.*). Boston: Kluwer Academic Publishers, pp. 547–570.

Turner, R.H. *et al.* (1986) *Warning for Disaster: Earthquake Watch in California*. Berkeley, CA: University of California Press.

Uitto, J.I. (1998) The geography of disaster vulnerability in megacities: a theoretical framework. *Applied Geography* **18** (1): 7–16.

White, G.F. (ed) (1974) *Natural Hazards: Local, National, Global*. New York: Oxford University Press.

Wisner, B. *et al.* (2004) *At Risk: Natural Hazards. People's Vulnerability and Disasters*. London: Routledge.

Witte, K. (1994) Fear control and danger control: the extended parallel response model (EPPM). *Communication Monographs* **61** (2): 113–134.

6
Disaster Cycles: Response and Recovery

In the aftermath of disasters, emergency managers and other emergency personnel involve themselves with response and recovery efforts to protect life and property of survivors, and return the affected community and individuals to pre-disaster or, preferably, improved conditions. Because it is impossible to eliminate all disasters, they need to participate in these two important phases of the disaster management cycle. Although there is certainly some overlap, the response and recovery phases are in sequential order and therefore they are treated in the same chapter. Salient features of response and recovery are presented below in a sequential order.

6.1 Response

Response refers to all actions taken immediately before, during, and after a disaster to save lives, minimize damage to property, and enhance the effectiveness of recovery at shortest possible time (Mileti, 1999). It is generally more narrowly defined to mean only the immediate actions (e.g., search and rescue operations for survivors, emergency medical care for the injured including mental health counseling, the proper identification and disposition of dead bodies, provision of temporary shelter, water, sanitation and food, repair utilities and key infrastructure, security for victims, protection of property, closure of roads and bridges, and attending to secondary hazards such as fire (in the case of earthquake) taken immediately after a disaster at various levels, such as individual, household, community, national, and international (Mitchell and Cutter, 1997).

Cleaning debris from disaster-affected areas and restoring roads and infrastructure if these are partially and/or completely destroyed by disaster are also

---

Environmental Hazards and Disasters: Contexts, Perspectives and Management, First Edition. Bimal Kanti Paul.
© 2011 John Wiley & Sons, Ltd. Published 2011 by John Wiley & Sons, Ltd.

activities belonging to emergency response. This is necessary to facilitate relief and rehabilitation operations, and provide emergency medical supplies and other disaster-related assistance in a more extensive manner. All actions are disaster-specific, and some action overlaps. For example, a tornado may not cause considerable damage to roads because it only affects a relatively small geographic area. But a tropical cyclone, tsunami, or winter storm generally devastates a very large geographic area and thus may cause severe damage to physical infrastructure such as roads, train tracks, bridges, as well as power and telephone lines and other mode of communications. For example, 2007 Cyclone Sidr that slammed coastal Bangladesh on November 15, 2007 totally destroyed 58 km of roads and partially damaged 1363 km. It caused power outages that resulted in a near-countrywide blackout for over 36 hours (*Natural Hazards Observer*, 2008).

In the United States, Hurricane Katrina, which struck Gulf Coast of Louisiana and Mississippi on August 28, 2005, brought a storm surge estimated at more than 20 feet (6 m) and overwhelmed the system of levees surrounding the city of New Orleans. As a result, more than 80% of the city went under water and much of its infrastructure was destroyed. This breakdown of infrastructure made it difficult for first responders to communicate with each other as well as to know the status of hurricane victims who were still in the city and to provide emergency assistance to them. This was one of the major reasons why the emergency response system failed in New Orleans (Eikenberry *et al.*, 2007).

Disaster response represents the early part of the relief phase of the disaster recovery cycle (Figure 6.1). Depending on the magnitude and type of the event, emergency response lasts for hours or days, perhaps up to one week (Fothergill and Peek, 2004). Inadequate or poor response creates a human-induced tragedy that exacerbates the plight of those already suffering the effects of a natural disaster. Such response also reflects a failure of government agencies and public administrators to provide adequate emergency support to disaster victims of a major event. Although this failure is very common in developing countries, it is not uncommon in developed countries like the United States. For example, the public response to Hurricane Katrina is widely considered an administrative failure (Eikenberry *et al.*, 2007).

Similar to inadequate and poor response, delayed response to an extreme event is also undesirable. Disaster experts maintain that the first 72 hours after a natural disaster are the most important, during which prompt mobilization of emergency resources can save many lives. Beyond individual and household levels, delayed response often leads to large-scale exodus of disaster victims from the affected communities. After Hurricane Andrew hit Homestead in 1992, many victims left the town because of delay in public response to the disaster. Homestead is a town of Florida with a population of 32 000 located in the Miami-Dade County metropolitan area. With a shrunken tax base, the city of Homestead went into a fiscal crisis several times in the 1990s that might have occurred less frequently had more people remained in the town (Franzino, 2004). Getting people back

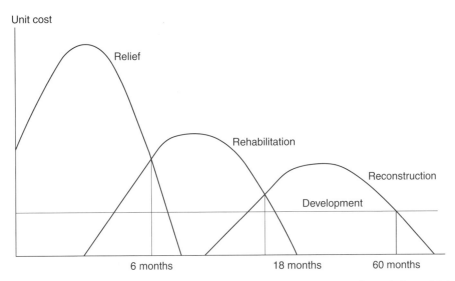

Figure 6.1 The disaster recovery cycle. *Source:* Frerks *et al.* (1995), p. 363.

into the town after the event as soon as possible is essential to reviving the economy of an area affected by disaster.

A plethora of private, public, humanitarian and nonprofit organizations, and individuals from both affected and non-affected areas/countries are involved in most major disaster responses. The number of organizations and individuals involved in response operations largely depends on the scale of the disaster. Irrespective of the scale, the neighbors and neighboring communities are generally first to respond. This is so because the community members have knowledge of socio-economic construction of disaster victims and of their multilayered local networks, such as family-kinship networks, networks of political parties, and religious networks. Without this knowledge, it is difficult to begin effective emergency response.

For high magnitude disasters, the national and foreign armed forces often participate in disaster responses. International nongovernmental organizations (INGOs), which operate across multiple countries or regions, typically respond to emergencies in developing countries. In the case of Katrina, however, more than a dozen INGOs, such as the International Rescue Committee, Oxfam, Save the Children, and UNICEF, provided significant humanitarian assistance and relief for the first time ever in the United States. INGOs responded to Katrina largely because they saw an overwhelming need that was not being met. It is noteworthy that some of these INGOs had to change or ignore their own organizational mandates to respond to a disaster in the United States (Eikenberry *et al.*, 2007).

Since numerous organizations and individuals are involved in responding to a disaster, there is a need for responders to coordinate and communicate ongoing situation assessment and resource mobilization during the emergency period.

Lack of communication and coordination often results in a waste of a sizable share of the resources flowing into the affected area. This also prolongs suffering of disaster victims and delays recovery efforts. Thus lack of coordination often becomes the primary reason for failure of response to a disaster. At the global level, the United Nations Office for the Coordination of Humanitarian Affairs (OCHA) typically acts as a coordinator of humanitarian emergency response by ensuring that an appropriate response mechanism is established and facilitating the smooth distribution of relief and other emergency needs to the disaster victims. A number of key activities associated with response are presented below.

6.1.1 Search and rescue operation

Search and rescue (SAR) operations consist of a number of activities undertaken immediately after a disaster to find victims who are injured and/or trapped under collapsed buildings, debris, or by moving water and remove them from danger or confinement. SAR procedures consist of three distinct but interrelated actions: locating victims; rescuing them from whatever condition has trapped them; and providing basic first-aid. The last action is necessary to stabilize victims so that they may be transported to regular emergency medical facilities for further treatment. Providing food and water to trapped victims is also the responsibility of SAR personnel.

Victims' friends, family members, neighbors, and others often participate in SAR operations. Some people in a disaster-affected community join rescue operations later, because they first have to rescue and/or attend to the needs of members of their own families, and then often to others, such as neighbors. In addition, local police, and fire department personnel, many individuals and organizations from neighboring towns, such as the National Guard, Department of Transportation personnel, the highway patrol, and the Salvation Army begin SAR missions soon quickly after a disaster. It has been found that about half of those rescued are rescued in the first 6 hours after a disaster occurs. The remaining half are rescued by trained and well-equipped SAR teams, known as emergent groups. Most disaster survivors are rescued within the first 24 hours (El-Tawil and Aguirre, 2010). However, some survivors are rescued days later due to persistent SAR efforts. This was the case after the Haiti 2010 earthquake.

Local responders who are usually the first to arrive on a disaster scene are not formally trained for SAR. Often they operate without appropriate and/or adequate equipment. As a result, they often place themselves in great danger. Despite rescuers being injured or killed in rescue attempts, many more lives are saved than lost. These responders use anything they can find to help them rescue trapped individuals and usually give preference to rescuing their friends and relatives over other victims. Empirical studies (e.g., Paul et al., 2007a) report that there is a strong association between SAR participation and their gender and age. In general women and the elderly are less willing to participate in such operations.

Formal SAR operations are conducted by trained professional teams who generally come from neighboring and/or more distant areas – even from foreign countries, and thus usually arrive comparatively late on the scene. For example, after the 2004 Indian Ocean tsunami, Thailand's navy and police were put into service for SAR operations. Rescue teams from throughout Asia, Europe, and North America soon joined these efforts. Similarly, nearly 2000 rescuers accompanied by 161 search dogs from Britain, Canada, the Dominican Republic, Argentina, the United States, China, Iceland, and many other countries came to Haiti two days after the country experienced a 7.0 magnitude earthquake on January 16, 2010. Members of these trained SAR teams are well equipped and as a result they can easily and quickly locate and save disaster victims. The type of equipment and tools needed depends on the situation. For example, cranes are usually needed to rescue people from high-rise buildings destroyed by earthquakes, whereas speed boats and ropes are needed to rescue people stranded by flooding.

One drawback to the deployment of foreign SAR teams is that it often takes days before members of these teams can begin searching for victims. Foreign teams need government approval from impacted countries which cause delays. They also usually need customs clearance for any equipment they may bring from their home countries. Often impacted countries do not allow rescuers or rescue equipment to cross borders from countries not on politically friendly terms with them. For example, the government of Pakistan required additional helicopters for rescuing survivors of the 2005 Kashmir earthquake. In addition to the rugged mountainous topography found in the Kashmir region, roads were blocked by landslides associated with this earthquake and its aftershocks. In this situation, helicopters were the only effective means to reach earthquake victims. Yet, despite this desperate situation, the Pakistani government refused an offer of Indian army helicopters (Ozerdem, 2006).

Neglected for many decades, only recently have public authorities and emergency response personnel made provisions for rescuing pets after an area is devastated by a disaster. In the absence of such provision, a considerable proportion of pet owners return to disaster areas to rescue their pets. For example, 80% of people who re-entered a flood evacuation site in Yuba County, California, in 1997, did so to rescue their pets (Heath *et al.*, 2000). A later study reported that 41% of pet-owning households that evacuated without their pets later returned the disaster-impacted area to rescue their cats, dogs, and other domestic pets (Beaver *et al.*, 2006). While no noticeable pet rescue effort was undertaken following Hurricane Katrina, animal rescuers did search for pets that had been left in backyards prior to the landfall of Hurricane Gastav on September 1, 2008 along the Louisiana coast. Similarly, more than 1000 animals were rescued from Galveston Island, Texas, in the weeks after Hurricane Ike made landfall in there on September 13, 2008 (Gallay, 2009).

6.1.2 Emergency medical care

Natural disasters are responsible for thousands of deaths and injuries each year world wide. Generally, more people are injured than die from direct impacts associated with an extreme event. For example, Hurricane Katrina killed some 2000 people, but injured 5698 – almost three times the number of fatalities. This hurricane made landfall along the southeast coastal portion of Louisiana, USA, as a category 3 storm on August 8, 2005. Cyclone Sidr, which made landfall along the Bangladesh coast on November 15, 2007, caused the death of 3406 people, but injured more than 55 000 (Paul, 2010). Most people injured in disasters need immediate medical attention in order to stabilize their condition; some suffering serious injuries need transport to a health facility where they can receive special medical services necessary to save their lives.

The nature of injuries incurred differs from one type of disasters to another as well as between developed and developing countries. Flooding generally produces hypothermia if victims are in the water for an extended period of time. Earthquakes, tornadoes, and hurricanes in general cause lacerations (from debris), fractures (from blunt trauma), and puncture wounds. In tornadoes and hurricanes, almost all body parts, such as feet, legs, arms, chest, thighs, eyes, head, neck, back, and ribs, are injured by falling/broken trees, flying/falling debris, or falling roofs and walls (Brenner and Noji, 1995; Brown et al., 2002; Paul, 2010). Fires are associated with burns and respiratory problems.

Before first aid and other emergency treatment can be given, it is necessary to locate injured victims as quickly as possible. Delay in finding and treating these victims can cause prolonged suffering and even deaths. Disaster victims often suffer from life-threatening injuries, including loss of blood and bodily fluids, damage to internal organs, and respiratory failure due to a lack of oxygen. It is important to mention that the general public are not the only ones at risk of injury; emergency responders, including police officers and firefighters also risk both death and injury while response and recovery operations are underway.

First responders are generally in a disaster-impacted area before other assistance arrives, and they provide emergency medical care to disaster victims. These personnel are usually followed by firefighters and ambulatory service staff, including emergency medical technicians, paramedics, nurses, and physicians. It is important for first responders to have basic first aid knowledge and skills. However, in addition to domestic medical teams, foreign countries also send mobile medical teams (MMTs), primarily to developing countries, to treat disaster victims with injuries. For example, Pakistan and the United States sent MMTs into Bangladesh after it experienced Cyclone Sidr in mid-November, 2007. The United States also sent an 18-member army medical team to operate a 250-bed hospital treating survivors from affected communities. Using a US Marine CH-46E Sea Knight helicopter, this MMT treated sick and injured persons in remote villages (Paul, 2010).

The United States also sent a navy hospital ship, the USNS *Comfort*, to Haiti after it was devastated by an earthquake in 2010. The ship was sent to provide emergency medical treatment, including surgery for people injured during the earthquake. Such ships are usually operated by military forces/navies of various countries around the world, as they are intended to provide a mobile, flexible, rapidly responsive afloat medical capability to treat combatant forces deployed in war or other operations. These ships' secondary mission is to provide full hospital services to support disaster relief and humanitarian operations.

The United States, Argentina, and several other countries also opened field hospitals in Haiti after the 2010 earthquake. These hospitals are temporary facilities constructed at or near the disaster-impacted areas to treat patients. In most cases, these hospitals are set up inside large tents or undamaged buildings, relying on equipment and staff transported in from hospitals from non-impacted areas and/or from foreign countries. Often hospitals of neighboring countries also treat people injured from disasters. For example, hospitals in Dominican Republic were made available to treat people injured by the Haiti earthquake. In addition, the Dominican Republic's emergency team assisted more than 2,000 injured people. The Dominican government sent eight mobile medical units along with 36 doctors, including orthopedic specialists, traumatologists, anesthetists, and surgeons. Other countries, such as Israel, Qatar, Iceland, and China either established field hospitals and/or sent medical teams to treat Haitian earthquake victims.

However, both first responders and professionals who volunteer their medical services must have sufficient qualifications. Malpractice lawsuits can result if a victim does not receive necessary medical treatment or if the treatment proves counterproductive. For a major disaster, which generally causes many injuries, hospitals of the impacted areas may soon be overwhelmed with patients. In such cases, there is an urgent need to set up field hospitals to reduce the burden on local hospitals. Another potential problem is that for some disasters, both domestic and foreign emergency medical teams may show up in large numbers, even in overabundance. This situation can create coordination problems, which may further cause delay in treating patients. Somewhat ironically, emergency medical services available are more than what is needed. For example, some survivors of the 2007 Greensburg, Kansas, tornado reported too many ambulances available immediately after the tornado hit the town and about half were not necessary. These ambulances came from neighboring towns (Paul *et al.*, 2007a).

6.1.2.1 Triage

One of the key steps that trained emergency medical personnel take to manage disaster first aid is triage. Triage is an initial assessment and separation of disaster victims (by attaching tags) into different groups for treatment based on the

Table 6.1 Triage tagging categories

Letter/code	Meaning
START	
D	Deceased
I	Immediate (victims needs advanced medical care within 1 hour)
DEL	Delayed (victim needs medical care, but treatment can be delayed)
M	Minor (victim can wait several hours before nonlife-threatening injuries are treated)
Advance triage	
Black	Expectant (victims' injuries are so severe they are expected to die)
Red	Immediate (victims are likely to survive their injuries, but only with immediate surgery or other life-saving treatment)
Yellow	Observation (victims are injured and need emergency medical care, but current condition is stable; must be monitored for change in condition)
Green	Wait (victims need medical care within several hours or even days, but will not die of their injuries if left untreated in the immediate future)
White	Dismiss (victims need little more than minor first aid treatment or basic care not requiring a doctor)

Source: Compiled from Coppola (2007), p. 257.

severity of their injuries. Separation of victims into groups helps paramedics and medical staff know which patients are in the most need of treatment and which patients can wait to be treated or for additional help to arrive. Obviously disaster-induced injuries vary in terms of seriousness. Some wounds are minor, while others are very serious or life threatening. By triaging injured disaster victims, first responders ensure that the highest priority cases are treated and transported to proper medical facilities before less serious ones.

Triage tagging involves marking injured disaster victims with a symbol on their forehead or a color-coded tag. There are two established systems of tagging – one for the situation when onsite medical resources are scarce and victims need to be transported to appropriate facilities, and the other for the situation when sufficient emergency medical care is available on site. The first system is known as START (Simple Triage and Rapid Transport) and the second system is called Advanced Triage. The categories included in both systems are listed in Table 6.1. Once patients are categorized, they may be physically moved to different locations. In transporting patients, care should be taken not to send people who do not require extra medical care. Also, it is important to ensure that triage decisions are only made by people with sufficient medical backgrounds.

6.1.2.2 Mental health counseling

Some disaster survivors often face post-traumatic stress and depression after experiencing death and injury among family members, friends, neighbors, or co-workers, loss of personal property, and/or witnessing damage and destruction. Disasters render some people homeless and jobless, with no apparent means to support their families. All cause psychological stresses of different types. One such stress is referred to as post-traumatic stress disorder (PTSD) or critical incident stress (CIS). PTSD is a type of anxiety disorder which occurs soon after a major trauma. As a response to a traumatic event, an individual may develop dissociative symptoms called acute stress disorder (ASD). First responders and other emergency management personnel may also become victims of PTSD/CIS or ASD. They are exposed to the emotional pain and suffering associated with death, injury, and destruction caused by disasters. They often have the added psychological pressure of feeling responsible for saving lives and protecting people at a time when both tasks are extremely challenging (Coppola, 2007).

Both disaster victims and responders may need mental health counseling, without which they may slip into behavioral health problems. Often, consequences of such depression include increase in crime rate, and rise in consumption of alcohol and drugs, a rise in the suicide rate and increased domestic violence. The nature and extent of stress-related illnesses generally differ according to age of survivors. Among all age groups, children are one of the most vulnerable populations because their neurophysiological systems are subject to permanent changes and their coping skills are rarely developed enough to manage catastrophic events (Baggerly and Exum, 2008). Typically, most children exhibit temporary stress-related symptoms both during and after disasters; yet most of these symptoms can be mitigated when parents and/or teachers provide emotional support and facilitate adaptive coping strategies. However, some children may experience serious clinical symptoms, which require interventions from professional counselors.

Much of the initial mental health disaster response involves two primary goals: (i) normalizing feelings: reassuring survivors that the strange and upsetting feelings they experience after a disaster are normal, and (ii) helping survivors find effective ways of coping with any ongoing stress. For this reason, there is a need for an adequate number of mental health counselors in major disaster-impacted areas. When there is an inadequate number of such counselors, survivors have to wait in order to (and many may not even) receive mental health counseling. However, such shortages are often compensated by family therapists in neighboring communities providing counseling services to disaster survivors.

Emotional stress, trauma, and other psychological impacts of disasters are also unevenly distributed across socio-economic classes. In general, high-income survivors suffer far fewer psychological impacts compared to low-income survivors, largely because most, if not all disasters exacerbate poverty (Mileti, 1999). Similarly, it has been found that women and girls suffer more emotional problems after disasters than do men. However, males experience increased rates of depression and alcohol abuse relative to females in developed nations (Lindell and Prater, 2003). Elderly persons are more likely to have symptoms of PTSD than young adults (Jia et al., 2010).

Mental health counseling for disaster survivors is still an alien concept in many developing countries. However, for major disasters, such counseling is often available from personnel responding from developed countries. After the 2004 Indian Ocean tsunami, developed countries not only dispatched military personnel and equipment to the tsunami-devastated countries to directly participate in relief efforts by distributing aid, removing corpses, and clearing rubble, they also sent health specialists to provide immediate medical attention, and offered psychological counseling. Developing countries need to pay more attention to providing health counseling to disaster survivors.

6.1.3 Identification and disposal of dead bodies

Following most natural disasters, particularly larger ones, there is fear among emergency managers, public health personnel, and members of the general populace that dead bodies will cause epidemics (de Goyet, 2007). This is one of the myths associated with disasters, one which may temporally direct the focus of emergency workers and responders away from the needs of survivors – towards combating false realities (Fischer, 1998; McEntire, 2007). However, due to the fear of an epidemic, authorities often quickly bury or cremate dead bodies without proper identification of victims, and spray "disinfectants." Burying disaster victims before identifying them or not following customary burial procedures anger and often cause mental health problems for family members and relatives of the victims. This practice also causes legal problems for the authorities. If disaster survivors do not touch dead bodies, the health risk for them is negligible. This is because most disaster victims die from trauma, drowning, or fire, and they are not likely to be sick with epidemic-causing infections. A few victims will have chronic infections in their blood, such as hepatitis, HIV, and tuberculosis. In a dead body diseases die within 48 hours, except HIV which can survive up to 15 days.

While the general public is relatively free from health risk, individuals handling dead bodies are at risk through contact with blood and feces. In addition, body recovery teams often work in hazardous environments (collapsed buildings and debris) and may also be at risk of injury and tetanus (transmitted via soil). It is important to note that body recovery is the first step in managing the dead and this phase of the response is usually chaotic and disorganized. Collecting bodies of human victims is seen as a priority activity because of the unpleasant nature of decomposing corpses, and the psychological burden it places on disaster survivors (Morgan et al., 2006).

A well-managed phase of body recovery should focus on a rapid retrieval of bodies for identification, which cannot occur without the use of modern technology. Whatever time it may take, body identification is important for victims' family members, relatives, and friends. Each body should have a record of the location, date of recovery, and unique reference number. Such record keeping plays an important role in reconstructing the event and identifying bodies and belongings. Belongings and body parts should be kept near the body or together at all times. Visual identification or photographs of fresh bodies are the simplest forms of identification and can minimize early nonforensic identification. Forensic procedures can take over after visual identification of bodies or photographs become impossible. Sooner is better for victim identification – decomposed bodies are much more difficult to identify and may require forensic expertise (Morgan et al., 2006).

After identification, a body should be released by the responsible authority to the victims' relatives for disposal according to local customs and religious practices. However, burial is the most practical method as it preserves evidence for possible future forensic investigation. For mass burial, careful thought must be given to the location of the burial site. Soil conditions, water table level, and available space must be considered. In addition, the proposed site should be acceptable to individuals and/or communities living near it. The World Health Organization (WHO) discourages mass burial to allow time for traditional religious funerary practices (Miller, 2005). In the first weeks after the 2004 Indian Ocean tsunami, some foreign groups buried bodies in mass graves in Sri Lanka and other tsunami-affected countries where many religious rituals are required before interment (Paul, 2007).

Long-term storage will be required for unidentified bodies and these bodies should be buried separately, with a funeral home or coroner keeping a record of their location. Identification and disposition of dead bodies necessitates the involvement of people of different professions, such as medical examiners, coroners, forensic investigators, morticians, mental health personnel, funeral home directors, and members of the clergy.

6.1.4 Debris removal

Many natural disasters generate large amounts of debris, causing considerable disposal challenges for public officials. Debris removal is the clearance, removal, and/or disposal of items damaged or destroyed by extreme events. It is necessary to: (i) allow the safe passage of emergency response vehicles in the aftermath of a disaster, (ii) eliminate public health and safety hazards, and (iii) ensure the economic recovery of the affected community to benefit the community-at-large. Delay in debris removal adversely impacts response and recovery operations. Thus, debris-removal crews must move quickly to help.

The volume and types of debris generated largely depends on the disaster type, spatial extent, and magnitude. Table 6.2 provides information on hazard-specific debris. Generally, tornadoes, hurricanes/cyclones, and earthquakes

Table 6.2 Hazard-specific debris

Type of hazard	Type of debris to expect
Flood	Mud, sediment, construction and demolition (C&D), appliances, tires, and hazardous waste.
Tornado	Buildings, roofing materials, cars, utility poles, traffic lights, homes, agricultural damage, roof, mobile homes/metal, and vegetation.
Hurricane	Sediment, trees, vegetation, personal property, construction, household items, and livestock carcasses.
Earthquake	Road and bridge materials, buildings, household and personal property, downed power lines, broken water pipes, and sediment from landslides.
Wildfire	Charred wood, ash, dead trees, and potentially mud and sediment from landslides due to denuded vegetation and loss of ground cover.
Hazardous materials	Hazardous materials can vary considerably in their amounts and toxicity. Debris removal must consider safety for the public and workers, determine the type of contamination and its effects on the environment, and develop appropriate identification, separation, removal, neutralization, interment, incineration, or other strategies for final disposal.
Terrorism	Buildings, vehicles, and hazardous materials.

Source: Compiled from EPA (1995).

produce more debris than floods and droughts (Figure 6.2). Among all disasters, hurricanes/cyclones account for the largest amount of debris from vegetation and man-made structures. For example, before Katrina, the event that left behind the greatest recorded amount of debris in the United States was Hurricane Andrew in 1992, which generated 43 million cubic yards (33 m^3) of debris in Florida's Metro-Dade County. Hurricane Katrina is believed to have generated more than 100 million cubic yards (76.5 m^3) of disaster debris (Luther, 2008). The volume of debris generated by Hurricane Katrina is unique because of the large area over which this storm caused damage. Parts of Alabama, Mississippi, and Louisiana, an area covering 90 000 square miles (233 000 square km), were declared a major disaster by President George W. Bush. Although debris was produced over the entire disaster area, the most significant property damage was concentrated within a 100-mile (150-km) radius of where the hurricane made landfall along the Gulf Coast (Luther, 2008).

Figure 6.2 Rubble created by a F3 tornado in downtown area of Pierce City, MO on May 4, 2003. *Source:* Author.

Debris produced in disasters can be classified into 10 groups. Table 6.3 shows that disaster-related debris may include many items, ranging from broken tree limbs, mud, rock, hazardous waste materials, broken glass and plywood, twisted metal, damaged property and possessions, automobiles, boats to dead animal carcasses. The excessive amount and types of debris generated by disasters must be dealt with by experienced and knowledgeable crews using heavy equipment to both clean and haul off these materials (Figure 6.3). Often there is a need to sort and store debris for disposal or recycling of some rubble, such as bricks, wood, and plastics. These materials can be used in the reconstruction process.

There are several debris-reduction methods: burning, chipping and grinding, recycling, and reuse. Burning reduces debris volume by 95% (Phillips, 2009). Open burning causes air pollution, and uncontrolled burning may cause wildfires. Controlled and open-air burning of woody debris is recommended because such burning produces minimal damage to the environment, but it should be limited to rural areas. The ash produced by burning can be used to increase soil fertility. Woody and vegetative debris can be reduced through various grinding and chipping methods. They can reduce woody debris by about 75% of its

Table 6.3 Types of disaster debris with examples

Type	Examples
Municipal solid waste (MSW)	Personal belongings and general household trash
Putrescibles	Rotten or spoiled fruits, vegetables, seafood, or meats
White goods	Refrigerators, freezers, washers, dryers, stoves, water heaters, dishwashers, and air conditioners
Household hazardous waste (HHW)	Oil, pesticides, paints, and cleaning agents
Construction and demolition (C&D) debris or aggregate debris	Asphalt, drywall, plaster, brick, metal, concrete, roofing materials, carpet, lumber, mattresses, and furniture
Vegetative debris	Trees, branches, shrubs, and logs
Automobile-related materials	Cars and trucks, fuel, motor oil, batteries, and tires
Electronic waste	Computers, televisions, printers, stereos, DVD players, and telephones
Dead animal carcasses	Livestock and wild animals

Source: Modified after Luther (2008), pp. 18–19.

original volume. One advantage of this reduction method is that reduced clean material can be reused and is thus economically beneficial as well as environmentally friendly (Phillips, 2009). In addition, abandoned furniture, retrieved siding, windows, shingles, and many other house construction materials can be reused and recycled.

Special care needs to be exercised to dispose of debris. Government agencies responsible for debris removal and disposal need to identify appropriate disposal sites and appropriate way to dispose of debris. Burying disaster rubble may adversely effect soil and water conditions. Thus, disposing of disaster rubble in inappropriate ways may have future impacts to human health. It is important to mention that governments often utilize private contractors to collect disaster debris for some communities because they may not be able to handle a large quantity of debris in a timely manner.

The duration of debris removal depends on the size of area impacted by a disaster and its magnitude. It often takes more than six months after a disaster to clear all debris. If it is not possible to completely remove all debris during the response phase, the quantity of debris should be reduced as quickly as possible. This can be accomplished in three ways. First, vegetative debris can be "chipped" which can reduce up to 75% of the total debris volume. Aggregates can be crushed and later used as road base. Finally, construction and demolition debris

Figure 6.3 A truck is carrying debris from Greensburg, Kansas after this small town hit by a category EF-5 tornado on the evening of May 4, 2007. *Source:* Mitchel Stimers.

may be recycled in some cases. Burning debris is another way to reduce debris, but must be carefully considered as previously mentioned.

All expenses associated with debris removal are generally paid by public authorities. In the United States, the most commonly used method for state and local governments to acquire assistance for debris removal is through the Federal Emergency Management Agency (FEMA) Public Assistance (PA) program. FEMA can either reimburse local governments for the cost of debris removal, or the local government can request direct federal assistance from FEMA, which then tasks the US Army Corps of engineers to complete the debris removal process (Luther, 2008). In order to be eligible for FEMA funding, debris removal work must be a direct result of a presidentially declared disaster, and must occur within the designated disaster area. The Environmental Protection Agency (EPA) ensures that debris management takes environmental protection into account.

Numerous public and private organizations, including survivors, their friends and relatives, and people from neighboring and even distant communities flock to the beleaguered community in the aftermath of a disaster to assist in clean-up operations. However, not all survivors participate in debris removal efforts. Ill-health and advanced age compel some survivors to forgo participation. Disaster survivors who participate in debris clearing operations remove their own debris first, and then join others to help clean streets and public places (Paul et al., 2007a). Time spent on clearing debris varies among individual volunteers. Because many organizations and individuals are involved in debris removal, there is a need to coordinate clean-up operations. There is also need for public outreach education efforts that explain where debris may be left for collection, what may be disposed of, and when it will be collected. Many houses and buildings may need to be demolished and survivors should be made aware of any such demolitions.

There are several problems with debris management. Debris or debris storage may attract rodents that have to be dealt with. Removal and transport of debris may create dust and noise that may annoy disaster survivors and others. Survivors often unload non-disaster-related items such as old mattresses, car doors, tires, and refrigerators at curbsides, believing that government agencies and/or private contractors will take these items away. Debris removal is also sometimes associated with scams by public officials and private constructors, who may often step over the line of legal activity in an effort to benefit financially from debris removal operations (Paul and Leven, 2002).

6.1.5 Post-disaster sheltering and housing

After a disaster in which people's homes are damaged or destroyed, families must seek alternative temporary housing until a permanent housing solution can be found (Johnson, 2007). Unlike survivors of hurricanes/cyclones or earthquakes, flood survivors may need alternative housing if their homes are submerged under flood water. The number of people requiring post-disaster shelter depends on the magnitude and type of disaster. In general, earthquakes, hurricanes/cyclones, and floods render more people homeless than do tornadoes or droughts.

The best choice for homeless disaster survivors for immediate shelter are facilities temporarily organized by both public and private agencies such as the American Red Cross and the Salvation Army within the community. Such private and public facilities should have been widely identified before the disaster, and they include covered stadiums, hospitals, schools, auditoriums, warehouses, and churches. Other options for temporary shelter include: (a) family, friends, relatives, or others outside the affected community for the short term, including hotels and motels; (b) camps, trailers, or other light housing set up privately or publicly to accommodate disaster survivors over a short to medium term; and (c) sturdy, but temporary, newly constructed houses for medium to long-term accommodation.

In constructing camps and sturdy temporary housing, relevant authorities must select sites in such a way that they meet the social needs of the affected population and they should be highly accessible. A selected site must be able to accommodate the number of displaced people that are planned to be located

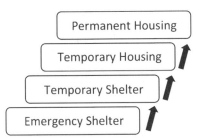

Figure 6.4 Sequence of return to housing. *Source:* Modified after Quarantelli (1982).

there. All sites should have access to water, adequate bathing and latrine facilities. Additionally, sites should allow for safe, healthy operation. When politics play a role in site selection, which is common in developing countries, disaster survivors may encounter challenges in just living at such a site. For example, Khan (1991) reported that the actual location decision and site selection of four flood shelters in four communities in northern Bangladesh was exclusively made by powerful local elites. By doing so, they secured their physical and social needs at the cost of actual flood victims.

Quarantelli (1995) offers a definition of the four distinct types of housing that may be used after a disaster by affected families (Figure 6.4). He makes the distinction between sheltering and housing in which the former denotes the activity of staying in a place during the height and immediate aftermath of a disaster, where regular daily routines are suspended, and the latter denotes the return to normal daily activities such as work, school, cooking at home, and shopping. Based on this distinction, the four types of housing are:

- Emergency shelter, which may take the form of a public shelter, refuge at a relative, friend, or neighbor's home, or under some form of shelter, including tents or even plastic sheets and is generally used for one night to a couple of nights during the emergency. Because of the short duration of stay, disaster victims do not normally resume their daily routines while staying in such shelters. Food is usually supplied either by relief agencies or by families and friends who have provided a place to stay.

- Temporary shelter, which may be a tent or a publicly designated mass shelter used for a few weeks following the disaster, and is also accompanied by the provision of food, water, and medical treatment.

- Temporary housing provides the means for a return to the daily activities of domestic life. Although disaster victims will be living in this temporary residence, most wish to move to permanent housing. Temporary housing can take the form of a rented apartment, a hotel room, or a small shack, depending on circumstances and available resources.

Figure 6.5 Cyclone Sidr survivors constructed makeshift temporary shelters in Char Khali, Patuakhali, Bangladesh. The cyclone made landfall on southwestern coast of Bangladesh on November 15, 2007. *Source:* Author.

- Permanent housing is the return to the former home after its reconstruction, or resettlement in a new home where the family plans to live on a permanent basis.

Those whose homes are affected by a disaster may or may not use all four types of shelters noted above. In developing countries, disaster victims usually use the last two types of housing during the post-disaster period. In such countries, temporary housing exclusively means small shacks built by the affected households without external assistance out of scrap materials (Figure 6.5). Living in such makeshift shelters, however, poses a serious threat to the health and well-being of disaster survivors. The stability and general quality of these shelters is typically far less than that of those destroyed or damaged by the disaster.

Temporary housing, however, no matter what form it takes, is the means by which affected families can begin to recover and reintegrate a sense of normalcy into their lives (Johnson, 2007). Depending on circumstances, it may take many months, or even years, to build and occupy permanent housing. Other factors that influence length of transition from temporary to permanent housing include the economic conditions of the disaster survivor, external assistance available, and the nature of new regulations imposed by authorities in affected communities. Strict regulations usually impose delays in repairing and/or rebuilding homes.

Post-disaster social ties influence where disaster survivors who have lost their homes seek alternative accommodations. In general, survivors who have small or weak social networks are more likely to use public shelters, while others choose family members, relatives, and friends (Mileti, 1999). Household income and resources, insurance, and access to affordable housing also have a significant impact on housing options after disasters (Johnson, 2007; Quarantelli, 1982). Existing literature (e.g., Paul *et al.*, 2007a) reveals that private shelters are more popular among disaster victims who have lost their homes than public shelters. This may be explained in a number of ways. For example, disaster survivors often perceive that they are safer and more familiar with their relatives and friends than with strangers encountered in public shelters.

In general, most survivors of disasters rebuild their homes in the same impacted community. Some survivors, however, relocate in neighboring or distant communities not affected by the disaster. During the post-disaster period, officials of affected communities often implement new zoning regulations which may prohibit some disaster survivors from rebuilding their homes in the affected communities. Because of the often slow pace of reconstruction, and confusion over reconstruction plans, some survivors decide to move to nearby towns. For example, when the 2007 Greensburg tornado destroyed 95% of this small Kansas town of 1400 people, more than two dozen former residents of Greensburg relocated themselves in nearby towns (Paul *et al.*, 2007a).

6.1.6 Repairing utilities and key infrastructure

Power, water, sewer, communication networks, and other infrastructure, such as transportation arteries, ports, airports, and rail lines, are often damaged or destroyed by major disasters. Therefore, there is a need to resume operation of at least critical services as quickly as possible. The kind of infrastructure and services interrupted by extreme events often depends on disaster type. For example, after the 1993 Midwest floods, the Mississippi waterway transporting cargo in the United States closed due to damage to bridges and ports. The May 2008 earthquake in Sichuan Province, China destroyed all bridges leading into the damaged area. As a result, rescue crews could not reach the injured and trapped persons immediately after the earthquake. This 7.9-magnitude earthquake also heavily damaged or blocked 70 of the roads (Phillips, 2009).

Unlike floods and droughts, winter or ice storms and hurricane or cyclone-associated windstorms damage or destroy electric poles, causing many people in affected areas to lose their electric power. Parts of five states (Colorado, Kansas, Nebraska, New Mexico, and Oklahoma) of the Great Plains region of the United States experienced severe winter storms on December 28–31, 2006.

More than 80 000 homes and businesses in the impacted areas were out of power and utility workers worked around the clock for more than a week to restore electricity. With phone lines down across much of the region and rural roads largely impassible, some residents were trapped. Ice and heavy snow also bent over electric towers and downed hundreds of miles of power lines (Paul et al., 2007b). About 2 million customers (homes and businesses) lost power after the windstorm-associated Hurricane Ike moved across Ohio, USA on September 14, 2008. This resulted in the most widespread weather-related blackout in the state's history. Like other windstorms and blizzards, this storm also damaged transmission towers (Schmidlin, 2011).

When power outages extend over large areas, repair crews of electric utility companies work round the clock. Often crews come from distant places to assist in the restoration of power. It sometimes takes several days to completely restore electricity to all customers. In addition to homes and businesses, power failures also force the closure of institutions and installations such as schools, park facilities, and hospitals. Some facilities can remain operational after storms with the aid of back-up electric power from generators. In general, public water systems, sewage treatment plants, and sewage pump stations continue to operate on generators after a power failure.

6.1.7 Safety and security

Although hazard researchers are divided on the issue of increase in antisocial behavior in the aftermath of a disaster, looting is often considered a security problem that follows a disaster, particularly in developed countries (Barsky, 2006). To eliminate the probability of looting, law enforcement authorities often impose curfews in disaster-affected areas, particularly at night. Security is also an issue within temporary shelters opened for people in disaster-impacted areas and resettlement camps. To provide safety and security for disaster survivors, local public safety departments add personnel and equipment during and after disasters to maintain order and protect public health and safety. Police departments frequently not only call off-duty police officers to help in keeping law and order under control, they also bring in outside law enforcement resources to ensure security. For similar reasons, some strategic roads in disaster-impacted areas are also closed to private traffic.

Response is by far the most complex of the four phases of the emergency management cycle, since it is conducted during periods of very high stress, in a highly time-constrained environment, and with limited information (Coppola, 2007). The skills of the people involved in response operations can be improved through exercises and training. Since numerous volunteers and emergent groups participate in disaster-response activities, efficient coordination is necessary to achieve success in such endeavors.

6.2 Recovery

The recovery phase of the disaster management cycle begins during and/or after the response phase, and its primary purpose is to reverse the damaging effects of disasters and restore survivors' lives. In disaster literature, the term "recovery" has been used interchangeably with rebuilding, reconstruction, restoration, rehabilitation, and post-disaster redevelopment (Quarantelli, 1998). Historically, the term recovery broadly implies putting a disaster-stricken community back together to its pre-disaster state (Mileti, 1999). It is a process of returning to "normalcy" after a disaster has occurred. The recovery process consists of a series of stages, steps, and sequences that disaster survivors, organizations, and communities move through at varying rates (Phillips, 2009). This post-disaster phase is also often used to increase safety and future disaster preparedness. Mileti (1999) views disaster recovery as a process of interaction and decision-making among a variety of groups and institutions, including households, businesses, and the community at large.

FEMA (2000) defines "recovery" as those non-emergency measures following a disaster whose purpose is to return all systems, both formal and informal, to as normal a state as possible. This definition of recovery encompasses repairing and reconstructing houses, commercial establishments, public buildings, lifelines, and infrastructure. However, FEMA's definition does not encompass returning to "pre-disaster" conditions (i.e., what would have been had the disaster not occurred), nor reaching a new, stable state. Recovery is thus a complex process, which overlaps with other three phases of emergency management cycle.

Both Coppola (2007) and Mileti (1999) claim that disaster recovery involves a host of activities that can be grouped as short-term and long-term activities. The former includes such activities as managing and dealing with volunteers and donated goods; conducting damage assessments; delivering emergency relief; rehabilitating the injured; providing temporary housing; and restoring and coordinating vital community services. These activities seek to stabilize the lives of disaster survivors in order to prepare them for transitioning toward rebuilding their future lives. Short-term activities tend to be temporary and usually do not contribute to the affected community's long-term development.

Repairing and reconstructing houses, commercial establishments, public buildings, and physical infrastructure are long-term activities. Other long-term activities include planning and implementing new construction, establishing social rehabilitation programs, and creating employment opportunities for disaster survivors. These activities seek to return life to "normal" (i.e., pre-disaster) or even improved levels. The long-term recovery period can be viewed as an opportunity to foster improvements in the built environment that reduce the impact of future disasters. In this chapter, most of the short-term activities listed above are treated in the response phase of the disaster management cycle. Emphasis here will be given to several long-term activities associated with the disaster recovery phase.

Activities associated with recovery are the most diverse of all disaster management functions. In relation to other functions, recovery is by far the most costly and more funding is dedicated to it than to any other disaster management cycle. The range of individuals, organizations, institutions, and groups involved in disaster recovery is also greater than in any of the other phases of disaster management cycle. The primary responsibility for most recovery efforts rests with the local government. Research has shown that locally based recovery approaches are most effective. In such an approach, the principal authority for carrying out recovery activities resides with community-based organizations, supplemented by technical and financial assistance from entities outside the community. This assistance can help strengthen local organizations and decision-making capacity as well as bring about the desired changes in rebuilding disaster-impacted communities (Mileti, 1999).

6.2.1 The disaster recovery cycle

Although time is by far the most compelling factor influencing local government recovery decisions, actions, and outcomes, the recovery from a disaster may take weeks, months, or years (Tierney et al., 2001). The magnitude of a disaster, community size, the extent of the physical damage, available resources, as well as recovery plans and goals are important determinants of the duration of the recovery period. Hazard researchers generally maintain that the disaster recovery period usually has no clear endpoint. Others (e.g., Wisner et al., 2004) contend that in some cases, the most vulnerable households and individuals never fully recover from the impacts of disasters. However, recovery period is often divided into four overlapping phases or cycles: development, relief, rehabilitation, and reconstruction (Figure 6.1).

As shown in Figure 6.1, the first and second cycles (i.e., development and relief) begin immediately following a disaster, but these two cycles differ markedly in terms of duration. While development encompasses the entire recovery period, relief generally lasts only for six months to a year, with a peak in the middle (Figure 6.1). In addition to providing disaster relief, this phase marks a period when disaster victims need clear information about entitlements and other opportunities associated with their assets and livelihoods. The duration of the third cycle (i.e., rehabilitation) is usually between 12 and about 20 months, and the last cycle from one to four years (Middleton and O'Keefe, 1998).

As can be seen from Figure 6.1., the points in the curves at which the last three phases overlap are the points at which the subsequent phase takes over – the final, gradual decline in the reconstruction phase leads back to post-disaster "normality." Although these cycles were developed based on the experiences of numerous disasters, the overall shape of the curve and the duration of each cycle vary considerably from disaster to disaster, from country to country, and between rural and urban areas (Figure 6.6). This is because of both physical

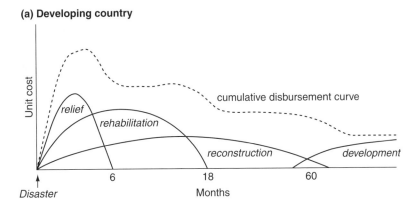
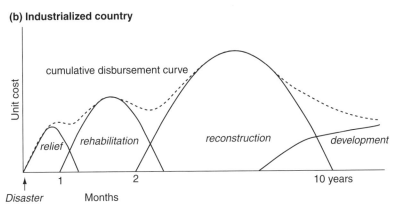

Figure 6.6 Stages in recovery from disaster: (a) for a developing country, and (b) for an industrialized nation. *Source:* Alexander (2000), p. 4.

and human factors, such as the extent of damage, available resources, existing infrastructure, and the cultural and political setting (Frerks *et al.*, 1995).

Although the primary interest of this chapter is not on the relief phase, it is important to mention that this phase of the disaster recovery cycle generally lasts longer for slow-onset disasters, such as floods and droughts, than for rapid-onset disasters, such as tornadoes, earthquakes, and flash floods. Because not all domestic and foreign agencies participate in the post-relief phase, many of these agencies begin to leave the disaster-affected area when they feel that humanitarian relief efforts are no longer needed. This marks the beginning of the rehabilitation phase of disaster recovery cycle. Only those agencies with resources and a practical or long-term mission continue working in the area during the recovery and/or reconstruction phases. For some disasters, such as earthquakes, the transition from the relief to the rehabilitation phase of the post-disaster intervention can be easily identified. Movement of the earthquake victims from tents to prefabricated temporary shelter marks this transition.

Communities that have suffered from a major natural disaster usually depend on external assistance for at least six months. If this dependence continues for more than six months, however, the impacted communities are liable to experience deep and lasting despair. This compounds their psychological trauma and can result in extremes of alienation and anger. According to the WHO (2006, p. 2), "the next three months are critical: the recovery must gain momentum in a way that reflects the extensive involvement of communities and attention to the needs of all, while, at the same time, maintaining relief efforts."

6.2.2 Special considerations in recovery

The process of disaster recovery involves several dimensions. For persons who have sustained physical injuries, recovery entails restoration of their physical well-being. Recovery also has a psychological dimension, as victims try to cope and come to terms with extreme stress, loss, and bereavement. Finally, recovery also has material and financial dimensions, involving the processes by which households, businesses, and communities attempt to restore property and compensate for financial losses. Our concern here is on the third dimension (i.e., material recovery), which is organized around at least two main themes: windows of opportunity and sustainable development.

6.2.2.1 *Windows of opportunity*

The post-disaster recovery phase provides a "window of opportunity" for disaster risk reduction (DRR). Local government entities, together with active support from emergency agencies at the state and/or federal level, generally introduce new hazard mitigation measures and enforce zoning regulations and building codes to better prepare for future disasters (Prater and Lindell, 2000; Tierney *et al.*, 2001). These measures seek to minimize the chance for loss of life, injury, and/or destruction of property as well as increase disaster resilience for communities affected by an extreme event (Mileti, 1999; Wisner *et al.*, 2004). Disaster survivors in general are more receptive to policy changes in the immediate aftermath of an extreme event and they tend to pay closer attention to things that were impacted the greatest (Mileti, 1999). Recency of disaster experience usually generates pledges of greater vigilance and increases responsiveness to safer behavior among disaster survivors. Personal awareness of disaster risk is generally raised, and commitment to future mitigation practices in light of a most recent disaster may appear strong (Tierney *et al.*, 2001).

Reconstruction funds, which are often provided more generously than those which are normally available to address risks, motivate survivors to adopt hazard-resistant land use and construction practices, and other mitigation measures. Economic incentives also reduce potential out-migration of disaster survivors to other communities. Often, official DRR goals that had been "forgotten" before the disaster gain renewed prominence in recovery policies, plans, and programs (Christoplos, 2006). Disasters also provide political opportunities and financial means for emergency managers, planners, and local government officials to implement DRR measures (Mileti, 1999). Such measures often make disaster-affected communities more disaster-resilient than their pre-disaster level.

However, support of disaster survivors to future DRR measures decreases over time and therefore the time available to take advantage of window of opportunity is limited. It may be as short as 30 days (Kingdon, 1995) to at most seven months following the disaster (Olshansky, 2006). This decrease of support is caused by a decrease in public awareness, lack of appropriate policy options, changing agency staff, and shifts in political resources to other issues (Kingdon, 1995). Shifting public opinion over time impacts the window of opportunity. Before a disaster, political support for hazard mitigation is balanced by political opposition, with most people neutral. But after a disaster, public opinion shifts in favor of hazard mitigation, which generally lasts up to six months after a disaster (Paul and Che, 2011). Available hazard literature suggests that the window of opportunity closes 18 months after the disaster (Prater and Lindell, 2000).

Many local government and concerned public emergency agencies in the United States and elsewhere have taken advantage of the "window of opportunity" disasters provide. In Chapter 5, it was mentioned that the city of Hoisington, Kansas took this opportunity by enforcing two new land use zoning regulations after part of the city experienced damage from an F-4 tornado on April 21, 2001. City officials encouraged residents to upgrade the quality of construction for new residential buildings to better withstand future tornadoes. Eighty homes have been rebuilt in the tornado-impacted area of Hoisington. An overwhelming majority of these home owners adopted new "tornado-resistant" technologies in the construction of their rebuilt homes. About 95% of those who rebuilt added safety features, such as tornado straps and foundation brackets; nearly 90% added a basement, and 30% of the rebuilt homes included safe rooms (Brock and Paul, 2003).

It is important to mention that FEMA, some state organizations, and many other private and public agencies provide financial support to disaster survivors to motivate adoption of hazard-resistant land use and construction practices, as well as other mitigation measures. Such support can take the form of subsidies, low-interest loans for retrofitting, tax credits, or cash. Incentive efforts may partially or fully compensate for losses incurred, encourage the adoption of adjustments by lowering the cost of implementation, and reduce individual and community vulnerability from future disasters. Society as a whole benefits from well-designed incentive programs because usually less tax money is spent on subsequent disaster assistance and recovery efforts (Mileti, 1999). For these reasons, residents and city authorities of Hoisington believe that in many ways, the 2001 tornado was a blessing because it led to the modernization of many buildings and homes that were in need of renovation (Brock and Paul, 2003).

Further, it is worth mentioning that the city of Hoisington bought floodplain zone properties with a generous grant provided by FEMA. Similarly, most of the buyout money in Cedar Falls, Iowa was provided by FEMA and state governmental agencies (Rahman, 2010). In addition, Cedar Falls received more than US$10 million in grants to build an industrial park and for construction of other public facilities damaged by the June flood of 2008. At the individual level, 93 homeowners received a total of over US$1.68 million to rebuild their homes or to purchase houses outside the flooded area using state funds. Forty small businesses also received over US$1 million in assistance (Rahman, 2010).

There is a danger, however, that redevelopment may increase inequality. Housing constructed to take advantage of the window of opportunity appears to be costly for a section of residents of disaster-impacted communities, particularly the poor and the elderly. For example, for economic reasons, many low-income households would not able to rent newly constructed houses/apartments and thus would be pushed out of their former place of residence. As many as 1500 people of "FEMA City" in Florida, USA had to live in 500 trailers more than a year after they were displaced by Hurricane Charley in August 2004 (Kaufman, 2005). They could not afford to live in the new houses rebuilt on the sites of their old homes. For the same reason, a considerable number of former residents of Greensburg, Kansas, USA had to build their houses in neighboring towns because building houses in Greensburg costs more as the city encourages residents to build eco-friendly houses after 95% of the town was destroyed by an EF-5 tornado (Paul and Che, 2011). This is discussed in more detail in the next subsection.

Making post-disaster city better and safer than ever often creates tension between city officials and planners, and the disaster victims. What victims want and need during the post-disaster period do not correspond exactly to what officials and planners envision for the areas impacted by extreme natural events. For example, Kobe, Japan experienced a powerful earthquake with a magnitude of 7.2 on the Richter scale on January 17, 1995. The Kobe earthquake survivors were displeased because they had little or no input in the redevelopment plans for earthquake-affected neighborhoods of Kobe. They protested against insensitive post-disaster planning and the ambitious reconstruction projects. The earthquake presented planners and city officials with a clean slate (Edgington, 2010). A similar tension between tornado survivors of Greensburg, Kansas, USA and city officials emerged, which is discussed in the next subsection of this chapter.

6.2.2.2 Sustainable development

Since the early 1980s, many hazard researchers (e.g., Alexander, 2000; Cuny, 1983; Mileti, 1999; Pelling, 2003; Wisner et al., 2004) have been urging the incorporation of elements of sustainable development into the process of rebuilding disaster-affected communities. Although many definitions of sustainable development exist, it is widely defined as "development that meets the needs of the present without compromising the ability of future generations to meet their own needs" (World Commission on Environment and Development, 1987, p. 188). The issues of sustainability are complex and oriented towards long-term management of resources in manners that are consistent with development and stability (Harrington, 2005). Nevertheless, a sustainable development approach can be seen as a heuristic device for bringing together the different human and physical elements that shape and are shaped by the social, economic, political, demographic, and environmental landscape (Pelling, 2003).

With increases in the frequency and severity of disasters in recent years, notions of sustainability, including sustainable development, have attracted serious attention among policymakers, emergency managers, and hazard researchers. Wisner *et al.* (2004) urge the inclusion of risk-reduction strategies in any sustainable development initiative, particularly those undertaken in the aftermath of a disaster. Mileti (1999, p. 29) maintained that "working toward sustainable communities (and, eventually, regions, nations, and the world) can go hand in hand with reducing losses from disasters." He further claims that disasters are likely to occur more frequently where unsustainable development has taken place and the occurrence of a disaster itself hinders movement toward sustainability because of the often resulting environmental degradation, ecological imbalance, negative socio-economic impacts, and lowered quality of life (Mileti, 1999). Therefore, actions designed to help communities mitigate disasters in a sustainable manner should strengthen overall resiliency to other social, economic, and environmental problems and vice versa.

If DRR is made a central concern in the design of sustainable development projects, then perhaps the upward trend of losses from disasters can be slowed or even reversed. In the context of DRR measures, components of sustainable development include: low-impact development practices that cause minimal impact on the environment, the use of recycled construction materials and materials that are highly insulated thereby reducing energy usage, and the use of alternative energy sources such as ground-source heat pumps (to capture and distribute geothermal energy), solar panels, and small wind turbines (US Department of Energy, 2009). These practices are also components of the concept of "going green," which has been increasingly used in recent years in conjunction with rising awareness of global climate change.

Going green as used here means to pursue knowledge and practices that can lead to more environmentally friendly and ecologically responsible decisions and lifestyles, which can help protect the environment and sustain its natural resources for current and future generations. Like the concept of sustainable development, going green emphasizes the future consequences of current practices. It entails a shift of perspective away from wasteful consumption of resources and toward activities that cause little or no damage to the environment. The concept of going green encompasses three major features (Osborn, 2010). The first is referred to as sustainable living, which involves limiting use of natural resources,

Figure 6.7 Billboard before entering Greensburg from east using state highway 54. *Source:* Che, D.

particularly nonrenewable ones, and increasing self-sufficiency. The second is use of environmental friendly (or green) products in our daily lives – such as recycled products. The third is the practice of recycling and reusing as many products as possible in order to reduce waste (Osborn, 2010). Essentially, the term "green" or "going green" embodies principles of sustainable development; however, the latter is a more encompassing concept than the former.

After devastation by a category EF-5 tornado on the evening of May 4, 2007, the mayor of Greensburg, a small town in southwestern Kansas with a population of 1400, announced the town was "going green" for its recovery (Phillips, 2009). The city council passed a resolution on December 17, 2007 requiring all publicly funded city buildings over 4000 square feet (370 m^2) to be built to the US Green Building Council's Leadership in Energy and Environmental Design (LEED) Platinum certification level. This level requires a reduction in energy consumption of 42% over standard building design (BNIM Architects, 2008). The city leaders saw, in this disaster situation, a great opportunity to create a community that has never been built: a sustainable or green community that will also have a reduced risk of destruction from future tornadoes (Figure 6.7). This means the new buildings will not only be sustainable, but also will be stronger in order to withstand the destructive power of a tornado. These new green structures in

Greensburg embody principles of sustainability, including resource conservation and recycling, along with best practices for community and building design, and construction that minimizes negative environmental impacts (US Department of Energy, 2009).

Greensburg city officials, including local residents, professional designers, and personnel from various federal agencies, began to form what is now called the "Greensburg GreenTown" movement – a nonprofit organization. This movement is making a concerted effort to rebuild Greensburg sustainably through alternative practices and energy sources. An important initiative that certifies sustainable building practices is the LEED program (US Department of Energy, 2009). Created by the US Green Building Council in 2000, the LEED program is essentially a green building certification system that designates any building type as a sustainable building as long as it meets specified criteria.

For the new construction/renovation rating system there are five overarching categories that must be followed in order for a building to obtain LEED certification: sustainable site, water efficiency, energy and atmosphere, material and resources, and indoor environmental quality. Each category contains a number of criteria that provide a total of 100 possible "base points" that a building could qualify for. If a building qualifies for a certain number of points, that building can be considered LEED certified. The break-down of the four different LEED certification levels are: Certified (40–49 points), Silver (50–59 points), Gold (60–79 points), and Platinum (80 points and above) (US Green Building Council, 2009). The green initiative put Greensburg in the national spotlight because only a very few communities in the United States have decided to embrace an eco-friendly community design. Despite many challenges, Greensburg is making a concerted effort to rebuild in a sustainable fashion.

There are, however, many challenges in order to rebuild a disaster-impacted community as an eco-friendly town. These challenges undeniably slow recovery efforts. In the case of Greensburg, many residents, particularly at the beginning, opposed the green project and confronted city officials and supporters of green initiatives. Many of them were simply not familiar with the concept and for some it took time to understand the concept, although they did eventually come round. Indeed, some residents who initially opposed the green initiative became green movement leaders (Cartlidge, 2010). Yet, many pre-tornado residents of Greensburg believed that they could not afford to live in a "green" city with increased upfront construction costs, even though these costs will be eventually offset by lower energy bills (Paul and Che, 2011).

Financial circumstances motivated some long-term Greensburg residents to purchase older, modestly priced homes in nearby communities rather than rebuilding in Greensburg. They used their insurance money to purchase homes in nearby towns at much lower price than the cost of rebuilding in Greensburg (Paul and Che, 2011). In addition, some former long-time Greensburg residents cited age, the physical/emotional challenges to rebuilding, and preference to rent as their reasons for leaving Greensburg. As indicated, for many Greensburg residents, a greater obstacle to rebuilding than the advised energy-efficiency initiatives was the gap between what they would receive in insurance payments for their demolished houses and the costs for constructing new ones (Paul and Che, 2011).

6.2.3 Recovery resources

Adequate funding is required at all levels to recover from the impact of disasters. Responsibility for reconstruction costs is divided between various sectors of the community. The local government is generally responsible for rebuilding public facilities and infrastructure, while the private sector, including industries, individuals, and households, will be responsible for rebuilding houses and businesses. The private sector, with help from governments at various levels, helps to restore overall economic vitality. The success in recovering from impact of a disaster in a timely manner depends, to a large extent, on how quickly and successfully affected community and disaster survivors are able to secure assistance from external resources (Coppola, 2007).

As indicated, a wide set of funds and programs are available to help recover from the impact of disasters. Some of these funds and programs are made specifically to individuals, and others go to states, nonprofit organizations, or other entities. Individuals and city emergency officials must be aware of these sources of funding. In the United States, funding may be obtained from federal, state, local, and private sources for disaster recovery. Other grants opportunities may be obtained through private corporations, foundations, organizations, and individuals. Numerous federal grants are available for disaster recovery and may be secured from the federal government after a presidential disaster declaration. In the absence of such a presidential disaster declaration, federal funds from agencies may still be available under relevant funding programs, such as low-income housing, or downtown revitalization programs (Phillips, 2009).

There are two distinct federal disaster assistance programs in the United States: individual and household assistance (IA) and public assistance (PA). The former program types are geared towards individual households and owners of small businesses. Some individual assistance programs are managed by the state government. IA is largely composed of six types of loan and grant programs: disaster loans for homes and individuals, disaster loans for businesses, economic injury disaster loans, farm service agency loans, individual and households program (IHP), and tax assistance. The last two are grant programs and individuals receiving these grants do not need to repay them. Some disasters generate need beyond the traditional 18 months of individual assistance that the federal government offers. Even after improvements over the past few decades, many disaster victims in the United States still find the application process for receiving federal disaster assistance is a very complex and lengthy process.

There are many other programs available in the United States and elsewhere to assist individuals, households, businesses, and even public and nonprofit entities recovering from the impact of disasters. Among others, these include insurance, government-based emergency relief funds, donations, and loans. To receive money from the insurance companies for repairing or rebuilding homes, there is a need to purchase such insurance prior to the disaster. Although insurance is not generally available in developing countries, insurance payments do not usually cover all expenses required to fully recover from the disaster impacts. For example, for many Greensburg residents, a challenge to rebuilding was the difference between what they would receive in insurance payment and the costs for constructing new homes. In addition, disaster insurance is disaster-specific, and often plays a small role in disaster recovery for many households and small businesses. Thus, even disaster insurance does not always shield disaster victims from large financial losses. In general, government-sponsored programs are the main mechanism through which disaster victims receive recovery assistance. These programs currently reach many more people than are covered by insurance.

Irrespective of development level, governments of both developed and developing countries have allocated in their annual budgets emergency relief funds which they use to cover anticipated expenses associated with disasters. Although such funds are primarily used to provide emergency aid to disaster victims of both domestic and foreign countries, a portion of these funds are often used for recovery purposes. The same is true for donations, which are philanthropic in origin and come from all over the world. Recovery donations do not only include cash and material resources, such as lumber, sheetrock, paint, major appliances as well as furniture like beds, tables, and chairs, but also volunteer time. Individuals, households, and businesses often need to secure loans to cover expenses associated with full recovery. Governments of the disaster-impacted countries can borrow money from international financial institutions, such as the World Bank, for rebuilding communities devastated by disasters.

For public assistance in the United States, the community affected by the disaster is able to make a formal request for federal disaster assistance within 30 days after the presidential disaster declaration (McEntire, 2007). FEMA, under a presidential disaster declaration, can assist state government agencies, local governments, tribal organizations and governments, and nonprofit institutions to engage in public assistance project. Eligible private or nonprofit institutions may include hospitals, educational facilities, museums, zoos, libraries, day-care, and adult-care centers. PA programs are generally designed to restore a facility to a functional condition and do not involve improvements (Phillips, 2009).

The PA program is extensive and requires that those involved become conversant with the procedures that are involved. In addition, grant writing skills are required to apply for public assistance grants. Successful grant writing also requires effective communication with people of various federal agencies. Entities seeking federal funds should work closely with state and federal partners to identify and secure appropriate funds. Grants must be managed carefully to ensure full compliance with all federal guidelines (Phillips, 2009).

6.2.4 Types of recovery

Damage and destruction caused by disasters disrupts the operational capacity of many entities including the federal and state governments to various degrees. However, four entities – individuals, households, businesses, and affected communities – are impacted the most. For this reason, only household, community, and business recovery are discussed below. Furthermore, because individuals are members of households, these two entities are treated together. Since the larger businesses have more resources, they tend to fare better than smaller businesses in attempting recover from disasters. In addition, larger businesses may also be part of a larger chain of stores or restaurants and can pull from those locations to replace damaged items or even borrow employees (Phillips, 2009). For this reason, recovery of smaller businesses is considered here.

6.2.4.1 Household recovery

Individuals and households that have experienced a disaster will ultimately recover from its impact at some point in time after occurrence of the event. This means that households will recover at different rates because of many factors. These rates are largely influenced by the amount of resources households own and how much financial assistance they receive from external sources. Both of these depend on many factors, such as socio-economic and demographic characteristics of individuals and household heads. In general, poorer families experience more difficulty recovering from disasters because they have few resources of their own and they often have difficulty in getting external assistance. Therefore, recovery takes a longer period for such families. Similarly, older individuals are less likely than younger applicants to receive disaster loans and grants, and when they are given money, they usually receive smaller amounts than they initially request. The same is true for households headed by a single mother.

Under and/or uninsured households will require longer to recover. Economic reserves and extended family support and assistance affect family recovery. Rich members of extended kin and social groups often provide recovery assistance to their poor relatives who have experienced damage and destruction caused by disasters. They provide their labor as well as financial assistance to help the poor relatives rebuild their homes. Because of this assistance, impoverished disaster victims often recover relatively quickly. All members in the social network of an impoverished disaster victim are not equally powerless. In general, poor relatives have one or more friends or relatives in their social network who have some degree of wealth and/or political power. In such cases, vulnerable members of a social network usually receive adequate aid, which facilitates a quick recovery from disaster. Furthermore, rural disaster survivors are more likely to use their kinship grouping as a source for recovery assistance than urban disaster victims. However, rural survivors are less likely to receive disaster recovery loans and grants than urban survivors (Eisenman *et al.*, 2007).

6.2.4.2 Community recovery

Disaster recovery cannot be approached solely on an individual or household basis, but rather is a community-wide concern. Thus, there are many components of community recovery – residential, commercial, industrial, social, and lifelines (Mileti, 1999). For example, even the if post-disaster rebuilding and repairing of houses is complete, the community may continue to suffer unless its social needs are adequately addressed. This means that, at the community level, there are varying degrees of recovery. Some aspects of community life, such as tax revenues, may take longer to return to normal than other aspects of community level recovery. As noted, it is not unusual for leaders of disaster-affected communities to put into place new codes, ordinances, regulations, and policies that govern building processes. Although such restrictions make these communities disaster-resilient, they generally delay the recovery process. Working out issues related to damage, rebuilding procedures, and funding programs may also delay community recovery.

How well a community recovers from a disaster tends to directly follow how well that community was doing economically before the disaster occurred. Successful communities are more likely to have the levels of civic pride and cohesion necessary to collectively move forward and even exceed pre-disaster prosperity levels. Successful recovery efforts also depend on horizontal and vertical integration into local, regional, and national networks (Mileti, 1999; Tobin and Montz, 1997). Strong horizontal integration reflects strong local government as well as cohesive system of public, private, and volunteer groups that are well integrated within the community. In such a community there is a strong local participation. Vertical integration means that a community has ties with higher levels of government and institutions that provide recovery assistance. Thus, the greater the vertical integration, the more readily recovery assistance can usually be obtained from external sources. The strong vertical integration is generally facilitated through strong local leaders and local government ability to take action on its own behalf.

As shown in Figure 6.8, vertical and horizontal integration are dichotomized either as strong or weak. This figure suggests there are at four types of communities with different capacities for recovery from a disaster. Community type 1 is characterized by both strong horizontal and vertical integration and this type of community has the greatest potential for effective disaster recovery in the shortest possible time compared with the other three types of communities.

	Strong	Weak
Strong	Type 1	Type 2
Weak	Type 3	Type 4

Figure 6.8 Type of community based on vertical and horizontal integration. *Source:* Berke (1993).

At the other extreme is community type 4 where both vertical and horizontal integration are weak. This type of community would likely face difficult time to recovery from the impacts of a natural disaster. It needs to be emphasized that size of the community (in terms of population) affects both horizontal and vertical linkages. As size increases, horizontal integration tends to decrease, but vertical integration tends to increase. For quick disaster recovery, such a community needs to encourage greater participation of community members and organizations in recovery process.

Tobin and Montz (1997) claim that three changes that occur in the post-disaster period facilitate community recovery. They maintain that conflict in disaster-impacted communities decreases because of the fact that most community members are preoccupied with the problem at hand. Specifically, there is a convergence of social values centering on cleanup and recovery that overrides pre-disaster differences. Individual and group interests tend to disappear or are subsumed for the community good. Finally, Tobin and Montz (1997) believe that immediately after a disaster many new but temporary organizations will emerge in the community to facilitate recovery. Impacts of these changes tend to decrease as the community returns to pre-disaster level. It is important to note that every disaster-impacted community does not experience similar magnitudes or intensities of change, nor for the same amount of time. Disasters create a new set of problems for communities, and community leaders and emergency managers must successfully address these problems.

Disaster-impacted communities often find that some portion of their cultural heritage has been damaged or destroyed along with historic buildings and other structures. Loss of such cultural components may result in a loss of community identity. Therefore, care should taken by community leaders and emergency officials to salvage/restore the unique historic and cultural character of their community. At the same time, rebuilding must take place without seriously damaging or degrading local environmental resources. Cultural recovery, which is closely associated with social recovery, usually comes from within the community, though outside assistance may be offered or even sought for such endeavors.

Table 6.4 Examples of community relocated in disaster recovery phase

Community	Country	Disaster (Year)
Dagara	India	Earthquake (2001)
Valmeyer, Illinois	USA	Flood (1993)
Babi Island	Indonesia	Earthquake and tsunami (1992)
Chernobyl	Ukraine	Nuclear accident (1986)
Gediz	Turkey	Earthquake (1970)
Valdez, Alaska	USA	Earthquake (1964)

Source: Compiled from various sources.

For efficient and quick recovery from disasters, communities should have recovery plans in place before a disaster occurs. However, the sad reality is that most communities develop recovery plans after they hit by disasters. Disaster recovery plans should involve all relevant departments and tasks must be clearly assigned to personnel in each department. Fundamentally, there is a need for training of department staff on what they need to do, how their work may change in an emergency context, and where they might be relocated to support critical services. Furthermore, recovery plans should address both the short- and long-term needs of the community. And, it is advisable to integrate recovery planning with comprehensive plans and land use planning. As indicated, disaster recovery is closely associated with rebuilding and repairing of destroyed and damaged houses. The recovery plans should emphasize the housing sector as well as ensuring equity. This is necessary because members of marginalized groups, such as the elderly, female-headed households, religious minorities, and the poor often bear a proportionally greater burden in terms of disaster consequences and usually experience more difficulty recovering than members of non-marginalized groups. Economic recovery should also receive priority in recovery plans so that local business activities resume in the shortest possible time so people can return to work.

In addition to those factors noted above, several other factors facilitate the disaster recovery process at community level. Prior disaster experience is directly associated with a rapid recovery process primarily because communities with prior experience usually have disaster plans and organizational arrangements for quick recovery from disasters. Effective leadership is another determinant of community recovery. As noted, often disaster-impacted communities (or portions thereof) are sometimes rebuilt in new locations to reduce future disaster risk. Table 6.4 lists several such communities which have relocated subsequent to a disaster. However, moving an entire community is a complicated undertaking because the costs associated with relocating are usually higher than the costs for rebuilding. Also disaster survivors are often reluctant to move to a new location due to strong place attachment.

6.2.4.3 Business recovery

Industries and business establishments, such as grocery stores, restaurants, cleaners, bookstores, fitness centers, day-care centers, and beauty salons, often sustain different degrees of damage from natural disasters. Because these are local economic drivers, loss of businesses not only negatively impact local economy, but also cause people to move out of the affected community. This considerably reduces the tax base of a local community at a time when it needs additional resources for long-term recovery. Yet, some businesses associated with post-disaster construction (e.g., lumber stores) generally experience growth related to recovery which helps the local economy.

Many of the damaged businesses remain closed for some time in the immediate aftermath of a disaster, while others experience closure. Both permanent and temporary business closures create unemployment and this hurts the economy of disaster-affected communities. Businesses that do reopen after closure often experience a reduction in the number of customers because many do not have the money to purchase items, or they may be unable to reach business facilities because roads are damaged by the disaster. If this occurs, business owners may be unable to pay employees or other bills.

It is vital for the local economy in the aftermath of a disaster that local businesses return to pre-disaster capacity as quickly as possible and take advantage of public and private recovery funding injected into the disaster-affected areas. Sources of funding discussed in the previous section also help small businesses recover from disasters. In addition to these funding opportunities, special funding is available for small businesses. For example, in the United States, business owners are eligible to apply for Small Business Administration (SBA) loans. The SBA offers physical disaster loans for permanent rebuilding and replacement of uninsured or underinsured disaster-damaged businesses. The maximum amount of money that the SBA can lend to qualified businesses under this loan is US$2.0 million. This amount can be used to repair and/or replace property, equipment, inventory, and fixtures. These physical disaster loans carry two rates: the low rate, which is 4%, and the high rate, 8% – given to applicants the SBA determines are able to obtain credit elsewhere. Businesses that receive the 4% interest rate have to repay within 30 years. Economic disaster injury loans (EIDL) from the SBA can provide up to US$1.5 million at an interest rate not to exceed 4% for a term of up to 30 years. These loans are intended to provide funds to cover operating expenses until the business recovers (Phillips, 2009).

Loans available from the SBA and other public sources cannot change problems that stem from losing customers or from regional economic downturns (Tierney, 2006). Many businesses rely on savings, including retirement funds, to preserve their livelihood. However, business recovery time varies according to type and size of businesses as well as owning versus renting business. Recovery for rental business takes longer compared to owner businesses. Furthermore, it is generally considered that businesses with past experience in disaster situations are more likely to prepare and thus more likely to recover faster. Location matters as well, because businesses that lose pedestrian traffic, parking, and/or rental spaces are more likely to experience a difficult recovery (Phillips, 2009). As noted, recovery also represents an opportune time to incorporate new features which may enhance the business-sector appearance and functionality, including green technology, such as streetscapes and use of alternative sources of energy.

Disaster recovery is usually a long and complex process. Considerable progress has been made in recent years in attempts to compensate and assist disaster victims, disaster-impacted communities, and businesses in the recovery process. Public funds for recovery and reconstruction tend to be a small but nonetheless important part of recovery financing. Apart from funding challenges, local politics may hinder the recovery process as well. At the community level, pressures to restore normalcy in response to victims' needs and desires are so strong that safety and community improvement goals are often altered, compromised, or even abandoned. Because numerous organizations are involved in the recovery process, extraordinary teamwork, cooperation, and coordination are required among all involved organizations, and among various local government departments. To make the recovery process more efficient and therefore faster, relevant public officials and staff of involved organizations should provide adequate education and recovery training to cope with recovery situations.

References

Alexander, D. (2000) *Confronting Catastrophe*. Oxford: Oxford University Press.
Baggerly, J. and Exum, H.A. (2008) Counseling children after natural disasters: guidance for family therapists. *The American Journal of Family Therapy* 36: 79–93.
Barsky, L.E. (2006) Disaster realities following Katrina: revisiting the looting myth. In *Learning from Catastrophe: Quick Response Research in the Wake of Hurricane Katrina* (ed. The Hazards Center). Boulder, CO: University of Colorado at Boulder, pp. 215–234.
Beaver, B. *et al.* (2006) Report of the 2006 National Animal Disaster Summit. *Vet Medicine Today: Disaster Medicine* 229 (6): 943–948.
BNIM Architects (2008) *Greensburg Sustainable Comprehensive Plan*. Kansas City, MO: BNIM Architects.
Berke, P.R. *et al.* (1993) Recovery after disasters: achieving sustainable development, mitigation, and equity. *Disasters* 17 (2): 93–109.
Brenner, S.A. and Noji, E.K. (1995) Tornado injuries related to housing in the Plainfield tornado. *International Journal of Epidemiology* 24 (3): 133–151.
Brock, V.T. and Paul, B.K. (2003) Public response to a tornado disaster: the case of Hoisington, Kansas. *Papers of the Applied Geography Conferences* 26: 343–351.

Brown, S. et al. (2002) Tornado-related deaths and injuries in Oklahoma due to the 3 May Tornadoes. *Weather Forecast* **17**: 343–353.

Cartlidge, M.R. (2010) There's no place like home: place attachment among the elderly in Greensburg, Kansas. MA thesis. Kansas State University, Mantahhtan, Kansas.

Christoplos, I. (2006) *The Elusive "Window of Opportunity" for Risk Reduction in Post-Disaster Recovery*. Bangkok: Provention Consortium.

Coppola, D.P. (2007) *Introduction to International Disaster Management*. Boston: Elsevier.

Cuny, F.C. (1983) *Disasters and Development*. New York: Oxford University Press.

de Goyet, C.V. (2007) Epidemics after natural disasters: a highly contagious myth. *Natural Hazards Observer* **31** (3): 4–6.

Edgington, D.W. (2010) *Reconstructing Kobe: The Geography of Crisis and Opportunity*. Vancouver: UBC Press.

Eikenberry, A.M. et al. (2007) Administrative failure and the international NGO response to Hurricane Katrina. *Public Administrative Review* December 2007: 160–170.

Eisenman, D.P. et al. (2007) Disaster planning and risk communication with vulnerable communities: lessons from Hurricane Katrina. *American Journal of Public Health* **97**: S109–S115.

El-Tawil, S. and Aguirre, B. (2010) Search and rescue in collapsed structures: engineering and social science aspects. *Disasters* **34** (4): 1084–1101.

EPA (Environmental Protection Agency) (1995) *Planning for Disaster Debris*. Washington, D.C.: EPA.

FEMA (Federal Emergency Management Authority) (2000) Hazards, disasters and the US emergency management system: an introduction, Section 6: *Fundamentals of US Emergency Management*. Washington, D.C.: FEMA.

Fischer, H.H. III. (1998) *Response to Disaster: Fact vs. Fiction and its Perpetuation: The Sociology of Disaster*. New York: University Press of America.

Fothergill, A. and Peek, L. (2004) Poverty and disasters in the United States: a review of recent sociological findings. *Natural Hazards* **32** (1): 89–110.

Franzino, R. (2004) *Homestead Florida: An Inventory and Analysis of Community and Economic Efforts*. Miami, FL: Department of Urban and Regional Planning, Florida Atlantic University

Frerks, G.E. et al. (1995) A disaster continuum? *Disasters* **19** (4): 362–366.

Gallay, C. (2009) Island vet Ken Diestler is honored for work after Ike. *Houston Chronicle* June 16.

Harrington, L.S.B. (2005) Vulnerability and sustainability concerns for the US High Plains. In *Rural Change and Sustainability: Agriculture, the Environment and Communities* (eds S.J. Essex, A.W. Gilg, and R. Yarwood). Cambridge, MA: CABI Publishing, pp. 169–184.

Heath, S.E. et al. (2000) A study of pet rescue in two disasters. *International Journal of Mass Emergencies and Disasters* **18** (3): 361–381.

Jia, Z. et al. (2010) Are the elderly more vulnerable to psychological impact of natural disaster? A population-based survey of adult survivors of the 2008 Sichuan Earthquake. *BMC Public Health* **10**: 172.

Johnson, C. (2007) Strategic planning for post-disaster temporary housing. *Disasters* **31** (4): 435–458.

Kaufman, M. (2005) FEMA's city of anxiety in Florida: many Hurricane Charley victims still unsure of next step. *The Washington Post* September 17.

Khan, M.M.I. (1991) The impact of local elites on disaster preparedness planning: the location of flood shelters in northern Bangladesh. *Disasters* 15 (4): 340–354.

Kingdon, J.W. (1995) *Agenda, Alternatives and Public Policy.* New York: Harper Collins.

Lindell, M.K. and Prater, C.S. (2003) Assessing community impacts of natural disasters. *Natural Hazards Review* 4 (4): 175–185.

Luther, L. (2008) *Disaster Debris Removal after Hurricane Katrina: Status and Associated Issues.* Washington, D.C.: Congressional Research Service.

McEntire, D.A. (2007) *Disaster Response and Recovery: Strategies and Tactics for Resilience.* Hoboken, NJ: Wiley.

Middleton, N. and O'Keefe, P. (1998) *Disaster and Development: the Politics of Humanitarian Aid.* London: Pluto Press.

Mileti, D.S. (1999) *Disaster by Design: A Reassessment of Natural Hazards in the United States.* Washington, D.C.: Joseph Henry Press.

Miller, G. (2005) The tsunami's psychological aftermath. *Science* 309: 5737.

Mitchell, J.T. and Cutter, S.L. (1997) *Global Change and Environmental Hazards: Is the World Becoming More Disastrous?* Washington, D.C.: The Association of American Geographers (AAG).

Morgan, O. et al. (2006) *Management of Dead Bodies after Disaster: A Field Manual for First Responders.* Washington, D.C.: PAHO.

Natural Hazards Observer (2008) Cyclone Sidr – Bangladesh. 32(3): 4.

Olshansky, R.B. (2006) Planning after Hurricane Katrina. *Journal of the American Planning Association* 72 (2): 147–153.

Osborn, S. (2010) Definition of going green. http://www.ehow.com/facts_4926406_defibition-going green.html (accessed October 2010).

Ozerdem, A. (2006) The mountain tsunami: afterthoughts on the Kashmir earthquake. *Third World Quarterly* 27 (3): 397–419.

Paul, B.K. (2007) 2004 Tsunami relief efforts: an overview. *Asian Profile* 35 (5): 467–478.

Paul, B.K. (2010) Human injuries caused by Bangladesh's Cyclone Sidr: an empirical study. *Natural Hazards* 54: 483–495.

Paul, B.K. and Che, D. (2011) Opportunities and challenges in rebuilding tornado-impacted Greensburg, Kansas as "Stronger, Better, and Greener." *GeoJournal* 76: 93–108.

Paul, B.K. et al. (2007a) Disaster in Kansas: the Tornado in Greensburg. Quick Response Report No. 196. The Natural Hazards Center, University of Colorado at Boulder, Boulder, CO.

Paul, B.K. et al. (2007b) Emergency Responses for High Plains Cattle Affected by the December 28–31, 2006 Blizzard. Quick Response Report No. 191. The Natural Hazards Center, University of Colorado at Boulder, Boulder, CO.

Paul, B.K. and Leven, J. (2002) Emergency Support Satisfaction among 2001 Hoisington, Kansas Tornado Victims. Quick Response Report No. 154. Natural Hazards Research and Applications Information Center, University of Colorado at Boulder, Boulder, CO.

Pelling, K. (ed.) (2003) *Natural Disasters and Development in a Globalizing World.* London: Routledge.

Phillips, B.D. (2009) *Disaster Recovery.* Boca Raton, FL: CRC Press.

Prater, C.S. and Lindell, M.K. (2000) Politics of hazard mitigation. *Natural Hazards Review* 1 (2): 73–82.

Quarantelli, E.L. (1982) General and particular observations on sheltering and housing in American disasters. *Disasters* 6: 277–281.

Quarantelli, E.L. (1995) Patterns of shelter and housing in US disasters. *Disaster Prevention and Management* **4** (3): 43–53.

Quarantelli, E.L. (1998) *What is a Disaster?* London: Routledge.

Rahman, M.K. (2010) Preparation and response to the flood of 2008 and windstorm of 2009 in Cedar Falls, Iowa. MA thesis, University of Northern Iowa, Cedar Falls, Iowa.

Schmidlin, T.W. (2011) Public health consequences of the Hurricane Ike windstorm in Ohio, USA. *Natural Hazards.* **58** (1): 235–249.

Tierney, K.L. et al. (2001) *Facing the Unexpected: Disaster Preparedness and Response in the United States.* Washington, D.C.: Joseph Henry Press

Tierney, K.L. (2006) Businesses and disasters: vulnerability, impacts, and recovery. In *Handbook of Disaster Research* (ed. E. Rodriguez et al.). New York: Springer.

Tobin, G.A. and Montz, B.E. (1997) *Natural Hazards: Explanation and Integration.* New York: The Guilford Press.

US Department of Energy (2009) *Rebuilding after Disaster: Going Green from the Ground Up.* Washington, D.C.: National Renewable Energy Laboratory.

US Green Building Council (2009) *Leadership in Energy and Environmental Design.* http://www.usgbc.org/DisplayPage.aspx?CategoryID=19 (accessed October 2010).

Wisner, B. et al. (2004) *At Risk: Natural Hazards, People's Vulnerability and Disasters.* London: Routledge.

World Commission on Environment and Development (1987) *Our Common Future.* New York: Oxford University Press.

WHO (World Health Organization) (2006) Update 95-SARS Chronology of a Hiller. http://www.who.int/csr/don/2003_07_04/en/index.html (accessed January 31, 2006).

7
Disaster Relief

In order to fully understand response and recovery operations of the disaster management cycle, there is a need to understand disaster relief efforts undertaken to provide emergency assistance to disaster victims. Information such as who are the providers and distributors of emergency assistance, how emergency aid flows from non-impacted to impacted areas, and what mechanisms are used to disburse emergency assistance may help emergency managers improve their performance in the post-disaster period as well as in subsequent disasters. Therefore, the first section of this chapter is devoted to a discussion of providers and distributors of disaster aid in both developed and developing countries, including the issue of participation of domestic and foreign military forces in disaster relief operations.

The second section of this chapter reviews disaster relief operations undertaken in the past in different countries of the world. It is important at the outset to mention that hazard researchers differ in their opinion over whether disasters should be addressed at all by relief measures. While some (e.g., Cupples, 2007; Paul, 2006) believe that emergency aid is essential for people who have lost homes and livelihoods, a number of authors (e.g., Bolin and Stanford, 1999; Hewitt, 1997; Susman *et al.*, 1983; Wisner *et al.*, 2004) have suggested that its arrival in a disaster-stricken area can have potentially harmful effects, replacing solidarity with self-interest or generating dependencies that reduce people's abilities to cope with future extreme natural events. Along with major criticisms of disaster aid provision, the second section also examines the role of nongovernmental organizations (NGOs) in emergency aid distribution,

7.1 Providers and distributors of disaster aid

For major disasters, governments of both developed and developing countries generally receive disaster aid from both domestic and foreign sources. This aid is then distributed among disaster victims. Before discussing sources of this disaster aid and entities involved in its distribution, it is important to address several concerns, such as how emergency aid flows from non-impacted to disaster-impacted countries? And who makes the appeal for external aid; as well as when is this appeal made? After a major disaster, the government of the impacted country usually drafts a needs assessment – often in conjunction with entities like the World Bank or a particular NGO. After completing such an assessment, it then makes a formal appeal, also known as a flash appeal, to the international community for assistance. Donor entities generally wait for this appeal before they make a pledge of assistance.

7.1.1 Flash appeal and pledge

A flash appeal is a process for identifying needs, coordinating a strategic response, publicizing funding need, and inventorying relief and early recovery projects. A pledge, on the other hand is a nonbinding promise that is often extremely general and could take the form of a simple ministerial statement (Batha, 2005). This may include both emergency relief and reconstruction aid. Donor entities cannot give away money or other forms of aid without a contract specifying its use. The critical step is when a donor makes a commitment – a legally binding, signed contract. At this point, the impacted country or relief agencies acting on its behalf may spend the money even if it has not actually arrived. When it comes to humanitarian aid, a donor will usually transfer money within couple of days of making a commitment (Batha, 2005). Some donor countries are slower than others in approving emergency aid. For example, Japan turns its pledges into commitments in just a few days, whereas the United States takes months to formally approve its aid (Batha, 2005). Delays usually relate to funds for reconstruction.

Whether or not the government of the disaster-affected country makes an appeal to help victims of an extreme event, a particular NGO may make such an appeal. For example, Bangladesh experienced a devastating flood in early August of 2007, which affected nearly 10 million people. Although the Bangladesh government made an appeal to all citizens of the country to come forward and join relief operation on August 5, 2007, it did not extend that appeal to include the international community. Observing this, the British NGO Oxfam made a formal appeal to the international community for assistance on August 8, 2007.

On November 18, 2007, donor nations and international agencies pledged over US$25 million in assistance to help Bangladesh meet the requirements arising in the aftermath of Cyclone Sidr, which impacted more than 8.5 million coastal residents on November 15, 2007 and killed over 3000 people (Paul, 2010). This assurance of help came before the government made any formal appeal for international assistance. The donor nations also urged the Bangladesh government to appeal for help from the international community. It is also not unusual for the United Nations (UN) to make an appeal on behalf of a disaster-affected country or countries. For example, immediately after the 2004 Indian Ocean tsunami, the UN Office for the Coordination of Humanitarian Affairs (OCHA) drafted a flash appeal in the amount of US$1.1 billion for emergency aid on January 6, 2005 (Valley, 2006).

The flash appeal, which was modified as new information came in, presented to the international community a unified assessment of needs and priorities of the Indian Ocean tsunami victims. In its first phase, the Indian Ocean tsunami flash appeal called for US$977 million to fund the work of some 40 UN agencies (e.g., the UN Children's Fund – UNICEF, the World Food Program – UNFP, and the UN Development Program – UNDP) and NGOs (e.g., Save the Children and CARE, Catholic Relief Service) in addressing the survival and recovery needs for an estimated 5 million people affected by the disaster. In April 2005, revised requirement was issued for US$1.1 billion (Schreurs, 2011).

There is, however, no central agency for channeling funds after a disaster. Donor governments generally disburse funds through the UN agencies, regional multilateral organizations, and/or domestic NGOs. The OCHA or the UNDP usually assumes the coordination role and manages the collection and disbursement of donated funds. The aforementioned organizations, in turn, provide services or deliver goods required to carry out emergency relief efforts (Coppola, 2007). Often bilateral aid is sent directly to governments of disaster-impacted countries. Japan, for example, transferred US$250 million directly to the government in several countries severely impacted by the 2004 Indian Ocean tsunami (Batha, 2005). Donor entities can also channel money through a multi-donor trust fund administered by the World Bank. This pooling of resources prevents the recipient country getting overwhelmed with multiple transactions involving dozens of donors all stipulating different conditions for spending their money (Batha, 2005). The flash appeal could also be likened to a trade fair where relief agencies lay out their needs in stalls and donor entities choose what they want to fund. Payment is made directly by the donor to the agency (Batha, 2005).

Donor governments provide emergency assistance in a myriad of ways, depending on the type of disaster, the needs of the affected country, and the capacity of the donor. These donations may come in the form of food, medicine, equipment, building materials, technical assistance, or debt relief. Donor entities provide this assistance in a variety of ways rather than just the direct transfer of cash or provision of supplies. Although monetary assistance is the easiest, and often the most needed form of assistance in a disaster's aftermath, donor governments are careful in disbursing cash to disaster-impacted countries. If the impacted country has a strong, credible and accountable government that is believed to be capable of carrying out emergency assistance program, the funds may be given directly to it (Coppola, 2007).

Donor entities know that frequently not all the money offered can always be spent by the governments of disaster-affected countries. There are several reasons for this. The capacity of a country to absorb emergency aid might be limited; furthermore, conditions are set by donor entities that sometimes delay or even prevent the distribution of aid. In addition, donor entity frustration may emerge, especially when governments transfer emergency funds to meet other needs. Moreover, sometimes donor countries feel pressure to pledge more funds than they are actually capable of providing, which is often termed a donor "beauty contest" (Telford and Cosgrave, 2007). During the first few days after the 2004 Indian Ocean tsunami, major donors were engaged in a veritable race to show who was most generous. Japan, for example, wanted to exceed the amounts offered by China and the United States – to politically show that Japan is economically more powerful than China. And as an Asian country Japan believed it needed to shoulder more of the relief burden than the United States since this disaster occurred in Asia (Tamamoto, 2005).

Donor countries also often need to increase their original pledged amount for several reasons. For example, the United States originally promised only US$15 million for the 2004 Indian Ocean tsunami, but in response to criticism from UN Undersecretary-General for Humanitarian Affairs Jan Egeland, the US government added another US$20 million to its original pledge made on December 18, 2004. Egeland categorized charitable contributions of rich countries as "stingy." As the death toll rose and contributions from Europe soared, the United States again increased its pledge to US$857 million (Wehrfritz and Cochrane, 2005).

As indicated, OCHA often issues a flash appeal for disasters that occur in developing countries, setting out the most pressing needs for up to six months. The OCHA heads the UN response to all emergency situations, including complex humanitarian emergencies. This assessment is made by the UN's top representative in the affected country with the help from UN agencies, NGOs, and the government. This appeal is generally divided into categories, such as food, shelter, water/sanitation, health, education, and agriculture. Other relief agencies, such as the International Federation of Red Cross and Red Crescent Society (IFRC and RCS) and Oxfam may make separate appeals. These agencies are generally much faster than the UN in getting their own appeals out, launching them within a day or two of a disaster and then revising them as necessary. The Red Cross may even make an appeal in advance of a disaster provided it knows it is about to impact an area (Batha, 2005).

It is worth mentioning that a gap generally exists between the amount of disaster aid pledged by donors and the amount of aid that actually is released. Table 7.1 illustrates examples of the gap between disaster aid pledged and aid delivered for four selected disasters occurring in four different countries. This table shows that the gaps range from US$200 million to US$6 billion. Such large gaps create a serious burden to the governments of disaster-affected countries and prolong the suffering of victims of extreme natural events.

It can further be noted that about half of the billions of dollars pledged by individuals, businesses, and governments around the globe for the 2004 Indian Ocean tsunami aid has not been sent – two years after the disaster. Quoting the BBC as a source, Tabor (2007) writes that a number of foreign governments

Table 7.1 Gap between disaster aid pledged and aid delivered (in US$)

Event	Pledged	Delivered	Gap
Bam earthquake (Iran), 2003	1.1 billion	17.5 million	1,082.5 million
Hurricane Mitch (Central America), 1998	9.0 billion	3.0 billion	6.0 billion
Mozambique flood, 2000	400.0 million	2000.0 million	200.0 million
Kashmir earthquake (Pakistan), 2005	530.0 million	85.8 million	445.2 million

Source: Compiled from various sources.

completely reneged on their promise to send aid. Others have only given a small percentage of what they promised. All told, some US$6.7 billion was pledged, but only US$3.4 billion had been sent at the time of the BBC report. China is one of many nations giving significantly less than it pledged. For example, China pledged US$301 million to Sri Lanka, but has actually given only a few million. Similarly, France, which promised US$79 million, has actually delivered just a little more than US$1 million, and Spain, which pledged US$60 million, has actually donated less than US$1 million. The BBC reports that the European Commission owes US$70 million in pledged funds, while Britain owes US$12 million. Also the United States has only donated about 38% of the dollars it promised (Tabor, 2007).

In both developed and developing countries scores of agencies are involved in disaster response. Coordination among agencies becomes necessary to avoid duplication and reduce overhead. This coordination speeds things up as well as promotes better utilization of resources. Such efforts may make the system more equitable as some sectors are habitually underfunded. Effective coordination is also needed to avoid waste and competition among relief agencies. The OCHA of the UN frequently takes responsibility for coordination of emergency relief. Many agencies, such the UN's World Food Program or the children's agency UNICEF, have pre-stocked warehouses around the world with food, clothing, and tents with the intent that these organizations can get such items to a disaster zone within a few hours of an extreme event occurrence. The World Food Program currently operates in 85 countries and maintains eight regional offices (Coppola, 2007). However, even without cash pledges, the UN agencies can respond immediately, using reserve funds, and recoup the money later (Batha, 2005).

International financial institutions, such as the International Monetary Fund (IMF), the World Bank, and the Asian Development Bank operate almost entirely bilaterally. Although these institutions usually do not fund disaster aid, they manage and channel disaster funds to appropriate organizations for disaster relief and reconstruction. They generally participate in medium- and long-term reconstruction and rehabilitation programs. Many contribute funds in the forms of grants and loans at a time when immediate relief operations terminate and

Table 7.2 Providers and distributors of disaster aid in developing countries

Category	External	Domestic
Providers	Donor governments	National/local governments
	Foreign NGOs/individuals	Local/non-local people
	Foreign corporate businesses	Private organizations (professional, business, and others)
	Expatriates	
Distributors	Foreign NGOs/individuals	Domestic NGOs
	Individuals/expatriates	National/local governments
	Foreign military forces	Domestic military force
		Individuals
		Others

infrastructure and livelihood projects begin in order to help disaster victims resume their lives.

As indicated, numerous public and private sources (both individuals and corporate or other private entities, such as religious, trade union and other associations) are involved in providing for and distributing emergency aid among disaster victims. It should be noted that all sources of disaster aid do not directly participate in emergency relief distribution and many distributors of relief goods depend exclusively on domestic and/or foreign sources for support. They usually begin their operations immediately after an event. Sources of disaster aid and entities involved in disbursing aid also differ between developed and developing countries. Aid providers and distributors are discussed separately in the following subsections.

7.1.2 Providers of disaster aid

Table 7.2 lists providers and distributors of disaster aid to developing countries. This table divides both providers and distributors into two broad categories: external and domestic. The former includes donor governments, foreign or northern NGOs, including faith-based organizations, foreign corporate businesses, individuals, and expatriates. Foreign private donations are primarily channeled through foreign NGOs and charities. These organizations collect donations in selected stores and offices, making other arrangements with selected stores, and/or through phone or online requests. When a charity like Oxfam knows it has raised US$1 million it informs its employees at a disaster scene they can spend up to US$1 million – the money does not have to be physically present. Figure 7.1

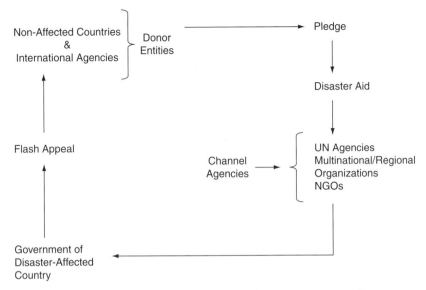

Figure 7.1 Channeling of disaster aid from donor entities to the government of disaster-affected country through channel agencies.

shows diagrammatically the flow of disaster relief from developed countries to disaster victims of developing countries.

Table 7.2 shows that domestic providers include national, state, and local governments, as well as fellow victims and neighbors. Commonly, not all members of a community are equally affected by a disaster. Less affected or unaffected community members, whom Taylor (1989) calls "secondary victims" of natural disasters, often provide assistance directly to disaster victims or indirectly through contributing donations requested by charities or other relief organizations.

For any given disaster, local people and agencies are always the first to respond to crisis. They show enormous solidarity with disaster victims by offering immediate assistance, such as food, shelter, clothing, cash, and labor and are both providers and distributors of disaster aid. In addition to physical and financial support, local people may also provide crucial psychological support – especially to those who experience the loss of family members (Paul, 1998). Domestic providers also include friends and relatives living outside affected communities, people in neighboring (or more distance places), professional and business associations, and other organizations in non-affected areas (Table 7.2).

In developed countries, external sources do not usually play much of a role in disaster relief operations. Because of their financial and technological advantage, they do not usually need the assistance of external providers in order to cope with loss and damage caused by natural disasters. Only in the case of large-scale disasters do some developed countries appeal for external assistance. In the United States, governments of disaster-impacted states along with the federal government can appeal for foreign help. After the Mississippi River swelled in

1993 and flooded nine states, flood-affected states requested and accepted assistance from all over the world. This flood caused property damage estimated at US$10–20 billion, and 20 million acres (80 000 km^2) of farmland were adversely affected, 8 million acres (32 000 km^2) of which were underwater for several days (Tobin and Montz, 1994). It is interesting to note that after the tornado in 1999, the Japanese government donated US$5000 for tornado victims of Kansas. This was unexpected help, because the Kansas government did not make a request for foreign assistance.

After Hurricane Katrina struck the Gulf Coast on August 28, 2005, the United States government was initially reluctant to accept disaster aid and assistance from foreign countries other than Canada and Britain. However, as the reports of the extreme damage surfaced, this policy was reversed. As many as 120 nations made pledges to provide more than US$700 million in cash and other aid to Katrina survivors (USAID, 2005). These countries included friends such as Britain and Germany and foes such as Cuba, Iran, and Venezuela. Katrina donors also included war-torn Afghanistan and countries impacted by the 2004 Indian Ocean tsunami such as India, Thailand, and Sri Lanka. Kuwait made the largest single pledge of US$500 million. Other large donations were made by Qatar, India, China, and Bangladesh.

International humanitarian agencies, such as Oxfam, generally do not respond to humanitarian crises in developed countries. As noted, governments of developed countries do not usually directly accept foreign donations. Such donations are normally channeled to domestic charities, such as the United Way or the American Red Cross (ARC) in the United States. It is worth noting that the Federal Emergency Management Authority (FEMA) is the principal government agency responsible for providing assistance to disaster victims in the United States, and ARC is the sole private organization, mandated by law, with providing relief in national emergencies. FEMA does not release funds to local governments until a plan for proposed funding is developed and approved. FEMA claims to have checks and balances in channeling disasters funds.

7.1.3 Distributors of disaster aid

In developing nations, emergency aid is generally distributed to disaster victims by local and national governments, and a host of other organizations including NGOs, businesses, and private citizens. As shown in Table 7.2, and mentioned previously, not all entities, particularly those from developed countries, involved in providing donations are engaged directly in distributing emergency relief to disaster victims of developing countries. Most NGOs in developing countries depend largely on external support for their disaster assistance programs and thus can be considered distributors (Table 7.2). With some exceptions, NGOs in developed countries usually support domestic NGOs in developing countries rather than engaging in the distribution of relief goods themselves (Roger et al., 1995).

Domestic NGOs are generally more familiar with disaster-affected areas, and their emergency relief efforts are usually more cost effective compared to foreign NGOs, particularly for those who have no personnel permanently stationed in disaster-affected countries. In addition, domestic NGOs traditionally stress the need to build on local capabilities. These organizations, however, also receive funding from bilateral and multilateral donors in emergency situations (Paul, 1998, 2003).

With growing immigrant populations in developed countries, expatriates have been playing an increasingly important role in disaster relief efforts. After each disaster, they take an active part in supporting disaster victims of their native land. These individuals collect donations either individually or through their organizations for the victims and send these donations directly or indirectly through the government or NGOs. For example, Bangladeshi Americans donated more than 10% of all the blankets distributed to victims of the extremely cold weather experienced in Bangladesh in January 2003. This event was the worst in nearly 35 years, and it killed around 750 people (Paul, 2005).

In developed countries, government agencies at different levels, along with some NGOs and charitable organizations, take part in disaster relief distribution among victims of extreme natural events. Local businesses also often participate directly in disbursing disaster relief. For example, after a tornado struck Fort Worth, Texas on March 28, 2000, local restaurants and grocery stores delivered hamburgers, pizza, fruits, and other food to tornado victims. Home improvement stores also distributed plywood to victims who needed to make quick repairs to protect homes (McEntire, 2007). However, the number of both providers and distributors of disaster relief is much higher in developing countries than in developed nations.

Since both foreign and domestic armed forces take part and play an important role in disaster relief distribution among victims of extreme natural events in both developed and developing countries, the advantages as well as the disadvantages of military participation will be discussed in some detail in the following paragraphs.

7.1.3.1 *Participation of foreign military forces in disaster relief distribution*

The most debated distributor of disaster aid is foreign military forces, which have a long tradition of involvement in disaster relief efforts (Palka, 2005). It has been widely reported that public distribution of disaster aid in developing countries suffers from mismanagement, personal and political interference, and corruption (Kennedy, 1999; Paul, 1998, 2003). As a result, disaster aid does not always reach the people who desperately need it. Many believe that disbursement of disaster aid will be fair and all disaster victims will have equal access to aid if foreign and/or domestic military forces take part in such an endeavor.

Disaster victims of developing countries often prefer developed country military involvement because of their perception that foreign military forces are well equipped and sufficiently trained to effectively handle most post-disaster situations. Because of the lack of modern equipment, military personnel in developing countries often cannot reach remote areas easily and quickly. Thus, intervention by foreign military forces in emergencies often results in saving lives. These forces are also able to repair roads, build houses and bridges, and carry relief goods to the disaster site very quickly (Paul, 2005; Walker, 1992).

The scale of foreign military involvement in disaster assistance varies from the use of a single aircraft to fly a rescue team in, to the large-scale involvement of all services over a period of weeks and months (Walker, 1992). Participation of foreign forces in disbursing disaster relief is often considered a serious threat to national sovereignty. Also, some may suspect that foreign military forces may not leave after disaster relief efforts are completed; then such deployment might be regarded as military intervention.

Foreign military forces often feel unwelcome by fellow disaster relief workers and by those they have come to help. Their past experience with an oppressive regime that has used military force to carry out their policies may cause them to be wary of working with or relying on military personnel (Cuny, 1983). However, military forces generally take charge of the entire spectrum of responsibilities when they are utilized. This may be a source of friction with other relief workers or the host nation as they may perceive the military has invaded and is taking over relief efforts, possibly in pursuit of other, not so humanitarian goals (Waldo, 2006).

There are also other serious issues: such as under whose command will foreign force(s) work in disaster-impacted areas? Who will initiate foreign military involvement in disaster relief? How is such military involvement to be monitored? Who will pay for the military? (Walker, 1992). In the context of monitoring military involvement in disaster-relief efforts, Walker (1992, p. 157) argues that "if the military is to be properly accountable then its command and control system must be integrated with that of the civilian relief operation. This may cause problems as the military style of management is very different from that adopted by most relief agencies." Military decision making is centralized and objective-driven, while relief agencies tend to maximize the delegation of decision-making, and are thus able to be more flexible to changing circumstances often encountered in disaster situations.

In addition, when a disaster occurs in an area of armed conflict, maintaining neutrality may also be a serious problem for participating foreign military forces. One of the important determinants of the success of a foreign military deployment in relief operations is whether or not the force is able to assume and maintain neutrality. Cuny (1983, p. 19) adds: "If at any time one or more parties of a conflict perceive that a foreign force has other than humanitarian objectives, either for itself or for another party, the operation in which it is involved will be regarded as a military intervention and the force will become engaged in a conflict."

In international disaster assistance, foreign military forces cannot participate in disaster relief efforts without an explicit request from the affected country. A formal arrangement must be made between the host government and the donor regarding the presence of foreign military in the disaster area. Frequently, though not always, foreign military personnel will be asked to work with the local military forces (Walker, 1992). The United States military takes its directions from the US State Department, which acts upon invitations from disaster-affected countries. The State Department also defines the extent and limitations of the support to be provided by the military (Gaydos and Luz, 1994).

Despite problems associated with the deployment of foreign military forces in disaster relief and rescue operations, foreign countries, irrespective of their level of economic development, often send their forces for such missions. United States forces have traditionally been involved in many disaster relief missions, including the responses to the Nicaraguan earthquake of 1972 and the Sri Lanka cyclone of 1978 (Gaydos and Luz, 1994). The United States also sent 7000 members of its naval forces to Bangladesh following the devastating cyclone of 1991, which killed some 150 000 people (Paul, 2005). These forces played a decisive role in transporting relief goods from Dhaka, the capital of Bangladesh, to those in need stranded on islands (Paul, 2003; Walker, 1992). This military intervention was welcomed by the people of Bangladesh and its government since the country had no resources to reach many cyclone victims quickly. Relief efforts of the US military were credited with saving as many as 200 000 lives. Walker (1992) claims that US military forces were a well-organized and appropriately equipped group that led to the success, not simply because of their military nature.

In addition to US military forces, three other external military forces (the United Kingdom, India, and Pakistan) were also involved in relief operations in Bangladesh in 1991. India provided six helicopters, Pakistan two, the United Kingdom six helicopters, and the United States 28 helicopters, six hovercraft, five C130s and a number of landing craft. Similarly, in the Caribbean, eight countries formed the Caribbean Disaster Relief Unit (CDRU) under the Caribbean Community and Common Market (CARICOM). Its purpose is to allow the military forces of each country to enter one another's territory in times of disaster, to render assistance to disaster victims (Walker, 1992). This implies that military resources may be deployed in friendly countries, but not in enemy countries or politically "unimportant" ones.

Immediately after the 2004 Indian Ocean tsunami, 34 countries, including Australia, Germany, India, Japan, New Zealand, Singapore, the United Kingdom, and the United States, dispatched military personnel and equipment to the tsunami-devastated areas in Indonesia, the Maldives, and Sri Lanka to directly participate in relief efforts by distributing aid, removing corpses, clearing rubble, providing immediate medical attention, and offering psychological counseling. The role of foreign military personnel was praised in several tsunami-devastated countries, particularly in Indonesia. The Indonesia National Military (or TNI) found that it lacked the equipment and capacity to provide assistance in a timely manner to the victims of the areas severely impacted by the 2004 Indian Ocean tsunami. This prompted the Indonesia government to accept foreign troops in the country. Similarly, military forces from several Western countries including the North Atlantic Treaty Organization (NATO) took part in disaster relief efforts in Pakistani-controlled Kashmir, which was struck by a 7.6 magnitude earthquake on October 8, 2005. These forces provided necessary helicopters for carrying the wounded to hospitals, transport search and rescue teams, and deliver relief assistance. At the same time, despite the desperate situation, the Pakistani government refused an offer of Indian army helicopters because India would not allow Pakistani pilots to fly them. Even the United States allowed foreign military forces from Canada, Mexico, Singapore, France, and Germany to take part in post-Katrina relief efforts in 2005. Canada sent a crew of 1600 on three naval ships and France dispatched supplies and troops to New Orleans.

7.1.3.2 Participation of domestic military forces in disaster relief distribution

As indicated, governments of both developed and developing countries often deploy military forces to distribute emergency relief among disaster victims (Fowler, 1990; Lewis, 1992; Paul, 2003). The military's command structure and ability to quickly deploy personnel, specialized equipment, and relief supplies under adverse conditions make it an ideal source of assistance during post-disaster operations (Gaydos and Luz, 1994). Members of these forces air drop relief materials, construct safe water resources, clean debris, bury the dead, repair damaged bridges, and provide many other essential services in disaster-impacted areas. Military forces are not only fast and efficient in providing such services to disaster victims, but are also widely considered nonpartisan and corruption free (Gaydos and Luz, 1994; Walker, 1992).

Sometimes the services provided by domestic military forces are also cheaper than commercial alternatives. The armed forces possess strategic and local mobility – essential in inaccessible areas – and a wide range of specialized equipment: helicopters, aircraft, earth-moving machinery, respirators, medical supplies, and power and lighting equipment. Most importantly, they are self-contained with their own rations and transport. They are also specialized in reconnaissance. However, a question that often arises is whether arm forces will work under civil authority. In some countries, such as the United States, the military acts in post-disaster situations only in support of civilian authorities – never as the leader, in accordance with US laws regarding civilian control over the military. Problems may arise when the military refuses to work under civilian authority.

Advocates of civil authority and most democratic governments, however, oppose military involvement in disasters and consider such involvement an erosion of citizen rights and responsibilities (Mitchell, 1999). Military involvement assumes a reduced capacity of individuals and social structures to cope with emergencies, implicitly expresses a distrust of individuals and structures concerned to make intelligent decisions in emergencies, and it creates a closed system intended to overcome the inherent weaknesses of "civil" society to deal with important emergencies. Some also oppose deployment of military forces in emergency relief distribution because such deployment often increases the demand for prostitution. Because of the limited choices available to economically challenged women and even children, especially in the aftermath of a major disaster, often they are the first ones forced into prostitution in order to survive (Ross et al., 1994).

Although the United States has a long tradition of employing the military in disaster assistance, both at home and internationally, some Americans oppose such deployment. These Americans believe that placing the burden of a heavy commitment to disaster-related missions on the military forces will markedly detract from training and preparation for their primary combat mission (Gaydos and Luz, 1994). Similarly, some military leaders also think that involvement of armed forces in disaster relief efforts may significantly detract from their primary mission of preparation for combat (Palka, 1995).

Others do not oppose military involvement, but they instead argue that the military must be given a well-defined mission in both domestic and international disaster relief operations. They maintain that a natural disaster is an opportunity to show a positive side of US foreign policy. Still others support the participation of US military forces in disaster-related roles if such involvement is in the best interests of the country, and such activities do not seriously detract from combat readiness (Gaydos and Luz, 1994). Palka (1995, p. 206) claims that US military personnel who have participated in disaster relief operations, particularly in foreign countries, have gained tremendous fulfillment and have found the experience to be both "personally and professionally rewarding."

Effective coordination and integration among military and all civilian organizations, including NGOs, can greatly relieve suffering due to extreme natural events in a disaster relief effort. All relief actors and, more importantly, all disasters victims, can immensely benefit from the efficiencies of their combined efforts. Military forces are expert at logistical planning; they have on-call medical, transportation, logistics, communications, engineering and security capabilities (Alexander, 2000; Cuny, 1983; Gaydos and Luz, 2004; Walker, 1992). They are also capable of bringing their own life support systems. In fact, the greatest argument for the involvement of military forces in disaster relief efforts is largely one of resources that can be rapidly deployed (Walker, 1992).

NGOs and other civilian relief agencies, on the other hand, are the disaster relief experts because their workers have participated in numerous relief efforts both in local and nonlocal communities. They are usually more familiar with an impacted area compared to military forces and thus have the trust of disaster victims. It is therefore necessary for all disaster relief actors to take actions to increase the efficiency of their own operation and to maximize their capabilities

Table 7.3 Comparison of 1999 India cyclone and 2000 Mozambique flood media coverage, and the volume of emergency assistance received

Event	Coverage type	Number
India cyclone, 1999	TV news spots on Danish national (Oct 15, 1999–Jan 15, 2000)	16
	Article in 23 newspapers (Oct 15, 1999–Jan 15, 2000)	91
Mozambique flood, 2000	TV news spots on Danish national (Feb 1, 2000–May 1, 2000)	87
	Article in 23 newspapers (Feb 1, 2000–May 1, 2000)	382
Emergency assistance received (in US$ million)		
India cyclone	23.10	
Mozambique flood	165.85	

Source: Compiled from Olsen *et al.* (2003), p. 114.

through coordinating with each other. Moore and his colleague (2005, p. 305) captured the essential role of coordination, citing that: "Improving the level of coordination among humanitarian aid organizations has been viewed as critical to optimize the flow of resources among agencies and increase the accountability, effectiveness and impact of aid operations."

While not actually providing or participating in disaster relief services, the national and international media plays a critical role in such an endeavor. The scenes of destruction, the dead and injured as portrayed by the media create great sympathy for disaster victims, and prompt substantial public and private response to disaster. The compelling images shown by the media are a large part of what incites people to donate their time and their money (Alexander, 2000). It is generally claimed that the official funding decisions are not taken on the basis of need, but rather in response to media and political pressure (Telford and Cosgrave, 2007). The correlation between the media's coverage of heart-rendering scenes of a disaster, and global response to such an event in the form of providing emergency assistance is commonly referred to as the "CNN effect" or "CNN factor" (Heeger, 2007).

Emergency relief organizations are some of the most conscious of the need for clear and abundant media footage to inspire donations. Jeremy Hobbs, executive director of Oxfam International rightly claims that a: "Disaster that does not get a high media profile – that's most of them – does not get the money" (quoted in Moszynski, 2005, p. 165). Information presented in Table 7.3 clearly shows that the volume of emergency assistance is directly associated with the extent of the media coverage. The table illustrates that TV coverage of the 2000 Mozambique flood was more than five times as extensive as the coverage of the 1999 India cyclone. A similar pattern is also found in the coverage of newspaper articles on these two events in Western Europe and the United States. As expected, Mozambique received more than seven times the amount of emergency aid that India received. Disaster experts have further observed that the outpouring of immediate assistance tends to diminish as the media coverage wanes.

7.1.4 Disaster aid provision: a review

At the outset, it should be noted that disaster relief is provided to victims of both developed and developing countries and it is considered to be a short-term (partial) damage reduction measure. Such aid is not intended to reduce the long-term vulnerability of disaster victims. Disaster aid does not and cannot compensate all losses caused by a disaster. Empirical studies reveal that disaster relief accounts for various proportions of the total loss incurred by a disaster. Paul and Leven (2002) reported that tornado victims of a Kansas town received disaster relief from various sources which amounted to about 70% of their total reported damage. In contrast, the amount of emergency assistance received by the 1998 flash flood victims in two Kansas cities – Augusta and Arkansas City – was only 26% of the total reported loss of property (Paul, 1999). Studies carried out in Bangladesh and India in recent years suggest that, in monetary terms, many disaster victims have received more emergency assistance (in monetary terms) than the monetary damages they incurred by a given disaster (Paul, 2006).

However, the goal of disaster aid is to save lives, reduce victim suffering, help to meet basic daily needs, and enhance the disaster recovery process (Tobin and Montz, 1997). Relief activities typically include providing food, medicine, safe drinking water, sanitation services, temporary housing, security, as well as psychological and social support. Disaster relief is not designed to mitigate or reduce the risk of future disasters, but to return victims' level of living to a pre-disaster level. Relief workers are interested only in dealing with the immediate needs of disaster victims.

Disaster relief also helps smooth the transition from necessary and immediate relief to long-term rehabilitation and redevelopment (Smith and Ward, 1998). A smooth transition is needed to quickly return survivors to (post-disaster) normal life. This implies that re-establishment to a post-disaster level of living and a return to normal life is not the same thing. The former refers meeting basic necessities, which can be achieved even while victims are staying in temporary housing. Returning to a "normal life" means much more. It may include owning a home and/or the means to rent a house, securing a permanent job, etc; in short, this means victims are no longer depending on any type of relief aid. Two points need to be mentioned here. First, not all disaster victims return to a "normal life," or fully recover from the impact of disaster. Secondly, a "normal life" to the most vulnerable people in a community, is often a continual struggle in which their situation may resemble those of disaster victims.

7.1.5 Criticisms of disaster relief provision

Many hazard researchers (e.g., Hewitt, 1983; Susman *et al.*, 1983; Wisner *et al.*, 2004) maintain that both relief and development aid is counterproductive; both can increase dependency on external sources; and create a "culture of dependence." These researchers claim that emergency and development aid not only reinforces the continued underdevelopment and marginalization of disaster victims, it also creates the pre-conditions for yet more disasters. Development aid and development programs are often to blame for raising human exposure to environmental risk. For example, development projects, such as dams, often increase risk and vulnerability, especially for marginalized groups. It is worth while mentioning that disasters are also viewed a consequence of development and industrialization. As noted, in Europe, experts believe that countries such as France and Germany are more adversely affected by floods today because major rivers, such as the Rhine, have been straightened to ease commercial traffic.

However, no empirical study to date has been conducted to support the contention that relief assistance contributes to cycles of underdevelopment. The amount of disaster aid received is generally closely associated with the degree of devastation caused by a particular extreme event. Local events usually attract aid from the immediate area, medium-scale events stimulate nation-wide contributions, and large-scale disasters usually involve assistance from all sources, including the international community. However, there is no doubt that disaster risk is much higher for more vulnerable groups than for less-vulnerable groups, because members of the former are more likely to be settled in marginal or disaster-prone areas (Wisner, 1993).

Critics of disaster relief (e.g., Sollis, 1994; Susman *et al.*, 1983; Wisner *et al.*, 2004) advocate integrating relief operations into policies, plans, and long-term development initiatives as well as empowerment by gender, race, and class. They believe that a long-term development perspective is critical, starting at the relief stage, in the allocation and use of resources in order to foster self-reliance, build capacity, avoid dependency, and reduce poverty (Yonder, 2005). These critics believe that emergency relief and recovery seek to return disaster victims to their pre-disaster "normal" life. However, they question what good it is to return to normal conditions that are characterized by injustice, extreme poverty, and disempowerment. To them, effective disaster relief without social and political change is not desirable (Mustafa, 2003). Critics advocate sustainable solutions to disaster problems instead of just repeating the same relief operations over and over again.

Relief assistance may cause more damage than a given disaster itself in the long run. Assistance can trigger societal changes that are almost as damaging as the disaster itself. For example, Goranson (2005) describes what happened in fishing villages in the Philippines following Typhoon Rita in 1978. The typhoon had wiped out the fleet of handmade wooden fishing boats in a group of sea-dependent Philippine communities, but with external support, these disaster victims were able to replace their destroyed or damaged boats with modern fiberglass versions with small petrol engines. The fishing economy rebounded quickly and flourished, but only for about 10 years. After that, the entire society collapsed because the fishermen no longer depended on boat builders, who had been the anchor of that society for over 2000 years. These craftsmen acted as priests, teachers, and judges; subsistence flowed according their goodwill and was supported by conventions of sharing and trust. This complex human balance was replaced by a cash economy in which those who could dole out precious petrol became the power brokers.

Critics have noted that international agencies and governments typically deliver disaster assistance in a top-down manner, often labeled as the "command and control model" of disaster response. This model includes incompetent and corrupt government bureaucracy at one end, and hazard victims at the other, and it makes the victims passive recipients of aid (Dynes and Tierney, 1994; Mustafa, 2003). It is widely held that direct participation of victims in setting priorities and distributing supplies, information, and work during the relief stage helps in restoring their self-confidence. Ignoring victim opinions and participation in emergency relief efforts often results in a waste of emergency funds and fosters victim dependency on external sources (Yonder, 2005).

Disaster relief critics urge decision-makers to encourage community involvement and cooperation, and to strengthen such capacities because local people and resources are the only ones who can bring effective help in the first few hours after a disaster occurs. For example, government and military forces, and international professional rescue teams arrived in the Marmara region of Turkey, which was devastated by an 7.4 magnitude earthquake on August 17, 1999, 72 hours after the event occurred. By that time, local people had rescued some 10 000 people from the rubble; the late-arriving professional teams rescued only about 500 more (Yonder, 2005). Some disaster experts also recommend using local medical teams, at least at the beginning of rescue operations, because these teams can be mobilized quickly and are culturally integrated with their patients (Smith and Ward, 1998). This underscores the importance of emergency preparedness training for local volunteers and medical personnel.

At this juncture, it is appropriate to mention that disaster victims in developing countries, particularly those who live in rural areas, have developed indigenous coping strategies – sometimes evolving over generations – to overcome the impacts of extreme natural events (Maskrey, 1989). Wisner et al. (2004) termed these strategies as the "people's science" that provides basis for much coping behavior of the disaster victims. More often than not, "official" relief pays little attention to what ordinary people do. As a consequence, resources are not only wasted, but such relief may destroy or undermine vernacular coping skills.

Although sometimes a local administration or an NGO has been able to build on such foundations, outside agencies must be informed about and understand these native coping strategies. Otherwise, some researchers (e.g., Wisner et al., 2004), claim "official" relief will create aid dependency and lead to unintended and detrimental outcomes.

Even when relief is recommended, there are many shortcomings in the provision of disaster relief. Throughout the relief stage, serious problems arise over what, when, and how disaster aid is delivered. These problems are discussed below under four headings: inequitable distribution of disaster aid, human rights violation, appropriateness of disaster aid, and other criticisms. Although there is considerable overlap, each relief effort is included under the most appropriate subheading.

7.1.5.1 Inequitable distribution of disaster aid

Disaster literature consistently emphasizes the need for appropriate and adequate amounts of emergency relief and frequently claims that the public distribution of aid suffers from widespread corruption and large-scale irregularities (e.g., Bolin and Stanford, 1999; Cupples, 2007; Lewis, 1999; Pelling, 2003a; Smith, 1992). Several approaches to environmental hazards, such as vulnerability and the political ecological paradigms, concede that public sources discriminate against marginal groups in the provision of disaster assistance. For example, it has been reported that after the 1972 earthquake in Guatemala, disaster assistance was directed mainly to middle-class people while the poor, who suffered the most, were largely ignored (Ozerdem, 2006). In Gujarat, India, after the earthquake in, 2001, relief workers believed that Muslims and Dalits (untouchables) did not receive an equitable share of disaster relief (Wisner et al., 2004).

During the 2010 floods in Pakistan, members of Ahmadiya community were not rescued from their homes because rescuers felt that Muslims must be given priority. Although Ahmadis claim to be Muslims, the Pakistani government designated them a non-Muslim minority in 1974. Ahmadis also complained that they were denied relief supplies and in some instances, they were forced to leave relief camps when their identity was disclosed (*Dawn*, 2010a). Members of Sikh community also complained of government apathy (*Dawn*, 2010b).

In the case of the Kashmir earthquake, there have been complaints that aid distribution in Pakistan-administered Kashmir was skewed – too much in some places, too little in others. Ozerdem (2006) reported that the slow and uncoordinated response put pressure on societal relationships between Kashmiris and Pakistan's population in general, as it increased the sense of alienation of some Kashmiris, who complained of discrimination. Kashmiris complained that relief was going faster to "Pakistanis" in earthquake survivors of the North West Frontier Province. Similarly, Katrina victims of New Orleans and Louisiana perceived that hurricane-affected families in Alabama and Mississippi not only received more emergency aid than them, they also received it faster than them. This might be associated with the fact that Congressmen and Senators of the latter two states occupy key posts on Capitol Hill.

Distribution of disaster aid by FEMA among Hurricane Katrina victims also suffered from serious problems. This organization initially handed out disaster relief debit cards worth US$2000 per family affected by the hurricane and provided rental assistance to the victims. In a report published in June 2006, the Government Accountability Office (GAO) noted that some payments went to people who had never lived in a hurricane-damage property. At least 1000 of them used prisoners' names and Social Security numbers, others gave false or nonexistent addresses. Some individuals received multiple payments and others received rental assistance even though they were living in hotels at government expenses. The GAO estimates that FEMA made "improper and potentially fraudulent" payments amounting to between US$600 million and US$1.4 billion (GAO, 2006). The report further claims that as much as 16% of the relief distributed by the FEMA was lost to fraud.

There are many reports of inequalities in the distribution of emergency aid to the 2004 Indian Ocean tsunami victims (Paul, 2007). Some affected communities in Thailand and in the Aceh province in Indonesia, were amply supplied with emergency aid, while others were largely ignored (IFRC and RCS, 2005). The greatest number of displaced people who found shelter in the homes of their relatives and friends received little assistance (Aspinall, 2005). Tsunami victims in Andaman Nicobar Islands, India, who originally settled there from mainland India over the last several decades, were largely ignored in relief efforts. Similarly, Myanmar refugees living in coastal Thailand were among the last to receive tsunami aid. In Thailand, most relief efforts were directed at tourist areas, and affected nontourist areas were largely ignored (Krauss, 2005). Because of inequitable and biased aid distribution, disaster victims belonging to vulnerable groups benefit least from emergency relief efforts and this further exacerbates the divide between vulnerable and nonvulnerable groups.

Disaster aid, particularly in developing countries, often does not reach the victims who need it. In the 1990s, for example, food aid was sent to drought-affected areas in Ethiopia and Sudan. But, because of the political instability and military turmoil, it was not possible to distribute this aid to drought victims (Tobin and Montz, 1997). In addition, it has been found in many instances that coverage by government disaster relief efforts was patchy and inadequate (Mustafa, 2003).

The disaster literature cautions that emergency aid provision is often perceived by victims as unequal and discriminatory in a number of situations when needs are great and while capacities to cope with them are limited. Such studies suggest that disaster response is carried out with little preparedness and in an untimely manner. Ozerdem and Jacoby (2006) further claim that even if the response is decided solely by the needs of disaster victims, this may also be perceived as favoritism of certain groups by communities who believe they are receiving less assistance than others. After the 2001 Bam earthquake in Gujarat, India, various religious and social groups perceived that the provision of emergency assistance was discriminatory.

Religious, caste, and sectarian discriminations have also been reported in earthquake-impacted areas of Kashmir. It has been claimed that the Swati castes were discriminated against by the Syeds, who received the bulk of relief in the Muzaffarabad district. A joint study by Church World Service Pakistan/Afghanistan and the Duryog Nivaran Secretariat reveals cases of religious-based discrimination against Christian families following this earthquake (IFRC and RCS, 2007).

7.1.5.2 Human rights violations

In developing countries, men are more likely to wait in relief line for goods, while women (as well as children and the elderly) become increasingly dependent on them for survival. This situation is exacerbated when a woman is a widow or single parent and is unable to compete for aid. Another well-documented event impacting women after a disaster occurs is the rise in violence against women, including rape, domestic violence, and other forms of abuse, which is discussed in the next chapter. Such violence and abuses were reported in all countries severely impacted by the 2004 Indian Ocean tsunami. Gender issues help explain why women are more likely than men to suffer from post-traumatic stress syndrome after disasters (Felten-Biermann, 2006).

Women help enormously in the aftermath of a disaster by managing displaced households and restoring family livelihoods. However, they also face a number of obstacles in applying for and qualifying for emergency aid, including limited literacy, limited access to information, limited experience dealing with bureaucracy, and sometimes, eligibility requirements that exclude them. Post-disaster aid efforts generally ignore this reality and target male-headed households as the primary claimants for government and other support and, thus, women may become doubly victimized – by the disaster itself and by the response to it. This situation substantially undermines prospects for household and community recovery. This suggests gender-sensitive programming/planning is essential during emergency relief (Yonder, 2005).

As indicated, the elderly, as a group, receive little or no special attention from relief agencies. Often, because of health conditions, the elderly find it difficult to travel to distribution sites and, even if they could make the journey, they may not have the strength to carry the relief goods back to their homes (IFRC and RCS, 2007). According to a report issued by HelpAge International in 2005, elderly victims in 2004 Indian Ocean tsunami-impacted countries were forced to seek out work to support themselves and their grandchildren, who had lost their parents in the tsunami. In Sri Lanka, there were reports that a high level of alcohol abuse among male victims was having an adverse impact on the well-being of older people (Paul, 2007). Elderly victims in all tsunami-impacted countries faced difficulties in resuming treatment for chronic illnesses, such as cancer and diabetes.

The HelpAge report also found in India that older people were virtually "invisible" during relief efforts. Researchers of this organization interviewed 856 older people in tsunami-impacted districts of southern India and concluded that they were not even recognized as a highly vulnerable group by aid agencies. The report further claims that there was no specific component in the relief operations for older people by any relief agency (Company News, 2005).

7.1.5.3 Appropriateness of disaster aid

Apart from distribution issues, relief goods are not always of the appropriate types. Hazard studies provide ample evidence of sending useless and/or redundant aid to disaster victims (e.g., Alexander, 2000; Gaydos and Luz, 1994). For example, after the 1976 earthquakes in Guatemala, some drugs sent by foreign donors were found to have an expiry date of August 1934! Most drugs were not only useless, they arrived unsorted, unlabeled, poorly packaged, outdated, or they were not intended for emergency use – often interfering with critical disaster relief functions (Alexander, 2000). Cupples (2007) reported that after Hurricane Mitch (1998) in Nicaragua relief agencies distributed beans among the survivors. However, these beans were so hard that they never cooked even using heaps of firewood.

The situation was no different in Armenia following the 1988 earthquake, which killed some 25 000 people. Autier *et al.* (1990) reported that after this earthquake in northern Armenia, foreign donors sent 5000 tons of drugs and consumable medical supplies. Less than a third of these medicines were immediately usable; some 11% of the remaining were inappropriate medicines, 8% had expired, and much of the remainder were inadequately labeled in as many as 21 different languages (Alexander, 2000). In addition, few of the medical supplies that were sent were appropriate for treating earthquake victims.

In the aftermath of the 1991 cyclone in Bangladesh that killed approximately 139 000 people, the country received a massive amount of drugs and medical supplies from foreign donors. Unfortunately, many of these supplies were not needed, while some needed drugs were not supplied (Rahman and Bennish, 1993). Often foreign military field hospitals are opened in disaster-impacted developing countries. All armies possess medical capabilities for dealing with battle injuries and such units are suitable for dealing with injuries caused by earthquakes. But for other disasters, such as floods or droughts, which do not normally cause extensive injuries, such field hospitals are not appropriate.

The German army opened a 200-bed field hospital in Iran in 1991 for Kurdish refugees, but the maximum number of patients in the hospital was never higher than 23 (Paul, 2007). The 2004 World Disaster Report (IFRC and RCS, 2005) claims that there were surpluses of physicians in several tsunami-impacted areas. For example, surgeons poured into Banda Aceh, Indonesia where 10 field hospitals were set up. None of these surgeons worked at full capacity, however, because of an insufficient number of patients. It was even reported that in the Indonesian town of Meulaboh, 20 surgeons competed for a single patient. In contrast, there was a dearth of midwives and nurses. As a result, some women had to give birth without medical assistance (IFRC and RCS, 2005).

A similar situation was observed in Pakistan-controlled Kashmir after the 2005 earthquake where medical teams from everywhere rushed in to help earthquake survivors. But army clinics in Muzaffarabad attending to the influx of injured people said they needed orthopedic implants and medical supplies, not more doctors (Ozerdem, 2006). Most health care needs after disasters and population displacements do not require complex equipment and specialized medical training. What is needed is appropriately trained local health staff, first aid providers, and preventive measures – such as the provision of safe drinking water, sanitation facilities, and adequate food.

After the 1974 flood in Bangladesh, which resulted a nation-wide famine that resulted in the loss of nearly 30 000 lives, several Western countries sent massive quantities of canned pork for distribution among flood victims, ignoring the fact that Bangladesh is an Islamic country, and this religion prohibits the consumption of pork. This was repeated after the 2004 Indian Ocean tsunami; meals containing pork were sent by the Red Cross (of Western countries) to Muslim tsunami victims of Aceh, Indonesia (Telford and Cosgrave, 2007). During the 2010 flood in Pakistan, relief camps personnel served beef to Hindus who were taking refuge there (*India Today*, 2010).

Shipments of large volumes of useless and redundant relief aid, such as outsized, filthy, or inappropriate garments have frequently been reported following disasters in developing countries. Gautham (2006) reported that after the 2004 Indian Ocean tsunami, NGOs flew in thousands of packets of sanitary napkins and distributed them to tribal women of the Nicobar Island who had never used them. Transporting relief goods from abroad often costs more than the value of goods themselves. Donating food is also problematic because verifying the safety and purity of donated canned foods is difficult and time consuming. Often relief goods are delayed in customs for inspection and/or clearance and never get to disaster victims. Storage space is yet another problem associated with useless and redundant relief goods. These inappropriate goods occupy space, which could be used to store genuinely needed disaster relief supplies. In general, space available for storing relief goods in developing countries is very limited. Alexander (2000) considers such practices by developed countries as a heartless and cynical form of dumping surpluses on victims of disaster, or an easy way of avoiding proper disposal at home while appearing to be generous.

Apart from space, there is another problem with the outpouring of donations, such as clothes, that inevitably follows a disaster. It takes time and volunteers to sort the items and dispose of things (often on roadsides) that are not wearable

(Moore, 2005). These disposed items block roads and may prove hazardous to livestock who often try to eat them. For this reason, the Red Cross in the United States does not accept donated clothes; however, it does accept cash so those in need can buy new and/or appropriate apparel (Moore, 2005).

Governments of developing countries affected by disasters have little control over this situation. Inappropriate and unneeded emergency relief items generally originate abroad and are often distributed to victims unilaterally by foreign agencies and private volunteers. These entities are not always familiar with the cultural and religious norms of the affected countries. Distribution of inappropriate relief items often disappoints disaster victims of developing countries, and can be detrimental to their dignity (UN, 2005).

7.1.5.4 Other criticisms

There are also other criticisms of disaster aid: timeliness of disaster aid, coordination of disaster relief efforts, and misuse and non-use of relief funds. Disaster aid can provide both an inefficient subsidy to risk-takers and can also reduce disaster victims' willingness to take remedial actions to minimize losses from future disasters (Haider et al., 1991). However, there is no empirical study available which shows to what extent the availability of disaster relief limits the adoption of hazard-reducing measures. Smith and Ward (1998) reported that an initial debate was triggered in the United States by the cost of direct property damage provided to the victims of Hurricane Andrew and the Great Midwestern floods of 1993.

Hurricane Andrew, which devastated Florida in 1992, caused property damage worth US$30 billion, while the corresponding damage caused by the Midwestern floods was estimated at US$10–12 billion (Tobin and Montz, 1994). Andrew victims received some US$18 million in direct property damage payments, but, in 1993, only US$1.7 billion was allocated for affected farmers in emergency payments by the US Department of Agriculture (USDA) for agricultural losses – in addition to crop insurance indemnities. Some critics regard such disaster aid as a government subsidy to floodplain agriculture which benefits individual farmers with taxpayer money.

In addition, an influx of donated food, deemed necessary for saving disaster victims from starvation, can also undermine market prices and, therefore, the incentives of local farmers to plant next season's crop (Albala-Bertrand, 1993; Tobin and Montz, 1997). Low food prices can also cause severe hardship to local farmers at a time when their livelihood has been disrupted by a disaster (Alexander, 2000). Therefore many disaster experts maintain that food aid should be reserved only as a last resort for victims of extreme events. Instead, they favor providing cash or vouchers to the disaster victims. This allows the victims to buy essentials, including food, soap, or whatever else they need most, even if they have lost their source(s) of earning a living.

Further, proponents of providing cash or vouchers to disaster victims claim that the prevailing food aid often advances petty interests of donors instead of the needs of disaster victims (Dewan, 2007). Rich countries often donate surplus food to subsidize their own farmers. For some reason, if they buy food items from the domestic market of the disaster-affected countries, it will intensify inflation on items already in short supply (Dewan, 2007).

Other criticisms include difficulty in delivering disaster aid in remote areas and in war-torn countries. Emergency aid, for example, is often used as a political weapon by powerful donor nations (Olsen *et al.*, 2003). International aid policies also are affected by domestic political leanings. Countries tend to favor allies so that more disaster aid generally flows to "friendly" nations. Moreover, some aid, as previously mentioned, comes with restrictions attached to its use. For example, the donor transportation system must be used, or aid may be linked to other activities. This creates obligations on the part of aid-receiving countries. In such cases, the needs of disaster victims are sometimes secondary to the political interests of donor and recipient countries (Tobin and Montz, 1997).

Disaster victims are generally ignored in the decision-making process and it is assumed that they benefit from the "trickle down" effect of public relief and reconstruction efforts. Victims are considered as "helpless recipients" who depend on whatever assistance that is given. They are not looked upon as those who can decide the relief needed, or as capable of planning and deciding how rebuilding is accomplished. In most disaster relief efforts, no attempt is generally made by governments to strengthen social protection measures for vulnerable people and communities.

Despite scores of criticisms, disaster victims of both developed and developing countries do not only expect and receive relief from outside sources to cope with losses caused by extreme events, they often become proactive within their communities in terms of building networks and providing assistance to other victims. Mustafa (2003) reported that the women victims of the 2001 flood disaster in the Rawalpindi–Islamabad conurbation in Pakistan challenged the government in doing its job of providing emergency relief to the affected areas. In one flood-affected neighborhood of Safdarabad, female victims collected cash to rent a bus to go to the district administrative offices and demand the relief aid that was promised to them. This group was vocal enough in their demands to be baton-charged by the police during one of their protest rallies.

7.1.6 Involvement of NGOs in disaster aid provision

Since the mid 1980s, when official donors began to channel both development and disaster aid to domestic NGOs in developing countries, many changes have occurred in the distribution of disaster relief (Paul, 2006). As a result, national governments have had decreasing success in drawing both emergency assistance and development aid from external sources (Townsend *et al.*, 2004). Such action was necessary because domestic NGOs have proven over the years that they are more efficient, effective, and impartial in the disbursement of relief than most governmental agencies (Fowler, 1990; Lewis, 1992; Paul, 1998, 2003, 2006). Because of participation by domestic NGOs in disaster relief efforts, relief distribution among disaster victims in some developing countries has been almost flawless. Government entities in developing countries, however, tend to downplay the role of the NGOs in relief efforts and stick to their traditional accusations of corruption and inefficiency among NGOs (Mustafa, 2002, 2003).

The fair and impartial distribution of emergency aid by domestic NGOs has forced governments of some developing countries, such as Bangladesh, to become more accountable and fair in emergency relief distribution (Paul, 1998, 2003, 2006). Unlike in the past, the mass media and concerned citizens of developing countries are now more vocal in criticizing improper distribution of emergency aid by national governments. However, governments of developing countries do not want to lose their authority and control over external disaster assistance. Paul (2006) maintains that if the Bangladesh government continues to provide disaster relief as efficiently as it did after the devastating flood in 1998, foreign donor agencies may rethink their positions and begin channeling a larger portion of disaster aid, and possibly development assistance, through the government. Before this flood, the Bangladesh government was not able to demonstrate its ability to properly manage and distribute relief goods to those affected by disasters that have occurred since 1971 – the year this country won its independence.

In addition to distribution issues, other changes in the provision of emergency relief have occurred in some developing countries, particularly in Bangladesh, a country that has made considerable advances in organizing disaster relief operations within the limits of available resources and which has more NGOs than any country of similar size in the world (Lewis, 1992). Several researchers (e.g., del Ninno and Dorosh, 2002) have reported that, in the aftermath of a disaster, food aid not only contributes to increased food availability but it also helps eliminate or reduce price-hikes in local markets – a common phenomenon in the post-disaster period. Empirical studies (e.g., Paul, 1998, 2003; Paul and Bhuiyan, 2004) have also provided evidence that, since the devastating cyclone of 1991, appropriate relief goods have been sent to disaster victims in Bangladesh. This is largely due to the increased involvement of NGO workers in emergency relief efforts, who know local culture and are aware of the actual needs of the people in the communities they serve.

With the involvement of domestic NGOs in development activities of developing countries, disaster relief has the potential to become increasingly integrated with development programs (Pelling, 2003a, 2003b). Because of these new opportunities to participate in development activities, many NGOs have moved away from a strictly relief orientation more toward a development and empowerment agenda (Elliott, 1987; Korten, 1990; Vakil, 1997; White, 1991). For example, in the 1970s, all NGOs in Bangladesh could be termed relief organizations. In the 1980s, many of these NGOs deliberately avoided relief operations in the belief that relief work causes disruption of normal development activities and often forces beneficiary groups into relief dependency (Paul, 2006).

By the early 1990s, many NGOs operating in Bangladesh, the Philippines, Turkey, and other developing countries realized that "development" is about reducing the vulnerability of the poor to both human-induced and natural hazards (Matin and Taher, 2001). As a result, it is increasingly difficult to separate disaster from development in any meaningful political, social or economic context (Benson et al., 2001; Cuny, 1983; Middleton and O'Keefe, 1998). In addition, NGOs are often involved in mobilizing support when they see that the government and other entities, such as private corporations, are involved in activities that increase the hazard vulnerability of the people and communities they serve (Luna, 2001).

Although the role played by NGOs in disaster relief and development programs varies remarkably from one developing country to another, their activities have been expanding over time. NGO membership is generally restricted to the poor with little or no resources. Empirical studies (e.g., Haque, 1993; Paul, 1998, 2003, 2006) conducted in Bangladesh indicate that NGO members frequently benefit more from the NGO's distribution of disaster aid than non-members. Members are also in a better position, compared with the pre-NGO period, and now are receiving more emergency aid and are thus better able to cope with the impacts of extreme natural events.

Moreover, NGOs have been trying to alleviate poverty by raising democratic consciousness, encouraging the poorest to articulate their social and economic needs, and by combining short-term relief efforts with long-term preparedness support. The contributions of NGOs in bringing vulnerable groups under the umbrella of development are widely acknowledged (Streeten, 1999). Alleviating poverty and helping people become more self-reliant are important elements of both social and economic development (Middleton and O'Keefe, 1998).

In the face of some criticism, poverty and vulnerability are increasingly considered identical, or at least highly correlated, and there is a belief in the NGO community as well as in governments of developing countries, that an increase in development brings disaster risk reduction (Wisner, 1993). Compared to the governments, NGOs are well placed to play a critical role in implementing development and disaster programs. They work with poor and vulnerable groups in society and, because of their participatory approach, NGOs can easily identify potential threats and vulnerability of these groups to hazards (Benson et al., 2001).

It is important, however, to mention that NGOs are not entirely free from problems. Almost all domestic NGOs are dependent on external sources for funding and they are not directly accountable to national governments or to the people. Financial irregularities in handling external funding and other allegations have also surfaced against some NGOs of developing countries. Although governments of these countries are not free from corruption and misappropriation of development funds, many perceive NGOs as a threat to their power and authority and carefully monitor the activities of NGOs (Matin and Taher, 2001). Pressure from donor countries and agencies also guide the proper use of external financial assistance received by NGOs.

7.2 Conclusion

An attempt has been made in this chapter to shed light on important aspects of emergency relief efforts undertaken in the aftermath of natural disasters. The post-disaster relief efforts generally pose enormous challenges for governments of the affected countries as well as others who participate in the relief efforts. Undoubtedly, such efforts face many challenges, some of which have been discussed in this chapter. Despite many flaws, the emergency relief phase aims at meeting basic needs of disaster survivors. Rapid mobilization of relief supplies helps meet the immediate needs of the survivors.

Evidence presented in this chapter clearly suggests that some of the criticisms regarding the provision of emergency relief for disaster victims of developing countries widely cited in hazard literature are still valid. These include: systematic biases in the disbursement of emergency relief, distribution of inappropriate and non-essential relief goods, lack of coordination among participating agencies, and human rights violations.

NGOs and other responding entities, should take local customs into account while developing their emergency assistance programs. Likewise, individuals and others who donate emergency aid should also take local traditions into serious consideration. Sending useless and unneeded items not only reduces valuable storage space, but can also often anger disaster victims. Clearly the preferred alternative is to donate cash for distribution. The direct donation of cash has several advantages: it does not occupy storage space, costs little to transport both short and long distances, and it is an item considered valuable by all cultures and societies. It is rapidly convertible and transferable. With today's technology, a cash donation can reach a relief agency in seconds, enabling the agency to purchase exactly what is needed on the ground while avoiding duplication and oversupply of items that are not needed. Using cash locally will stimulate the economy of the affected region. In addition, to improve coordination, donating aid is better than personally participating in disaster relief efforts.

Discussion in the last section of this chapter suggests that some of the arguments made in the hazard literature against the provision of emergency relief for disaster victims are no longer appropriate in some developing countries – at least to the extent that they were two or three decades ago. Since NGOs are increasingly considered by donors as a viable – even preferable – alternative to national governments for channeling disaster as well as development assistance, the condition of the rural poor in developing countries will likely further improve over time. During the last three decades, many NGOs have been working in developing countries to address poverty, rural development, gender equity, environmental conservation, disaster management, human rights, and other social issues. To empower women and the poor in social and economic terms, NGOs have vastly broadened their activities to include the provision of microcredit, education, health care, nutrition, family planning and welfare, as well as legal aid. They are largely responsible for reducing rural poverty in many developing countries. These and other changes need to be included in a recurrent debate among hazard researchers over whether natural disasters should be addressed by relief measures.

References

Albala-Bertrand, J.M. (1993) *The Political Economy of Large Natural Disasters, with Special Reference to Developing Countries*. Oxford: Oxford University Press.

Alexander, D. (2000) *Confronting Catastrophe*. Oxford: Oxford University Press.

Aspinall, F. (2005) Indonesia after the tsunami. *Current History* **104** (6): 105–109.

Autier, P. et al. (1990) Drug supply in the aftermath of the 1988 Armenian earthquake. *The Lancet* **335**: 1388–1390.

Batha, E. (2005) How does aid flow after a disaster? http://alertnet.org/thefacts/reliefresources/112230013428.htm (accessed January 6, 2006).

Benson, C. et al. (2001) NGO initiatives in risk reduction: an overview. *Disasters* **23** (3): 199–215.

Bolin, R., and Stanford, L. (1999) Constructing vulnerability in the first world: the Northridge Earthquake in Southern California. In *The Angry Earth: Disaster in Anthropological Perspective* (eds A. Oliver-Smith and S.M. Hoffman). London: Routledge, pp. 89–112.

Company News (2005) Elderly "invisible" in tsunami aid efforts, says HelpAge. http://agency.facts.com/news/company-news/Corporate/5078.html (accessed January 6, 2006).

Coppola, D.P. (2007) *Introduction to International Disaster Management*. Amsterdam: Elsevier.

Cuny, F.C. (1983) *Disasters and Development*. London: Oxford University Press.

Cupples, J. (2007) Gender and Hurricane Mitch: reconstructing subjectivities after disaster. *Disasters* 155–175.

Dawn (2010a) HRCP condemn denial of relief to Ahmadis. 21 August.

Dawn (2010b) Govt. accused of neglecting Sikh. 10 August.

del Ninno, C. and Dorosh, P.A. (2002) Maintaining food security in the wake of natural disaster: policy and household response to the 1998 floods in Bangladesh. *Journal of Bangladesh Studies* **4** (1): 12–24.

Dewan, A.A. (2007) Disaster aid and anti-inflationary recovery. *The Daily Star* December 3.

Dynes, R.R. and Tierney, K.J. (1994) *Disasters, Collective Behavior and Social Organization*. Newark: University of Delaware Press.

Elliott, C. (1987) Some aspects of relations between the north and south in the NGO sector. *World Development* **15** (1): 57–68.

Felten-Biermann, C. (2006) Gender and natural disaster: sexualized violence and the tsunami. *Development* **49** (3): 82–85.

Fowler, A. (1990) Doing it better? Where and how NGOs have a comparative advantage in facilitating development. *AERDD Bulletin* 28.

Gautham, S. (2006) Teach the girls to swim tsunami, survival and gender dimension. http://www.countercurrents.org/gen-gautham100806.htm (accessed February 4, 2007).

Gaydos, J.C. and Luz, G.A. (1994) Military participation in emergency humanitarian assistance. *Disasters* 18: 48–57.

Goranson, H.T. (2005) Local science for large disasters. *Jordan Times*, September 10.

GAO (Government Accounting Office) (2006) Expected assistance for victims of Hurricanes Katrina and Rita: FEMA's control weaknesses exposed the government to significant fraud and abuse. United States Government Accountability Office, Washington, D.C.

Haider, R. et al. (1991) *Cyclone 91: An Environmental and Perceptual Study*. Dhaka: BCAS.

Haque, E.C. (1993) Flood prevention and mitigation actions in Bangladesh: the "sustainable floodplain development" approach. *Impact Assessment* 11: 367–390.

Heeger, B. (2007) Natural disasters and CNN: the importance of TV news coverage for provoking private donations for disaster relief. Natural Hazards Center, University of Colorado at Boulder, CO.

Hewitt, K. (ed.) (1983) *Interpretations of Calamity*. Boston: Allen and Unwin.

Hewitt, K. (1997) *Regions of Risk: A Geographical Introduction to Disasters*. London: Longman.

India Today (2010) Uproar as Pak Hindu are served beef in flood camp. *India Today* 24 August.

IFRC (International Federation of Red Cross) and RCS (Red Crescent Societies). (2007) *World Disaster Report*. Geneva: IFRC.

IFRC (International Federation of Red Cross) and RCS (Red Crescent Societies). (2005) *World Disaster Report*. Geneva: IFRC and RCS.

Kennedy, C.H. (1999) Reconsidering the relationship between the state, donors, and NGOs in Bangladesh. *The Pakistan Development Review* 38: 489–510.

Korten, D.C. (1990) *Getting to the 21st Century: Voluntary Action and the Global Agenda*. West Hartford: Kumarian Press.

Krauss, E. (2005) *Waves of Destruction: The Stories of Four Families and History's Deadliest Tsunami*. Emmaus, PA: Rodale Press.

Lewis, D.J. (1992) *Catalysts for Change? NGOs, Agricultural Technology and the State in Bangladesh*. London: ODI.

Lewis, J. (1999) *Development in Disaster-Prone Places: Studies of Vulnerability*. London: Intermediate Technology Publications.

Luna, E.M. (2001) Disaster mitigation and preparedness: the case of NGOs in the Philippines. *Disasters* 25 (3): 216–226.

Maskrey, A. (1989) *Disaster Mitigation – A Community Based Approach: Approaches to Mitigation*. Oxford: Oxfam.

Matin, N. and Taher, M. (2001) The changing emphasis of disasters in Bangladesh NGOs. *Disasters* 25 (3): 227–239.

McEntire, D.A. (2007) *Disaster Response and Recovery: Strategies and Tactics for Resilience*. Hoboken, NJ: John Wiley & Sons, Inc.

Middleton, N. and O'Keefe, P. (1998) *Disaster and Development: The Politics of Humanitarian Aid*. London: Pluto Press.

Mitchell, J.K. (1999) Natural disasters in the context of mega-cities. In *Crucibles of Hazard: Mega-Cities and Disasters in Transition* (ed. J.K. Mitchell). Tokyo: United Nations University Press, pp. 15–55.

Moore, M.T. et al. (2005) Cities bursting of seams with excess used clothes. *USA Today* September 23: 3A.

Moszynski, P. (2005) Generosity after tsunami could threaten neglected crisis. *British Medical Journal* 330 (1): 155–165.

Mustafa, D. (2002) Linking access and vulnerability: perceptions of irrigation and flood management in Pakistan. *The Professional Geographer* 54 (1): 94–105.

Mustafa, D. (2003) Reinforcing vulnerability? Disaster relief, recovery, and responses to the 2001 Flood in Rawalpindi, Pakistan. *Environmental Hazards* 5 (1): 71–82.

Olsen, M.R. et al. (2003) Humanitarian crises: what determines the level of emergency assistance? Media coverage, donor interests and the aid business. *Disasters* 27 (2): 109–126.

Ozerdem, A. (2006) The mountain tsunami: afterthoughts on the Kashmir Earthquake. *Third World Quarterly* 27 (3): 397–419.

Ozerdem, A. and Jacoby, T. (2006) *Disaster Management and Civil Society: Earthquake Relief in Japan, Turkey and India*. London: IB Tauris.

Palka, E. (1995) The US army in operations other than war: a time to rivert military geography. *Geojournal* 37 (2): 201–208.

Palka, E. (2005) Decades of instability and uncertainty: mission diversity in the SASO environment. In *Military Geography from Peace to War* (eds E. Palka and F. Galgano). Boston: McGraw-Hill, pp. 87–214.

Paul, B.K. (1998) Coping with the 1996 tornado in Tangail, Bangladesh: an analysis of field data. *The Professional Geographer* 50 (3): 287–301.

Paul, B.K. (1999) Flash Flooding in Kansas: a Study of Emergency and Victims' Perceptions. Quick Response Report No. 118. Natural Hazards Research and Applications Information Center, University of Colorado, Boulder, CO.

Paul, B.K. (2003) Relief assistance to 1998 flood victims: a comparison of the performance of the government and NGOs. *The Geographical Journal* 169 (1): 75–89.

Paul, B.K. (2005) Evidence against disaster-induced migration: the case of the 2004 tornado in north-central Bangladesh. *Disasters* 29 (4): 370–385.

Paul, B.K. (2006) Disaster relief efforts: an update. *Progress in Development Studies* 6 (3): 211–223.

Paul, B.K. (2007) 2004 Tsunami relief efforts: an overview. *Asian Profile* 35 (5): 467–478.

Paul, B.K. (2010) Human injuries caused by Bangladesh's Cyclone Sidr: an empirical study. *Natural Hazards* 54: 483–495.

Paul, B.K. and Bhuiyan, R.H. (2004) The April 2004 Tornado in North-Central Bangladesh: A Case for Introducing Tornado Forecasting and Warning Systems. Quick Response Research Report No. 169. Natural Hazards Research and Applications Information Center, University of Colorado at Boulder, Boulder, CO.

Paul, B.K. and Leven, J. (2002) Emergency Support Satisfaction among 2001 Hoisington, Kansas Tornado Victims. Quick Response Report No. 154. Natural Hazards Research and Applications Information Center, University of Colorado at Boulder, Boulder, CO.

Pelling, M. (2003a) Disaster risk and development planning: the case of integration. *International Development Planning Review* 25: i–ix.

Pelling, M. (2003b) *Natural Disasters and Development in a Globalizing World*. London: Routledge.

Rahman, M.O. and Bennish, M. (1993) Health related response to natural disasters: the case of the Bangladesh cyclone of 1991. *Social Science and Medicine* **36** (7): 903–914.

Roger, R.C. et al. (1995) *Non-Governmental Organization and Rural Poverty Alleviation*. New York: Oxford University Press.

Ross, J. et al. (1994) *Linking Relief and Development*. Sussex: Institute of Development Studies.

Schreurs, M.A. (2011) Improving governance structures for natural disaster response: lessons from the Indian Ocean Tsunami. In *The Indian Ocean Tsunami: The Global Response to a Natural Disasters* (eds P.P. Karan and S.P. Subbiah). Lexington: The University Press of Kentucky, pp. 261–280.

Smith, K. (1992) *Environmental Hazards: Assessing Risk and Reducing Disaster*. London: Routledge.

Smith, K. (2001) *Environmental Hazards: Assessing Risk and Reducing Disaster*. London: Routledge.

Smith, K. and Ward, R. (1998) *Floods: Physical Processes and Human Impacts*. New York: John Wiley & Sons Inc.

Sollis, P. (1994) The relief-development continuum: some notes on rethinking assistance for civilian victims of conflict. *Journal of International Affairs* **47**: 451–471.

Streeten, P. (1999) Globalization and its impact on development co-operation. *Development* **42** (1): 9–15.

Susman, P. et al. (1983) Global disasters: a radical interpretation. In *Interpretations of Calamities* (ed. K. Hewitt). London: Allen & Unwin, pp. 274–276.

Tabor, N. (2007) A tsunami of a problem. http://www.theconservativevoice.com/article/21753.html (accessed June 1, 2007).

Tamamoto, M. (2005) After the tsunami, how Japan can lead. *The Far Eastern Economic Review* **168** (2): 10–18.

Taylor, A.J. (1989) *Disasters and Disaster Stress*. New York: AMS Press.

Telford, J. and Cosgrave, J. (2007) The international humanitarian system and the 2004 Indian Ocean earthquake and tsunamis. *Disasters* **31** (1): 1–28.

Tobin, G.A. and Montz, B.E. (1994) *The Great Midwestern Floods of 1993*. Fort Worth: Saunders College Publishing.

Tobin, G.A. and Montz, B.E. (1997) *Natural Hazards: Exploration and Mitigation*. New York: The Guilford Press.

Townsend, J.G. et al. (2004) Creating spaces of resistance: development NGOs and their clients in Ghana, India and Mexico: *Antipode* **36**: 871–889.

Vakil, A.C. (1997) Confronting the classification problem: toward a taxonomy of NGOs. *World Development* **25**: 2057–2070.

Valley, P. (2006) Review of the year: disaster – nature's assaults have shaped a new reality for mankind. *The Independent* **11** (14). March 13.

UN (United Nations) (2005) Regional Workshop on Lessons Learned and Best Practices in the Response to the Indian Ocean Tsunami: Report and Summary of Main Conclusions. Medan, Indonesia.

USAID (United States Assistance for International Development) (2005) Agency channels foreign aid for Hurricane Katrina victims. *Front Lines* October, pp. 1–2.

Walker, P. (1992) Foreign military resources for disaster relief: an NGO perspective. *Disasters* **16** (1): 152–159.

Waldo, B. (2006) The rise of the relief-and-reconstruction complex. *Journal of International Affairs* **59** (2): 281–296.

Wehrfritz, G. and Cochrane, J. (2005) Charity and chaos. *Newsweek* January 17: 30–32.

White, S.C. (1991) *An Evaluation of NGO Effectiveness in Raising the Economic Status of the Rural Poor: Bangladesh Country Study*. London: Overseas Development Administration.

Wisner, B. (1993) Disaster vulnerability: scale, power, and daily life. *GeoJournal* **30**: 127–140.

Wisner, B. *et al.* (2004) *At Risk: Natural Hazards, People's Vulnerability and Disasters*. London: Routledge.

Yonder, A. (2005) *Women's Participation in Disaster Relief and Recovery*. New York: Population Council.

8
Hazards/Disasters: Special Topics

Hazard researchers widely believe that gender perspectives are not well incorporated into disaster research as well as disaster planning and management. However, relevant studies in both developed and developing countries have noted differences between male and female behavior in time of emergencies. As a group, particularly in a male-dominated, patriarchal society, women encounter a set of problems that are quite different from those encountered by their male counterparts when both are affected by natural disasters. Gender responses and coping mechanisms are also different in the wake of a disaster. The first half of this chapter will highlight some of these issues along with the ways in which women are often more vulnerable not only in post-disaster situations, but also during and in pre-disaster periods. The second half of this chapter will cover salient features of global warming and global climate change – a topic that has recently received much attention by scientists and non-scientists alike.

8.1 Gender and natural disasters

Much of the gender and hazard/disaster literature calls for more gender-sensitive disaster response by focusing on the ways in which women are more vulnerable to disasters. Available studies (e.g., Cupples, 2007; Enarson and Morrow, 1998; Fothergill, 1998; Ikeda, 1995; Neumayer and Plumper, 2007) clearly show that the vulnerabilities and impacts of natural disasters are not distributed evenly by gender. Gender issues are often overlooked or underutilized in disaster planning and emergency management strategies. This section will pay close attention to gender identity and subjectivity as these are constructed and reworked through the disaster process to highlight the complexities and contradictions associated

---

Environmental Hazards and Disasters: Contexts, Perspectives and Management, First Edition. Bimal Kanti Paul.
© 2011 John Wiley & Sons, Ltd. Published 2011 by John Wiley & Sons, Ltd.

with female response to disasters. This will be performed by critically reviewing existing studies focusing on women, and their experiences, needs, and responses to different types of disasters.

As noted, men and women are impacted in different ways by natural disasters. After observing this difference, Enarson (2000) claimed more than 10 years ago that there exist "gendered disaster vulnerabilities" and, therefore, a "gendered terrain of disasters." The situation has changed little since then. Many developing countries are still struggling with gender issues in the face of a noticeable recent increase in the number of natural disasters and the number of people affected by these extreme events. Female vulnerability to natural disasters is closely associated with gender discrimination as it exists throughout the world and begins long before a disaster occurs.

Women in male-dominated societies of most developing countries have less access to education, health, food, and other resources. They are less mobile, possess less decision-making capacity, experience more inequality in law, in public and private arenas, and enjoy an overall lower level of human rights than men. Together, these socially determined inequalities combine to form a complex web of female vulnerability to natural disasters. Thus, women in such societies are relatively more at risk than men from the effects of natural disasters from the onset. They also lack resiliency and agency in the aftermath of a disaster due to the pre-existing burden of gender roles, inequalities, and power deficits associated with male-dominated societies and/or cultures (Neumayer and Plumper, 2007).

8.1.1 Mortality

Gender differences in disaster impacts are most evident in the context of deaths caused by these events. Based on an analysis of over 100 papers on natural disasters, Fothergill (1998) concluded that women are more likely to die from natural disasters than are men. All studies (e.g., Chowdhury et al., 1993; Ikeda, 1995; Sommer and Mosley, 1972) conducted in Bangladesh after the devastating cyclone in 1991 reported much higher mortality rates among women compared to men. Furthermore, these studies found that the age cohort of 20–49 had female mortality rates averaging four to five times more than that of males in the same cohort. This finding is consistent with the study of Neumayer and Plumper (2007) who claim that natural disasters lower the life expectancy of women more than that of men. In other words, in the natural disasters they studied that occurred between 1981 and 2002, on average, more women were killed than men or more women were killed at an earlier age than men.

As indicated in Chapter 4, as many as four times more women than men were killed in 2004 tsunami-affected areas of Indonesia, India, and Sri Lanka (McDonald, 2005). In the four villages surveyed by Oxfam in the Aceh Besar province, Indonesia, only 189 (28%) of 676 survivors were female. In the worst affected village, Kuala Cangkoy, for every male who died, there were four

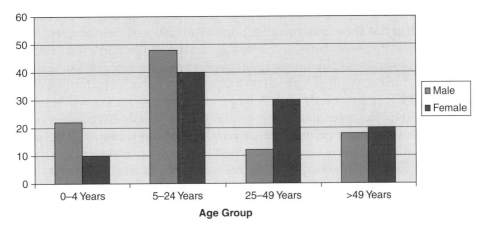

Figure 8.1 2005 Earthquake fatalities in Pakistan by gender and age group. *Source:* Khan and Mustafa (2007), p. 210.

females. In Cuddalore in Tamil Nadu, India, almost three times as many women were killed as men, with 391 female deaths, compared with 146 men (Oxfam, 2005).

Many women in Indonesia, Sri Lanka, and India were waiting on the shoreline for their husbands to return from sea with their catch. It is these women who traditionally sort the catch and take it to the market to sell. Other women were looking after their children at home located close to the beach. A considerable number of them died because they did not know how to swim, climb a tree, or were slow in running against rushing water from the sea. Many women also lost their lives in their attempts to save their children and elderly relatives who were with them at home (Oxfam, 2005). In the Batticaloa District of Sri Lanka, the tsunami hit at the time when women who live along the coast usually bathe in the sea (McDonald, 2005).

In contrast to the 2004 tsunami, the 2005 earthquake in Pakistan killed almost the same number of males and females according to Khan and Mustafa (2007). However, the mortality differs by age between male and female victims of this earthquake (Figure 8.1). Among pre-school children 68% of the fatalities were male. Among grade school to college-age going children and young adults, 55% of the fatalities were male. More male deaths relative to female deaths in the 0–24 age group was clearly associated with a higher proportion of male enrollment in schools. As noted in Chapter 4, the 2005 earthquake casualties were high in Pakistan because the event occurred when students were attending classes. The earthquake caused the collapse of many school buildings. In the married and middle-aged group of 25–49 years, 69% of the fatalities were female. This figure suggests that gender roles in Pakistani society keep women indoors and hence they are more exposed to building collapse (Khan and Mustafa, 2007). This finding is also confirmed by Mahmood (2007) who reported that 53% of the earthquake orphans living in refugee camps had lost their mother, as opposed to 43% who had lost their fathers, and 4% who had lost both their parents.

Mortality differentials between males and females can be partly explained in terms of biological and physiological differences between men and women. The relatively greater strength of males allows them to run faster, swim farther, climb higher, and hold on to steady objects longer than most women can (Ahsan and Khatun, 2004; Neumayer and Plumper, 2007). These disadvantages are exacerbated when women are pregnant or carrying one or more infants or children. Also women are relatively smaller and weigh less than men and therefore would seem to be more easily swept away by rushing water and/or impacted by high velocity winds (Ikeda, 1995). However, biological differences do not fully account for the drastic differences in mortality rates between males and females. Religious and cultural factors are also responsible for some differences observed, particularly in Muslim countries where women's spatial mobility is restricted by the norm of *purdah* (veil) (Ikeda, 1995). Purdah not only marks women's sphere as private and men's as public, but it also correlates to their overall lack of social power and lack of autonomy and basic rights, which further hinders their ability to cope in a disaster (Ahsan and Khatun, 2004).

The lesser social and economic value of females in male-dominated societies places them in vulnerable situations in disasters. Women generally receive less food, health care, and education, which place them at additional risk, even of death, when a disaster occurs. Females in such societies suffer more from malnutrition, undernutrition, and ill-health than do males; therefore it is harder for them to endure the impacts of storms and, in some cases, the surge of water (Ikeda, 1995). In addition, women in developing countries possess fewer resources and capital than men; thus, they have fewer means of survival when disaster occurs. Single-head households headed by females are also disproportionately affected by natural disasters in terms of mortality (Ollenburger and Tobin, 1998). Other social determinants of higher female deaths in South and Southeast Asian countries observed during cyclones and in the 2004 Indian Ocean tsunami have been linked with their traditional garment known as the "Sari" in South Asia and the "Kain Sarong" in Southeast Asia as well as their long hair. Such clothing and long hair hampers quick movement in water and during high winds associated with tropical cyclones and associated storm surges.

8.1.2 Access to disaster information and evacuation differentials

Cultural and religious practices of purdah, primarily found in Muslim and male-dominated societies, not only create differential mortality rates by sex, but also restrict women's access to information on impeding extreme events and their willingness to seek safe shelter from tropical cyclones and flood events. Women have fewer opportunities to obtain information on their risk and how to minimize it. Because males' social and spatial networks are larger than most females, men are able to access more information on impending natural disasters. For example, while many males in coastal Bangladesh knew that Cyclone Sidr was approaching, most coastal women learned of it from their husbands and/or close male relatives, yet there were some women who were not aware of it at all. Most of these women's husbands were away fishing, and they did not receive the cyclone warnings because such warnings were given in the marketplace which, in most cases, was distant from their homes (Government of Bangladesh, 2008). One female survivor suggests that: "if we could make recommendations to the government, we would tell them to ... make sure that warnings are announced on microphones in all villages, not just in the market, so that the women can also hear them in their homes" (Government of Bangladesh, 2008, p. 151).

Because of the fear of the social consequences that may arise, women in Bangladesh hesitate to leave their homes for safer places, including public cyclone shelters, prior to landfall of tropical cyclones or during flooding. Ikeda (1995) also notes that the practice of purdah is stronger in the more conservative coastal areas than elsewhere in Bangladesh. However, Paul (2010) suggests that while Ideka's observation is generally true for the eastern coast of Bangladesh, it is not so for the southwestern coast which was affected by Cyclone Sidr in 2007. Although the practice of purdah seems to be decreasing over time in Bangladesh, many male heads of households in coastal areas are still reluctant to allow females to leave homes for safer shelter. In contrast, one NGO reported that some people, such as sex workers, were denied access to public shelters prior to the landfall of Cyclone Sidr (Government of Bangladesh, 2008).

As indicated, the traditional female role as caretaker of the family often increases women's risk and vulnerability for hazards and/or disasters relative to men. In the case of evacuation, women need to look for their children, and carry and/or hold them, thus they are often delayed in reaching shelter. In such a situation, even the smallest amount of time may mean the difference between life and death. Protecting children at the time of a disaster discourages women from going to a safer place. With reference to the women who were at home in Aceh, Indonesia at the time of 2004 tsunami, Felten-Biermann (2006, p. 82) claims that: "the first thing they did when the wave reached was to try to save the lives of children and old people."

8.1.3 Sexual and gender-based violence and other gender issues

Women's lower social status compared to men in many developing countries makes them more vulnerable during the post-disaster period. Although sexual and gender-based violence are associated with all major natural disasters (Cupples, 2007; Enarson, 2000; Pikul, 2005), examples are drawn here primarily from the 2004 Indian Ocean tsunami and Hurricane Katrina in 2005. A report (UN, 2006) published by the UN entitled: "Tsunami Response: A Human Rights Assessment," claims that allegations of human rights violations in tsunami-impacted areas were rampant in the 2004 Indian Ocean tsunami event. These include discrimination in aid distribution, and sexual and gender-based violence. Gender-sensitive disaster relief was absent after the 2004 Indian Ocean tsunami. Basic humanitarian aid was refused to women, particularly widows and unmarried women (McDonald, 2005).

Shortly after the 2004 tsunami, many women were raped in Sri Lanka in temporary shelters by their rescuers or by other men who were taking advantage from their defenseless situation (Felten-Biermann, 2006). Quite a large number of sexual harassment claims were also reported in Aceh refugee camps (Felten-Biermann, 2006). Also medical services were not available for women in most of the camps built after tsunami in Indonesia, India, and Sri Lanka. When medical services were available, there was often no opportunity for women to speak privately with medical practitioners, and there was a general lack of trust between women and their physicians (Rees *et al.*, 2005). Moreover, women in these temporary camps were often verbally and physically harassed by men (McDonald, 2005). Sexual and gender-based violence in both temporary and also in permanent shelters is a major cause of injury and disease world-wide, and it also frequently erodes self-esteem as well as reduces a woman's capacity to care for herself and others (Krug *et al.*, 2002).

Fortunately, in several camps and temporary shelters constructed after the 2004 tsunami in India, women doctors and police officers were posted. At another extreme, NGOs flew in thousands of packets of sanitary napkins and distributed them to tribal women in the Nicobar islands, India who have never used them. They were first used as toilet paper because there was a water crisis, and then as pillows. This only illustrates that even when women-centric actions take place, they can be misplaced, if the women themselves are not allowed to articulate what they want (Gautham, 2006).

Living conditions in the temporary shelters and relief camps set up after the 2004 Indian Ocean tsunami were far below minimum standards set by the UN. Overcrowding, inadequate lighting, lack of toilets and running water, exposure to extremes of heat and cold, and an utter disrespect for privacy and security of women have been experienced and witnessed in all tsunami-impacted regions. Most of the temporary shelters built to house tsunami survivors contained only one room – this did not provide women and girls with the necessary space or privacy to change their clothes. Flimsy and often broken partitions between structures further encroached on women's privacy. In some cases, the location of tents close to sidewalks also exposed women to public scrutiny. The presence of military personnel in the same camps where tsunami survivors were living caused serious concern for the physical safety of female survivors. It was also difficult for women to use bathrooms at relief camps because they often had gaps in the doors, making it possible for everyone to see in from outside. Further, it was difficult to use latrines and bathrooms at night because of the distance from a shelter and limited lighting (UN, 2006).

In both India and Sri Lanka, women complained that toilets built close to temporary shelters lack water or that they had to walk long distances to use toilet/latrine facilities. In Tamil Nadu, India, where temporary structures were built with tar sheeting, women were left with the difficult choice of sleeping outside along with other members of the community – putting themselves at risk of abuse – or risk heat-related health problems by sleeping inside the inappropriate, suffocating shelters. As indicated, this was compounded by an absence of adequate health services (UN, 2006). The temporary shelters and camps in the 2004 Indian Ocean tsunami served as microcosms of dysfunctional gender relations in Indonesia, India, and Sri Lanka and underscore not only the importance of repositioning women centrally in disaster situation, but also endowing women with equal status and power within the broader society (Rees *et al.*, 2005).

Even women who went to public shelters prior to the landfall of Hurricane Katrina in 2005 experienced sexual and gender-biased violence. Security of women and girls was a great concern for 2005 Katrina survivors, particularly those who took shelter in the Superdome or in the New Orleans Convention Center. In the days following this hurricane, major media providers such as CNN and the Chicago Tribune reported rapes were taking place in the Superdome and at the Convention Center. The New Orleans' Police Authority quickly denied any incidence of rape and promised that it would investigate rape cases if anyone came forward (Marshall, 2005). It was not really surprising that the police in New Orleans did not know of any rape reports in as much as government officials were ignorant for some time that there were people even taking refuge in the Convention Center. Most obviously, the usual channels through which rapes are reported were unavailable. Furthermore, there were no functioning rape crisis centers, law enforcement offices, or medical facilities at the height of the crisis in New Orleans where such cases could be reported (Marshall, 2005).

The Louisiana Coalition Against Domestic Violence (LCADV), an organization serving sexual assault survivors, reported that all direct services to survivors of domestic violence had been suspended but that LCADV was receiving reports of women being battered by the their partners in emergency shelters. According to Captain Jeffrey Winn of the New Orleans Police SWAT team, "police officers on the scene at the Convention Center told him that a number of women had been gang-raped. Similar reports were also made by emergency personnel and National Guard troops" (Marshall, 2005, p. 15). Women often were unable to report rapes until after being transported to other jurisdictions. While it is common practice for jurisdiction to take "courtesy reports" of crimes committed in other locales, until September 13, police in major evacuation centers, including Houston, did not take reports of crimes that happened in New Orleans (Hafiz, 2005). Increases in the sexual abuse of women generally mean increases in unwanted pregnancies and sexually transmitted diseases, including HIV/AIDS (Ross, 2005).

In many developing countries, fulfillment of traditional female gender-based roles in the overwhelmingly patriarchal society becomes more difficult subsequent to a disaster (Paul, 1999). For example, during flooding, women cannot easily perform their domestic activities. Cooking areas may be inundated and flooding often makes the collection of firewood more difficult, if not impossible. Women must also travel greater distances to obtain fresh water during flood seasons. Furthermore, women not only have to cope with their own physical and emotional reactions to disasters, but are also called on to support both the physical and psychological needs of their families. The price of staple food grains usually increases while the demand for labor decreases during and/or after the occurrence of most natural hazards. In order to earn money to buy food for their family, many men, particularly those from the poorest social groups, migrate to nearby large cities in search of employment. Women remaining at home must take sole responsibility for their family and must also accomplish the work for which the migrant men from their households were responsible (Paul, 1999).

Unfortunately, far too often migrant husbands fail to send a portion of their earnings home to support their family. For whatever reason, non-receipt of such funds causes their wives to face difficulties in managing normal household affairs (Tichagwa, 1994). Failure to financially support their spouse serves as an early warning for the wives that their husbands may have deserted them. Husbands who do not leave home in search of employment may also desert their wives (Cupples, 2007). Destitute and without any means of employment, women are sometimes compelled to migrate elsewhere and face the most acute conditions of physical and social insecurity. Often an entire family is uprooted by a disaster and must migrate to an urban center. Young girls may remain unmarried because their families are unable to provide the kind of (or any) dowry customarily needed for marriage. Sometimes powerful and relatively prosperous villagers seize this opportunity to take young girls as their second and third wives; desperate parents may become the victims of such circumstances. After the 2004 tsunami there were many reports of forced marriages of girls and young women from India, Sri Lanka, and Indonesia. These girls and women were often married to much older tsunami widower males and were pressured to have children quicker and closer together, with significant implications for their reproductive health. Sometimes these girls were not more than 13 years old. Others were married merely for their husbands to receive state subsidies for marrying and starting a family (Felten-Biermann, 2006; Rees *et al.*, 2005).

Women displaced and/or abandoned during the post-disaster period are often lured away by professional gangs or pimps with promises of obtaining jobs elsewhere. Many of these women are likely to be forced into prostitution in towns within the country or even abroad by human traffickers, and other exploitative individuals and/or networks (Ahsan and Hossain, 2004). Other women and girls willingly accept prostitution in an effort to earn income to support their families. Older destitute women often become beggars and live on charity (Paul, 1997). In addition, women who survive a disaster may be at greater risk for adverse reproductive health outcomes, including infections, rashes, and menstruation problems due to unsanitary post-disaster conditions, compounded by the lack of normal feminine products and conditions in temporary shelters (WHO, 2005).

In describing the experience of female adolescents during the 1998 floods in Bangladesh, Rasid and Michaud (2000) reported that with the floods, most of these adolescents found difficulty in being appropriately "secluded." They claim that:

> Many were unable to sleep, bathe or get access to latrines in privacy because so many houses and latrines were under the water. Some of the girls who had begun menstruation were distressed at not being able to keep themselves clean. Strong social taboos associated with menstruation and the dirty water that surround them made it impossible for the girls to wash their menstrual clothes privately or change them frequently enough. (Rasid and Michaud, 2000, p. 54)

The situation is even worse for female adolescents who take refuge in the unfamiliar environment of flood shelters and relief camps. In such shelters and camps they are unable to maintain their "space" and privacy from male strangers and thus a flood situation is a uniquely vulnerable time for them (Rasid and Michaud, 2000).

Evidence produced here clearly suggests that women are more vulnerable than men before, during, and after disasters and that biological factors associated with being female fail to fully explain significant gender differentials expressed in mortality rates due to natural disasters. This further suggests that natural disasters do not exist in isolation from the social, cultural, and religious constructs that already marginalize women and place them at risk of violence. Sufferings of women increase in the wake of disaster and these sufferings are associated with gender inequality and the limited representation of women in disaster response. Thus, extreme natural events have negative impact on women's health, identity and family relations, and these factors are often ignored in hazard management as well as in disaster relief and recovery operations. If a disaster process is not managed carefully, it can reproduce or reinforce gender inequalities, particularly in developing countries.

8.2 Global warming and climate change

Global warming is the increase in the average temperature of the Earth's atmosphere and oceans since the mid-twentieth century and its projected continuation. According to the 2007 Fourth Assessment Report by the Intergovernmental Panel on Climate Change (IPCC), the average global surface temperature has increased $1.33 \pm 0.32°F$ ($0.74 \pm 0.18°C$) during the twentieth century, and the average surface temperature will likely rise by $1.60–3.4°C$ over the next century; even possibly more than $9°F$ ($5°C$) in some places (IPCC, 2007). This increase in surface temperature in recent decades has been higher at high latitudes in the Arctic. Specifically, in the past 90 years, temperatures in the Arctic region have increased more than 14°F (about 8°C). Although the Earth is continually undergoing climate change, the observed warming of the earth's surface in recent decades is extraordinary because it appears to be primarily anthropogenic and may be happening more quickly than climate changes in the past.

Global warming is closely associated with an increase in the concentration of greenhouse gases (i.e., gases in the atmosphere that absorb and emit radiation within the thermal infrared range). The main greenhouse gases in the Earth's atmosphere are water vapor, carbon dioxide, methane, nitrous oxide, and ozone. Global warming is the main cause of an increase in the amount of thermal radiation near the Earth's surface. Many scientists agree that the enhanced greenhouse effect is leading to rising global temperatures which drive climate change. However, global warming and climate change are topics that will continue to be the center of passionate dialogue between scientists and politicians.

A scientific survey conducted among 10 257 earth scientists and others in the United States in 2008 revealed that 52% of the respondents surveyed agreed that the Earth has been warming in recent years, and 47% believed that human activities are a major cause of that warning (Doran and Zimmerman, 2009). This contradicts Oreskes' (2004) findings in which 928 abstracts from peer-reviewed research papers were compared. He found that more than 75% of these authors either explicitly or implicitly accepted the consensus view that Earth's climate is being affected by human activities. While both studies suffer shortcomings, the bottom line is that global warming and climate change have emerged as major issues of common concern to the world community. An exhaustive National Research Council report published in early 2011 urges to stop debating the reality of climate change and asks to focus on the "pressing need for substantial action," including an extreme reduction in greenhouse gas emissions and more aggressive investment in alternative energy sources (BASC 2011).

After discussing the main causes, consequences, and mitigation measures involved with global warming, steps to reduce impacts of global warming and associated climate change will be covered in detail. The last part of this chapter will present the impacts of global climate change in Bangladesh and related issues with special emphasis on climate change-induced migration. Bangladesh has been selected as a case study because it is recognized as one of the most vulnerable countries in the world to the impacts associated with global climate change.

8.2.1 Causes of global warming

The natural greenhouse effect provides the Earth with a warm atmosphere envelope that supports all life, including human life. This warmth comes from trapping of incoming and outgoing solar radiation by an array of natural greenhouse gases in the atmospheric zone closest to Earth, the troposphere. Without the greenhouse effect, the Earth would be virtually uninhabitable with an average temperature well below freezing. Houghton (2004) claims that without the greenhouse effect, the Earth's average temperature would be at least 20°C cooler than its current temperature. Although the amount of greenhouse gases present in the atmosphere has varied somewhat over long periods, they were relatively stable until very recently. Only since the industrial revolution has the amount of those greenhouse gases increased dramatically, primarily from the human-induced burning of fossil fuels such as coal, natural gas (methane), and petroleum. This has caused a remarkable increase in the average Earth surface temperature over the past few decades. Carbon dioxide, methane, and water vapor all occur naturally in the environment. For example, most atmospheric water vapor evaporates from the oceans. Carbon dioxide and methane are emitted by erupting volcanoes, animals, decaying vegetation, and forest fires.

Common sources of anthropogenic greenhouse gases are cars and coal power plants. Coal-fired power plants emit carbon dioxide as well as other gases that pollute the environment. Power plants that use coal as a source of energy are the largest contributors of carbon dioxide into the Earth's relatively thin atmosphere, more than gasoline-powered cars and trucks combined (Hyndman and Hyndman, 2011). Apart from the above major contributors, other activities associated with greenhouse gases are deforestation and large-scale agriculture involving grazing animals, including beef and dairy cows. These farm animals, as well as the decomposition of agricultural materials, release methane and nitrous oxide. Decaying landfill also generates methane.

Other sources of greenhouse gases are various refrigerants used in air-conditioning units and refrigerators. Another contributor to global warming is land use change. Vast amounts of forests and grasslands are being destroyed and used for residential and other purposes. Forests take in carbon dioxide from the atmosphere, releasing the oxygen, and store the carbon in their biomass. As more trees are cut down and their wood used for fuel, more carbon enters the atmosphere because less is taken out, and less is also stored as biomass. It has been recently reported that 30% of the buildup of carbon dioxide in the atmosphere is a result of the loss of trees and forests (Pulsipher and Pulsipher, 2011). Furthermore, any increase in greenhouse gases has a cumulative effect, because almost everything expelled into the atmosphere generally stays there, often for decades or even centuries.

Four major greenhouse gases account for the bulk of the increase in Earth surface temperatures. Carbon dioxide (CO_2) contributes more than 50% of the human-generated greenhouse gases by volume (Figure 8.2). As indicated, the recent increase in atmospheric carbon dioxide is primarily a result of the combustion of fossil fuels used for transportation, industrial, and residential purposes. Atmospheric carbon dioxide was measured at 280 parts per million (ppm) in 1860, and had increased to about 388 ppm by 2010 (Rowntree et al., 2011). Although initially some scientists argued that the current increase in atmospheric carbon dioxide was merely natural variation, the vast majority of scientists now agree that this increase is a result of human activities. Furthermore, carbon dioxide is projected to increase to more than 450 ppm by 2050 (Rowntree et al., 2011).

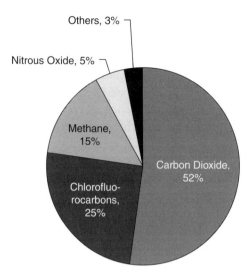

Figure 8.2 Four major greenhouse gases and their contribution to the global warming. *Source:* Rowntree *et al.* (2011), pp. 38–39.

Chlorofluorocarbons (CFCs) account for nearly 25% of the human-induced greenhouse gases. These come mainly from widespread use of aerosol sprays and refrigeration systems, including air conditioning (Figure 8.2). Although CFCs have been banned for years in developed countries, their use continues in many developing countries. Importantly, these gases are highly stable and may reside in the atmosphere for extended periods of time, perhaps as long as 100 years or more. In addition, a molecule of CFC absorbs 1000 times more infrared radiation from the Earth than a molecule of carbon dioxide; as a result, the role of CFCs in global warming is highly significant.

Since 1750 concentrations of methane have increased 151% in the atmosphere as a result of burning vegetation associated with rainforest clearance, anaerobic activity in flooded rice fields, cattle and sheep effluent, and leakage of pipelines and refineries connected with natural gas production. Landfill, sewage, oil and gas drilling, coal mines, and feedlots also release methane. Currently, methane accounts for some 15% of anthropogenic greenhouse gases by volume. Nitrous oxide accounts for over 5% of the human-generated greenhouse gases and results primarily from the increased use of chemical fertilizers (Figure 8.2).

The current public discourse on global warming is dominated by an assumption that it is entirely or largely the result of anthropogenic greenhouse gases (Rashid, 2011). However, the increase in the global temperature of the atmosphere near the Earth's surface may also caused by natural factors. Among other natural factors, volcanic eruptions and variation in solar output have been linked to changes in global average temperature (IPCC, 2007). More significant changes in the temperature have been attributed to variations in the earth's orbit around the sun (Archer, 2007). However, the long geologic time cycles within which these operate are not comparable to the relatively short historical horizons within which the current global warming has been occurring (Rashid, 2011).

8.2.2 Effects of global warming and climate change

The full extent of changes that will occur due to global warming is still unknown. One obvious consequence of global warming is changing global climate patterns. Although such changes are currently thought to bring greater variability of weather with more extremes of temperature, stronger winds, and precipitation, these and other effects will not be uniform across the globe. There are a chain of consequences due to global climate change which could have devastating consequences for the planet and the people of many countries of the world. Some of these consequences are discussed below.

8.2.2.1 Precipitation changes

As noted, global warming will impact on precipitation, including the likelihood of greater extremes of wet and dry seasons, as well as more intense storms. Higher atmospheric temperatures means more evaporation from both land and water bodies, and more water dissolved into the atmosphere. This situation will probably cause wetter climates in polar and equatorial regions, and a warmer and drier climate in mid-latitude regions. As a result, the areas that are now largely dry, such as Mexico, will get drier, while many wet areas, such as Bangladesh, are expected to be wetter, particularly during summer. Summer precipitation across most of the United States is predicted to not change much, but the Pacific Northwest region is expected to be 20–30% drier (Hyndman and Hyndman, 2011).

If atmospheric carbon dioxide concentrations in the atmosphere rises from its current levels to 450–600 ppm before the year 2100, long-term dry season precipitation is expected in some areas to drop to conditions similar to those of the American "dust bowl" of the 1930s. The southwestern United States and southern Europe would likely become more arid; a long-term decrease in rainfall has already been documented in these regions. However, the northern plains of the United States, Canada (east of the Rockies), and eastern Asia north of China, are expected to become wetter (Hyndman and Hyndman, 2011). Some areas with colder climates, such as northern parts of Russia, will likely experience a warmer climate, and these areas will be subjected to both flooding and droughts. More precipitation will probably increase drought because of increased evaporation. Higher temperatures cause more evaporation, which may have a larger effect than the expected increase in precipitation.

Rising temperatures also mean that future floods are likely to occur earlier in the season with more frequency and severity. This is particularly true for countries located in higher and in the mid-latitudes. These countries experience snowfall and because of warmer temperatures, this snow will melt faster and earlier in spring. This will almost inevitably lead to earlier and higher floods in these countries. Higher floodwaters threaten levees not built for those levels. Other consequences include more severe mudflows, and coastal and riverbank erosion. Higher temperatures mean that more precipitation falls as rain in the countries located in higher and mid-latitudes than snow. This will likely lead to the precipitation washing away quickly rather than slowly seeping into the ground as the snow melts. Decreased rainfall in already dry areas, on the other hand, could severely reduce the water supply for drinking and other uses. Climate change is thus likely to increase water scarcity in areas that are already experiencing water stress. Droughts will probably become more frequent and their continuous reoccurrence may overcome the coping mechanisms of individual households and communities (Warner *et al.*, 2010).

8.2.2.2 Glacial and ice cap melting

Warmer temperatures are causing Arctic ice and glaciers to melt at a faster rate than ever measured before. This melting can cause flooding downstream of the glacier and could displace people and animals. Melting of Arctic ice has a feedback mechanism that will accelerate melting. White ice reflects 90% of the sun's energy back into space. As more Arctic ice melts and more open water is exposed, the darker surface of the ocean absorbs approximately nine times the solar energy relative to snow and ice, which causes more melting (Hyndman and Hyndman, 2011). It is important to mention that mountain glaciers are an important source of fresh water for downstream populations – especially during any dry season. Melting of mountain glaciers will have severe impacts for freshwater systems that supply water for millions of people.

The melting ice caps will also cause the salinity levels of the oceans to drop. This could have huge impacts on underwater ecosystems and may change ocean currents. Some ocean organisms have to have a certain salinity level to survive. Furthermore, a change in the ocean currents resulting from changes in salinity would affect weather systems throughout the world. The warm, salty Gulf Stream, which moves northeast along the east coast of North America and across to Europe, might change its circulation pattern, creating devastating effects on the climate of Western Europe (Pulsipher and Pulsipher, 2011). Currently, westerly winds carry warm moisture across the Atlantic and keep much of Western Europe warm in winter, in spite of its northern latitude.

Any change in the flow of the warm Gulf Stream to the north could cool Europe and some other northern hemisphere climates. As a result of this change, some areas may receive, on average, the same amount of precipitation, but not in the same way. For example, such areas might receive precipitation in more intense forms with longer durations of little or no precipitation between events. This will increase flash flooding and the severity of droughts. Besides predicted precipitation changes as a result of changes in ocean currents, there may be important temperature changes as well. Obviously, with the melting of polar ice caps sea level will rise too. Currently, projections are for a global rise in sea level of at least 3.3 feet (1 m) by the year 2100. Although this may not seem like great, even a small increase will endanger low-lying island nations throughout the world, as well as coastal areas in Europe, Asia, and North America. Global warming leads to sea level rise from two primary factors – about half from water added due to melting ice and half from the physical heating and resultant thermal expansion of sea water.

Rising sea levels present a very real threat to people living along coastal areas. Although they constitute only 2% of the total land surface of the earth, coastal areas are home to some 10% of the current world population and 13% of the urban population (Warner et al., 2010). Many of the world's major population centers, such as London, New York, Amsterdam, Tokyo, Mumbai, and Venice, are located on the coast. Sea level rise will also displace millions of people, particularly in developing countries, and will undoubtedly cause large-scale migrations of people from coastal to inland areas as well as from one country to another. Increasing temperatures also mean thawing of the permafrost which currently traps in the soil a huge amount of methane. Melting of the permafrost in tundra region will create marshes and peat bogs. This environmental condition, along with decomposition of organic materials, would release a significant quantity of methane gas into the atmosphere, exacerbating the global warming. This is believed to be happening now in the tundra regions of Siberia and Canada (Environmental Protection Agency, 2009).

8.2.2.3 Warming ocean

As the oceans are slowly soaking up heat from the warming atmosphere, ocean waters are inevitably becoming warmer. The average change in ocean temperature over the past 40 years was 0.3°C in the top 300 m and 0.06°C in the top 3.5 km (Hyndman and Hyndman, 2011). Warmer ocean water is predicted to be one of the main ingredients in forming more intense tropical storms and hurricanes/cyclones. Warmer oceans will likely cause these events to become stronger and to occur more frequently in the future. As indicated, water expands as it warms and as a result, sea level will rise via global warming even if no glaciers and/or ice sheets melt. Global warming will also cause more evaporation of water from the oceans. This will probably increase water vapor content in the atmosphere and act as a positive feedback loop, causing still more global warming, although the increased water vapor could result in greater cloud formation, releasing some of the excess water vapor.

8.2.2.4 Effects on environment and human population

Global warming will affect ecosystems throughout the world. Increased temperatures may mean that certain plants/organisms can no longer survive in current locations, allowing new species to move in. There will, in all probability, be several migrations of organisms moving to areas that are more suited for them, changing existing landscapes. Global warming will also cause extinction of some species. For example, the polar bear and arctic fox have already become endangered species due to the polar ice caps melting. They both depend on the cold and the ice to hunt and survive.

Agricultural activities will be severely impacted by global warming. Global climate change will cause a major shift in agricultural areas. For example, the "wheat belt" in the United States might receive less rainfall and become warmer and drier, endangering this staple grain production. In addition, southern parts of the United States and Europe can expect a warmer and drier climate that will demand even more irrigation for crop production. There may also be problems with dairy cows as a higher quality of milk is generally produced by cows in colder climates. While more northern countries, such as Canada and Russia, might experience a longer growing season because of global warming, the soils in these two countries and other high latitude countries are generally not as fertile as soils of mid-latitude nations, such as the United States. As a result, world grain production may decrease, resulting in increasing malnutrition and under-nutrition. Public health would also suffer from the spread of insect-borne diseases such as malaria, cholera, and dengue fever brought to higher latitudes and elevations due to global temperature increase.

Nevertheless, some sectors of national economies would benefit from global warming. One such sector is the construction industry. Some coastal cites and towns will need to move inland because of sea level rise. This means there will likely be a lot of rebuilding. However, increased construction could increase global warming if places do not use existing building materials and sustainable practices. If not, more forest resources will be consumed and this will also cause an increase of carbon dioxide in the atmosphere (Environmental Protection Agency, 2009). Furthermore, rapid industrialization and growth of urbanization will diminish forest areas. Less plant life means less recycled carbon dioxide, which will lead to build-up of carbon dioxide in the atmosphere, exacerbating the "greenhouse effect." Economic disruption, population migration, political upheaval and conflicts over resources, to name but a few, seem more likely under most predicted scenarios associated with global warming.

8.2.3 Mitigation of global warming and climate change

Given the existing levels of greenhouse gases in the atmosphere and the worldwide trend in their increase, there is an immediate need to slow or stop global warming. This can be accomplished in a number of ways, some of which are discussed below.

8.2.3.1 Reduction of energy consumption

One of the important ways to mitigate the impact of global warming and associated climate change is through reducing energy consumption. This can be achieved by slowing population growth and/or reducing per capita energy and/or resource consumption. There is a direct relationship between population increase and consumption of energy. Although currently the citizens of developing countries do not consume as much energy as people of developed countries, most of the world's population is now residing in developing countries. In addition, the rate of natural increase of population is much higher in developing nations than in developed nations. As a result, the demand for energy is increasing in the former countries at a much greater rate than found in most developed countries. The total annual carbon dioxide emissions from fossil fuels in developing countries have now surpassed those of all developed nations, and the rate of increase is much faster for developing nations. For example, increase in US energy demand is between 1% and 2% per year, but in China energy demand increases by 10% per year (Hyndman and Hyndman, 2011).

Per capita energy consumption reduction can be achieved through conservation and greater efficiency. Energy conservation is a simple, effective, and inexpensive way to save large amounts of electric power, thereby reducing carbon output used in power production. Buildings use a large amount of energy for heating and cooling purposes. Conservation mechanisms for buildings include insulation to minimize loss of heat, maintaining cooler environments, more efficient generation of light, as well as efficiently circulating and filtering air. As noted in Chapter 6, using green technologies, a building can realize reductions in energy consumption of at least 40% over standard building designs. The design of more energy-efficient household electronics could also significantly reduce emissions of carbon dioxide. In addition, the use of smaller cars and mass transit such as buses and railroads can significantly reduce energy consumption. Cars, especially the large ones, consume relatively large amounts of fuel. The use of hybrid cars, which consume alternative fuels, can also reduce gasoline consumption.

8.2.3.2 Use of alternative energies

Increased use of alternative energy sources, such as hydro, wind, solar, and nuclear energy (both at the household level and in industrial sectors) can significantly lower greenhouse gases emissions into the atmosphere. Because wind power produces no carbon dioxide or other pollutants, its use is usually viewed

Figure 8.3 Wind farm in Kansas. *Source:* Mitchel Stimers.

favorably by both public and electric power generation companies. In 2008, wind accounted for only about 1% of total electricity generation in the United States. However since then, about 35% of all new electricity generation was from wind-driven systems (Hyndman and Hyndman, 2011). Realizing the significant drop in the cost of production of wind-generated electricity, wind farms are rapidly spreading from the pioneer adopting states of Texas, California, Oregon, and Washington to other states such as Kansas, Nebraska, Iowa, Oklahoma, Montana, Wyoming, and the Dakotas (Figure 8.3). However, there are a number of challenging issues associated with using wind as an energy source. Giant wind turbines can and do kill birds and bats. Although noise was initially a concern, modern wind turbines are very quiet. Another challenge for wind energy is that it may cost as much as US$100 billon to build transmission lines from suitably windy locations to the primary transmission grids so that the power can be delivered to large east and/or west coast cities in the United States.

Solar power has great potential as an alternative source of clean energy. Producing energy from this source requires converting sunlight into electricity, either directly using photovoltaics (PV) or indirectly using concentrated solar power (CSP) systems. Concentrated solar power systems use lenses or mirrors and tracking systems to focus a large area of sunlight into a small beam. Domestic and commercial uses of solar power are not yet widespread because solar energy is too expensive to use on large scales or too complicated to install. The payback might take 8–14 years. In order for solar energy to reach the average consumer it has to be simple to maintain and it has to be cheap. New thin-film solar materials, which can be used to coat windows and roofs, use little or no silicon, are much cheaper to manufacture, and promise to reduce costs significantly. Simple and effective roof-top solar collectors or solar panels that heat water for domestic use are currently in use, particularly in the summer, in some countries with above average annual days of sunshine, such as Greece, Turkey, Israel, Australia, Japan, Austria, and China. Even on the southwestern and west coasts of the United States roof-top panels are a common sight, and US tax codes are changing, accelerating their adoption (Hyndman and Hyndman, 2011).

Another alternative, and cleaner and cheaper than gasoline or diesel energy sources is natural gas, including methane. Natural gas is already widely used for heating homes and driving power plants. Although methane is a greenhouse gas, burning it generates much less carbon dioxide than other fossil fuels. The vast majority of new power plants being built in the United States are designed to run on natural gas. Many global warming experts suggest that cars should also be modified to use it. Although vast reserves of methane have been found (in frozen form) on most seafloors, the technologies needed to tap these offshore reserves are not presently economically feasible. It is worth mentioning that unburned methane is 20 times more potent than carbon dioxide as a contributor to atmospheric warming (Hyndman and Hyndman, 2011). However, unlike carbon dioxide, which remains in the atmosphere for hundreds of years, methane does so for only about a decade. Thus cutting the amount of methane released into the atmosphere by burning or other means has an almost immediate and very significant effect in reducing global warming.

Some experts recommend using "clean coal," which is a term used to describe technologies that may reduce emissions of carbon dioxide and other greenhouse gas that arise from the burning of "dirty coal" for electrical power. Typically, clean coal has been used by coal companies in reference to carbon capture and sequestration, which pumps and stores carbon dioxide emissions underground, and to plants using an integrated gasification combined cycle which gasifies coal to reduce carbon dioxide emissions. Although coal gasification plants are in the experimental stage, these plants should reduce pollution, use less water, and more easily collect carbon dioxide for underground disposal (Hyndman and Hyndman, 2011).

Given the numerous and serious concerns over emissions of greenhouse gases, use of nuclear energy is being considered as one option for an alternative energy source. Safety and disposal of nuclear waste is still a serious and contentious issue, but such concerns now appear to be less of an issue than the consequences associated with further increases in greenhouse gas emissions. Countries with little fossil fuel resources already generate significant amounts of nuclear energy. France, for example, generates about 78% of its power from nuclear fission – the breakup of uranium atoms – and it recycles much of the nuclear waste into additional fuels. Nuclear power presently accounts for about 16% of worldwide energy supply (Hyndman and Hyndman, 2011). Other clean energy sources include use of tidal and geothermal power to generate energy and using hydrogen fuels, whose primary byproduct is water. Tidal power systems, as currently envisioned, are pollution free, but expensive to build.

Biofuels, such as ethanol and biodiesel, are gaining increased public and scientific attention in the United States because of their role in slowing global warming, improving national energy security, and reducing oil price hikes. Bioethanol is an alcohol made from corn and sugar components of selected plants. With advanced technologies being developed, cellulosic biomass, such as trees and grasses, can also be used for ethanol production. Ethanol can be used as a fuel for vehicles in its pure form, but it is usually used as a gasoline additive in order to increase octane ratings and improve vehicle emissions. However, researchers question the efficacy of ethanol as a fuel, pointing to increased greenhouse gas emissions during the production process. Although ethanol burns cleaner than gasoline, the number of miles per gallon of fuel consumed is lower. In addition, the use of corn, soybeans, and other crops in the production of ethanol has increased their prices and reduced their availability as feed and food crops. One estimate suggests that it would take about 43% of all crop land in the United States to replace only 10% of all gasoline and diesel fuel consumed currently (Hyndman and Hyndman, 2011).

8.2.3.3 Carbon trading

Carbon trading is a market-based way to lower greenhouse gases in the atmosphere. Coal and natural gas are the least expensive among all currently used fuel sources. Therefore, one way to reduce the use of carbon and natural gas is to raise the cost of fossil fuels by taxing carbon emissions. Such a policy would make alternative fuels more affordable as well as encourage conservation of energy. Although a few countries have already implemented taxes based on carbon use, this policy may not be politically viable in the United States. Most countries of the European Union try to limit emissions by encouraging carbon trading. In these countries, companies are limited to a certain level of carbon emissions based on past emission levels. This policy encourages reduction of greenhouse gas emissions without hindering economic growth (Hyndman and Hyndman, 2011).

As an alternative measure, a cap-and-trade law could be implemented to limit carbon emissions. Under this scheme, an explicit national limit or cap is set on greenhouse gas emissions. That cap is then divided into tradable permits, which are then distributed to all regulated firms for a fee, through auctions, for free, or through some combination of these methods. A notable example includes the Clean Air Act of 1990, which implemented a successful sulfur dioxide and nitrous oxide trading system in the United States. Within five years, participating power plants reduced emissions by more than 20% at a cost of less than one-third of the lowest estimates originally made (Hyndman and Hyndman, 2011). Cap-and-trade, currently practiced in Europe, commits nations to set responsible limits on global warming emissions and gradually reduces those limits over time. Setting commonsense rules, cap-and-trade motivates the competitiveness and ingenuity of the marketplace to reduce emissions as smoothly, efficiently, and cost-effectively as possible.

To reduce global average temperatures, scientists have proposed various engineering interventions which can be classified into two categories: temperature treatment and carbon management (Rashid, 2011). The former is aimed at moderating heat by blocking or reflecting a small portion of the sunlight hitting the earth. This can be achieved by placing solar reflectors in the upper atmosphere or injecting sulfur dioxide particles in the stratosphere for reducing reflection of parts of the incoming solar radiation back to the space (Dyer, 2008). The most radical geo-engineering proposal is a call by the IPCC for the scientists and engineers to investigate how the North and South Poles could be cooled to end sea level rise (McDonald, 2009). Most of these proposals seem to be academic in nature. Carbon management, on the other hand, refers to gradually remove large amounts of carbon from the atmosphere. While some progress has been achieved in reducing global carbon emissions, the overall progress is exceedingly slow.

8.2.4 Global cooperation

In its First Assessment Report, the IPCC stated the need to reduce current emission of greenhouse gases by 60–80% if concentrations are to be kept at the 1990 level (IPCC, 1990). Considering this assessment and recognizing the severe consequences of global warming and climate changes, the United Nations organized the first Earth Summit in 1992 in Rio de Jeneiro, Brazil. In this summit the UN presented an agreement for reducing global warming. Signatories were bound by international law to reduce their greenhouse gas emissions by agreed-upon target dates. Within a year, 167 countries had signed. However, this agreement did little to actually reduce emissions.

A similar summit was held in Kyoto, Japan in 1997 and another agreement was proposed to reduce carbon emissions, known as the Kyoto Protocol. In this meeting, with the notable exceptions of the United States and Russia, 38 industrialized countries agreed to reduce their emissions of greenhouse gases by 5.2% below their 1990 levels by 2008–2012, the so-called "first commitment period" (Sari, 2008). Russia signed the protocol in 2004, bringing the percentage of industrialized nation signatories to 55, thus finally meeting the treaty's threshold for activation. However, this protocol faced numerous obstacles between 1997 and early 2005, when the agreement finally became international law. As noted, the first commitment period of the Kyoto Protocol will expire in 2012, and negotiations continue over the overall global emissions limitations and reduction commitments in the second commitment period (Sari, 2008).

The United States was the largest emitter of greenhouse gases in the world until very recently. In 2006, the leading position was assumed by China which emitted 6017 million metric tons (MMT) of carbon dioxide into the atmosphere in that year (Rowntree et al., 2011). Unfortunately, the United States did not (and has not, to date) sign the Kyoto Protocol, fearing that emissions controls might constrain business, thereby slowing the economy. Another reason for not signing the protocol was that large developing countries, primarily China and India, are not seen by the United States as seriously committed to reducing greenhouse gas emissions. As a result, the United States is the only industrialized country that has not ratified the Kyoto Protocol. Recently, President Obama has pledged that the United States will become a full member of the protocol and reduce its greenhouse gas emissions accordingly. However, this goal remains politically and economically problematic. Although the Kyoto Protocol will not solve the global warming problem, it is an important first step toward serious international cooperation in mitigating climate change. It laid the groundwork for what the world needed to do. Hopefully, a second step to be taken in 2012 will see increased international commitments and cooperation in reducing greenhouse gas emissions when the second phase of the Kyoto Protocol emission reduction process begins. However, questions remain unanswered as to how to get the United States back into the treaty and how to include key developing countries (Sari, 2008).

There have been, perhaps not surprisingly, tensions between developed and developing countries over implementation of measures recommended in the Kyoto Protocol. It is obvious that developed countries have created the bulk of the global warming problem, and these countries even today contribute about half of the total human-caused greenhouse gas emissions. As a result, many argue that the developed countries should be required not only to take stringent steps to curb their own emissions, but should also subsidize and underwrite some of the costs of emission controls in developing countries. Understandably, the developing nations, particularly China and India, are reluctant to sign any emission control agreement that they deem will constrain their economic future. However, without the cooperation of large emitters like China and India, the prospect for significant reduction of emissions is bleak. China still generates far less greenhouse gases than the United States on a per capita basis, but its production is rapidly increasing.

Coal is the primary source of energy both in China and India. Although it is the cheapest among all currently proven energy sources, coal generates the most greenhouse gases. Due to the widespread poverty, people in China and India can hardly afford to uses alternative sources of energy, which are costlier than coal. China and India are determined to expand access to electricity, grow their economies, and reduce poverty. Showing great foresight, China did pass laws in late 2009 requiring its power companies to purchase electricity from renewable

Table 8.1 Emission reduction pledges by industrialized countries except Ukraine

Country	2020 (%)	Reference year
Australia	5–15	2000
Belarus	5–10	1990
Canada	17	2005
Croatia	−5	1990
EU-27	20–30	1990
Iceland	15–30	1990
Japan	25	1990
Kazakhtan	15	1992
Liechtenstein	20–30	1990
Monaco	30	1990
New Zealand	10–20	1990
Norway	30–40	1990
Russia	15–25	1990
Switzerland	20–30	1990

Source: Dutt (2010), p. 33.

sources – even if such costs are more expensive than coal-fired electricity. Two years before, in 2007, 21 Asia-Pacific countries agreed to reduce their greenhouse gas emissions by 25% by the year 2030. In 2009, G8 country members also committed to reducing their carbon emissions by an ambitious 80% by the year 2050 (Dutt, 2010).

In the 2009 Copenhagen climate change conference of the United Nations, most participating countries pledged to reduce greenhouse gas emissions between 5 and 40% by 2020 (Table 8.1). The lower values in Table 8.1 correspond to what countries are willing to do unilaterally, while the upper end represents what each could do, provided there was a broad agreement involving other parties. To illustrate, Australia pledges to reduce its 2020 emissions by 5% compared to its 2000 emissions, but could pledge as much as 15% if other countries become committed to legally binding emissions reductions. Unfortunately, missing from this list of proactive emission reduction committed nations is the United States. It offered a pledge as a series of commitments for different future years (Table 8.2). The countries that have made a pledge to reduce or limit their greenhouse gas emissions account for at least 80% of the world's total emissions (Dutt, 2010). Although the Copenhagen climate conference was widely considered unsuccessful in achieving its goal, it did achieve partial success in reducing deforestation

Table 8.2 US pledges on emission reductions (for different future years, with respect to 2005 emissions)

Year	Emissions reduction (%)
2020	17
2025	30
2030	43
2050	83

Source: Dutt (2010), p. 33.

from forest degradation (Dutt, 2010). A number of developed countries assigned substantial funds to support these activities in developing countries. Two countries (Brazil and Indonesia) responsible for the bulk of deforestation made strong statements to reduce deforestation in their respective countries.

Unfortunately, the scientific community considers that the current pledges for reduction of greenhouse gases fall far short of what is needed for stabilizing the global climate. Also, pledges by well-meaning nations do not always translate into reality. Moreover, developing countries expect financial support from developed countries for efforts in reducing their emissions as well as for adapting to climate change. The Copenhagen Accord suggested amounts for this financing, but no consensus was reached on who would pay for what, or who would receive how much (Dutt, 2010).

8.2.5 Climate change and sea level rise: a case study of Bangladesh

Concern over anticipated sea level rise associated with global climate change first surfaced in Bangladesh in the late 1980s. At that time this concern was primarily confined to a limited number of Bangladeshi scientists and researchers, all of whom were in disagreement as to the occurrence of and/or manifestations of this possible change. After two back-to-back floods and one powerful cyclone that severely impacted the country in 2007, the topic of climate change came to the forefront, this time not only among Bangladeshi climate specialists and related scientists, but also among concerned citizens, politicians, and the print and electronic media. It is now widely accepted in the country that Bangladesh faces grave challenges from impacts associated with climate change.

Climate change is going to affect Bangladesh adversely in many ways. It will exacerbate the poverty situation, increase environmental migration, and ultimately hinder the sustainable development process in the country. The most damaging effects of climate change are increased flood frequency, severity, duration and areal extent, salt water intrusion, and droughts, all of which drastically reduce

Table 8.3 Affected area and population due to sea level rise in Bangladesh

	Flooded area		Affected population	
Sea level rise (projected year)	Area (sq. km)	Percentage	Pop. (millions)	Percentage
10 cm (2020)	3 150	2.0	1.8	1.3
25 cm (2050)	6 300	4.0	3.5	2.5
50 cm (2075)	16 130	11.2	6.6	5.5
1 m (2100)	26 350	18.3	30.0	20.4

Sources: World Bank (2000, 2009).

crop productivity. Climate change-induced challenges are scarcity of fresh water due to less rain, drainage congestion due to the rise in sea level, riverbank erosion, frequent floods and cyclones, prolonged and widespread drought, greater salinity in the surface water, and an increase in incidence of diseases, especially water- and vector-borne diseases.

Sea level rise is another critical factor that determines the vulnerability of Bangladesh to climate change impacts. A portion of the country's coastal zone will likely be permanently inundated. The coastal zone of Bangladesh covers 47 201 km^2, which is 32% of the total landmass of the country (Sarwar and Khan, 2007). In 2002, the total population living in the coastal zone was 35.1 million (i.e., 28% of the total population of the country) (Bangladesh Bureau of Statistics, 2003). According to the World Development Report 2010 (World Bank, 2009), about 18% of all Bangladesh land will be submerged following a local sea level rise of 1 m (Figure 8.4). Should this occur, it will result in the displacement of almost 30 million of its current 147 million people (Table 8.3). Scientists, including those of the United Nations' Intergovernmental Panel on Climate Change (IPCC), maintain that the number of the displaced or "climate refuges" will increase in Bangladesh during the coming years as rising sea levels devour low-lying coastal areas of this delta country (World Bank, 2009). In such a scenario, the Bangladesh government would have to relocate this huge number of internally displaced people with few available resources. Among other aspects, this section will focus on issues related to these displaced people, who are interchangeably referred to as "environmental refugees" or "climate refugees."

Although linkages between environmental change and human migration have long been debated by population researchers and other scholars (e.g., Biermann and Boas, 2007, 2008; Black, 2001; McGregor, 1993; Myers, 2005), it is obvious that coastal residents will be forced to migrate if their homes are permanently inundated due to a rise in sea level (Myers, 1995, 1997, 2002). As late as 2004, environmental refugees were a peripheral concern for many in the United Nations (McNamara, 2007). However, the December 2004 Indian Ocean tsunami and

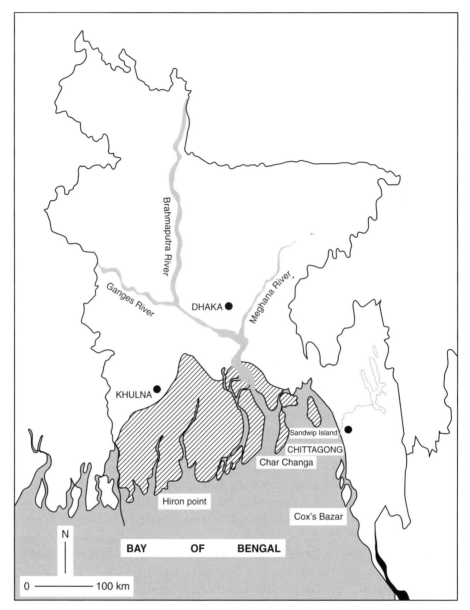

Figure 8.4 Area likely to be affected by 1 m sea level rise (shaded part) along Bangladesh coast. *Source:* Compiled from various sources.

Hurricane Katrina in August 2005 in the United States, the subsequent release of the Stern (2006) report as well as the IPCC's 4th Assessment Report (IPCC, 2007) have all provided crucial evidence that global climate change has been taking place. Globally, this has led to a wider acceptance of and openness towards the notion that people can be displaced due to climate change. Leading climate change scholars maintain that migration may be the only response and a coping strategy for people whose livelihoods are undermined by climate change (Barnett, 2001; Barnett and Adger, 2007; Warner *et al.*, 2010; World Bank, 2009).

Since the end of 2007, not only have attitudes changed regarding acceptance of the reality of climate change at the global level, but there has also been heightened interest and concern in the relationship between climate/environmental changes and migration. There is a growing recognition that greater research and policy attention needs to be given to the environmental causes of migration. Increasing emphasis is also being placed on recognizing the potential of migration as an adaptation strategy for communities directly impacted by climate change (e.g., Barnett and Webber, 2009). This shift has already shaped how the issue of climate change and population movement is viewed in Bangladesh by both its citizens and its government. For example, in terms of government strategies to deal with climate change-induced migration, Bangladesh began seeking the cooperation of wealthy countries in relocation planning for its citizens likely to be displaced by sea level rise.

At the 2009 Copenhagen climate change conference, the Finance Minister of Bangladesh urged the United Kingdom and other wealthy nations to accept millions of Bangladeshi climate refugees (*Daily Star*, 2009). He called on the UN to redefine international law to give climate refugees the same protection as people fleeing political repression. Although the Finance Minister's proposal reflects the gravity of the problem, not all, both within and beyond Bangladesh, support his notion that the developed countries have moral responsibility to accept climate refugees from Bangladesh and other developing countries susceptible to climate change-induced sea level rise.

Bangladeshi climate specialists and members of civil society support the idea that richer nations should accept climate refugees as land becomes scarce in this densely populated South Asian nation. Based on reports published recently in Dhaka-based daily newspapers, and formal and informal conversations with officials of Bangladeshi environmental organizations, it seems that the Bangladesh government has not yet outlined any policy regarding the mechanics of climate refugee migration to foreign countries. Without a clear policy in place, many concerned citizens of Bangladesh suspect that the most vulnerable – poor and largely illiterate coastal residents – will never be able to migrate to the United States, Canada, Australia, or other developed countries. Rather, many suspect politically powerful, and/or highly educated urban residents will take this opportunity to migrate to developed countries as climate refugees.

Even if the government does develop a comprehensive policy on climate-induced migration, many suspect that it will fail to implement such a policy because of rampant corruption within the government. In addition, the Bangladesh government is not transparent and not accountable to the public. It will be evident from information presented below that the international community has reservations about the relocation of climate refugees in developed countries and it advocates that climate-induced migrants be resettled within the national boundaries of the impacted countries (World Bank, 2009).

8.2.5.1 Climate refugee discourse: a review

While the phenomenon of climate change has been receiving increasing acceptance on a global scale, several issues regarding environmental or climate refugees are still unresolved. First, there is semantic debate over the terminology of "environmental" or "climate" refugee (Bates, 2002; Leighton et al., 2009). Refugee has a specific legal meaning in the context of the 1951 Geneva Convention Relating to the Status of Refugees. This convention defines refugees as people with a well-founded fear of being persecuted for reasons of race, religion, nationality, membership in a particular social group or for political opinion (Bates, 2002). People who will likely be displaced due to sea level rise are not legally protected in any way by this convention. Therefore, the term refugees should not be used; "environmental migrants," "environmentally motivated migrants," or "climate migrants" are more appropriate terms. Even terms such as "climigrant" and "climate exile" have been coined (Leighton et al., 2009).

Along with sea level rise, climate change is projected to worsen the intensity, duration, and frequency of chronic or sudden onset hazards. After carefully reviewing the available literature on disaster migration and environmental degradation, Raleigh et al. (2009) concluded that environmental or climate change-induced large-scale migration is unlikely because the poor would be the primary victims of such an event. Evidence from developing countries suggests that the poor often find it impossible to move, as they lack the necessary funds and social support. Those who can migrate are more likely to make short-term, short-distance movements than permanent long-term ones. Raleigh et al. (2009), however, acknowledge that disasters vary considerably in their potential to instigate migration. Individual, community, and national vulnerabilities shape responses as much as disaster effects do. In studying tornado and flood victims in Bangladesh, Paul (2003, 2005) found that where the provision of disaster aid was well managed and without irregularities, people do not move from affected areas.

Given the above evidence, it may not be appropriate to use either "environmental migrants" or "climate migrants" as terms for people at risk of being displaced by sea level rise. If it occurs as projected, people of most island countries (e.g., the Maldives), along with people of low-lying coastal areas of vulnerable countries like Bangladesh have no other alternative but to migrate internally or internationally (Kothari, 2002). A more appropriate and specific term for such migrants is "sea level rise (SLR) migrants." Climate migrant is a subcategory of environmental migrant and sea level rise migrant is a subcategory of climate migrant. It is, however, unclear what form such migration will take. Will it be a managed retreat or the progressive abandonment of land and structures? This will depend on the nature of rising sea levels and the depth of water in coastal communities. Will the rise be gradual or abrupt? There is as yet no consensus on these questions.

The second debate concerns the links between environmental change and migration. Can environmental factors be recognized as a root cause of migration? Scholars such as Biermann and Boas (2007, 2008), Conisbee and Simms (2003), Jacobson (1988), and Myers (1995, 1997, 2002, 2005) claim that if lands become degraded or other environmental problems surface, such as disappearing forests or rising seas, then people will be forced to flee their homeland. Some of these scholars believe that environmental or climate refugees have, or will soon become, the largest class of displaced persons in the world. However, other scholars, including geographers (e.g., Black, 1998, 2001; Hulme, 2008; Kibreab, 1997; McGregor, 1993; Wood, 2001) question the very concept of environmental or climate refugees. These critics argue that migration as well as the decision to migrate is often multifaceted; to single out environmental factors as the reason for this migration is misleading and oversimplifies the causal relationship between environmental degradation and migrations.

8.2.5.2 The Bangladesh government's position on climate refugees

In the context of global rise in sea level, more important issues than the debate over definitions or the root cause of migration are: who is responsible for such rise in sea level and who is responsible for resettlement of people displaced by sea level rise? Unlike deforestation, national governments of vulnerable countries are not directly responsible for rise in sea level. Therefore, all types of land degradation and sea level rise should not be treated in a similar fashion. As noted, the official position of the Bangladesh government is that it is a moral obligation of industrialized countries to accept displaced people because they are largely responsible for the greenhouse gas emissions that are the root cause of sea level rise. Furthermore, Bangladesh claims it is one of the least responsible nations for the problem, yet it is one of the most at risk from the consequences of the impending climate change, specifically rising sea levels. This position of the Bangladesh government is consistent with a call for "climate justice" advocated by environmental activists across the globe (*Daily Star*, 2009). Climate justice is a moral position, based on the fact that those people least responsible for climate change are often the most affected by it, and have the least means to cope with it.

Like Bangladesh, low-lying coastal portions of countries in Europe, North America, and Australia may themselves face a similar refugee problem. International agencies, including the World Bank, hope that most people forced into migration due to climate change can be resettled within their own countries (World Bank, 2009). There is a great reluctance on the part of the UN High

Commissioner for Refugees (UNHCR) regarding revision of the definition of refugees to include climate refugees. A UNHCR report warned in August 2009 that given the current political environment, such a revision could result in decreasing protection standards for nonclimate refugees and even undermine the international refugee protection system altogether (UNHCR, 2009). However, global civil society groups, such as the International Campaign on Climate Refugees' Rights (ICCR), and environmental activists have been consistently demanding a legal and institutional framework for protecting the rights of and rehabilitating "climate refugees" that may be displaced due to climate change. The ICCR has suggested adoption of a safeguard protocol to ensure the political, social, cultural, and economic rights of climate refugees are not violated. Concerned groups have further demanded a review of the 1951 Geneva Convention on Refugees by taking climate refugees into consideration, and called for a separate international framework to ensure that climate refugees are treated with dignity and respect (*Daily Star*, 2009).

Given current reluctance on the part of most developed countries to accept climate refugees, some Bangladeshis feel that the Bangladesh government should also approach Middle Eastern countries for providing resettlement options for Bangladeshi climate refugees. They claim that wealthy Middle Eastern Muslim countries have an obligation to accept displaced Bangladeshi Muslims.

Regardless of whether Western or Middle Eastern countries may become possible destinations for Bangladeshi climate refugees, experts maintain that after their homes are inundated due to sea level rise, affected coastal residents of Bangladesh will first move within the country – most likely to larger cities located inland, such as Dhaka (Islam, 2009). These experts further claim that such movements have already started against a background of rapid global urbanization, in which the rapid urban population growth far outpaces jobs and infrastructure. According to them, the next step in this predicted migration pattern is illegally crossing national borders to neighboring countries, where resources may be only slightly less scarce (Islam, 2009). Nag (2008) postulates that about 75 million people from Bangladesh would migrate to India as climate refugees.

Cross-border migration will almost certainly become a great political issue and will cause political tension (Gaan, 2005). India claims that there are millions of illegal Bangladeshis already living in India. Further increase in such "illegal" migration will surely lead to conflict. It is likely that illegal migration from Bangladesh will extend beyond neighboring countries in the advent of climate-induced sea level rise. It is probable that some climate refugees will be victims of those who are in the trade of trafficking women and children, which has been occurring for more than three decades in Bangladesh (Paul and Hasnath, 2000).

Relocating people from vulnerable areas is an example of adaptation to sea level rise and focuses on coping ex post or reactive adaptation, to the adverse impacts of climate change, after they have occurred (Barnett, 2001; Shalizi and Lecocq, 2009). Lal and Aggarwal (2000) argue that it is usually better to react to climate change as it unfolds, rather than alter activities in anticipation of climate change. However, they list a number of possible problems with reactive adaptation strategies. They claim that the consequences of climate change, even after reactive adaptation, may be unexpectedly severe. Moreover, reactive policies run the risk of misallocating resources by adopting short-term incremental approaches and failing to subsidize and/or address possible greater changes in the future.

Given uncertainty regarding the acceptance of climate (or sea level rise) refugees by any country, the Bangladesh government should also focus on other (proactive) adaptations, such as modifying zoning laws on coasts in anticipation of stronger sea surges, planning for large areas of forest in flood-prone areas along rivers and coastlines, perhaps building embankments in some areas to cope with rising waters, along with shifting to crops that are more resistant to drought and saline water. The goal of these and other proactive adaptation measures is to reduce the impacts of climate change by reducing vulnerability to many of its direct effects (Lal and Aggarwal, 2000). Proactive adaptation, like mitigation, uses resources now to prevent possible crises in the future.

From an economic point of view, the costs of proactive adaptation are generally lower than the costs of reactive adaptation strategies (Athukorala and Resosudarmo, 2005; Burby, 2006). A common problem with most adaptive strategies is that many if not all impacts of climate change will not be visible until the next few decades. In addition, because of uncertainties among climate experts regarding changes in regional climates, the specific local impacts of climate change (even at national levels) are also uncertain (Lal and Aggarwal, 2000). Despite these limitations, the research on this issue finds that past sea level rise has not led to displaced coastal populations; instead people coped through a variety of different adaptations (Black, 2001; Perch-Nielsen, 2004; Raleigh *et al.*, 2009).

In the context of climate change, mitigation consists of reducing emissions (or removing greenhouse gases from the atmosphere), shifting from coal to natural gas-fired power plants, developing renewable energy, and reducing deforestation and associated emissions of carbon dioxide (IPCC, 2001). All these measures can substantially reduce the impacts expected with ongoing climate change. Although there is some disagreement among climate specialists, mitigation measures are generally more appropriate responses to climate change for developed countries (Shalizi and Lecocq, 2009). The Bangladesh government feels it has little responsibility for mitigation measures since its contribution to greenhouse gas emissions that cause global warming is insignificant. Bangladesh emits an estimated 34.0 million tonnes of carbon dioxide annually, an amount that represents about 0.1% of the global total (Alam and Rahman, 2008).

Experts, however, recommend that the country should consider both mitigation and adaptation options, even though mitigation involves global efforts to be effective and adaptation is more varied and local. Remarkably, the Bangladesh government has already incorporated both adaptation and mitigation strategies into the country's overall development strategy (Alam and Rahman, 2008). Surprisingly, no one has yet advocated reduction in population growth rate as a response to what appears to be imminent sea level rise. Bangladesh has a natural population growth rate of 1.7%, which is higher than the world average growth rate of 1.2% (Population Reference Bureau, 2009).

8.2.5.3 Bangladesh government response and policy regarding climate change

Considering what could be catastrophic effects of climate change, the Bangladesh government has devised a strategy and has action plans in place to address climate change. As one of the initial steps, the government set up a national climate committee in 1994 to provide policy guidance and oversee the implementation of obligations under the UN Framework Convention on Climate Change (UNFCCC). The Bangladesh government itself initiated a study in 2002 which presented national data on greenhouse gas inventory and emissions, vulnerability, adaptation and mitigation, and climate change response strategies (Alam and Rahman, 2008). Data from this study led the government to develop the National Adaptation Program of Action (NAPA) in 2005 to address impacts of climate change. NAPA outlined strategic guidelines for climate change-related threats by economic sector. NAPA put major emphasis on the following actions (see Islam, 2008):

- developing saline-tolerant crop varieties for coastal areas,
- constructing new flood shelters and assistance centers to cope with enhanced recurrent floods,
- promoting adaptation of coastal fisheries through culture of salt-tolerant fish species,
- adapting fisheries in areas prone to enhanced flooding,
- reducing the impact of climate change hazards (such as increased storm surge) through coastal afforestation project with increased local community participation,
- providing drinking water to coastal communities to combat enhanced salinity due to sea level rise, and
- enhancing the resilience of urban infrastructure and industry in coping with the impacts of climate change.

While a significant step in the right direction, these NAPA recommendations fail to foresee climate change-related problems as a multi-scalar (Eakin, 2005) and do not make it clear how vulnerable communities will be made more resilient with their resources. In 2008, the Bangladesh government formed a fund known as the Multi-Donor Trust Fund (MDTF), and asked developed countries for financial assistance to fight climate change. Britain and Denmark have already pledged more than US$115 million to this fund. At the local level, Bangladesh has initiated the Knowledge Network on Climate Change (KNOCC) and the Global Initiative on Community-based Adaptation to Climate Change (GIBACC) programs.

In 2009, the Bangladesh government formed the All Party Parliamentary Group (APPG) on Climate Change and Environment. The APPG considers that climate change brings new opportunities as well as the challenges and risks (*New Nation*, 2009a). This group aims to:

- facilitate and develop cross-party consensus building, recognizing the importance of the environment as a whole and climate change in particular in formulating policy initiatives/guidelines that will be followed even when there is a change in government;

- network with regional and international forums and groups on climate change and establish strategic alliances and common positions and standards;

- inform and educate the people at large on issues related to climate change and environmental consequences, and build awareness thereof;

- promote political leadership and stewardship of the issues and resources impacted by climate change and environment agendas;

- develop linkages between lawmakers, local government representatives, opinion builders, business communities, and NGOs to promote an advanced level of understanding and cooperation between important stakeholders in coordinated responses and actions to promote adaptation and sustainable environmental governance;

- advocate specific policy initiatives to mainstream climate change and environmental concerns;

- encourage and promote use of renewable/alternative energy throughout Bangladesh.

The Bangladesh government has signed and ratified both the UNFCCC and the Kyoto Protocol; it has also signed other multilateral environmental agreements (MEAs) and is participating in international environment/climate meetings and conferences. Among these forums the government has played a proactive role in the UNFCCC COPs. At the 2009 summit in Copenhagen, Denmark, Bangladesh not only urged developed countries to accept climate migrants, it also demanded creation of a climate change adaptation fund in addition to their usual overseas development assistance fund (*New Nation*, 2009b). Furthermore, Bangladesh stated in the summit that allocations from this adaptation fund should be awarded in proportion to the percentage of its population exposed to climate change.

It is noteworthy that both international and national NGOs are also playing an important role, and are expanding their activities in disaster management, and livelihood improvement in Bangladesh. The Climate Change and Development Forum (CCDF) for example, are working in Bangladesh to share information on climate change and how it can be integrated into ongoing and future activities. There are also community-level activities to address development and climate change, with a special focus on disaster management (Alam and Rahman, 2008).

It is important to mention that Bangladesh was among the first countries to prepare and submit its NAPA with the UNFCCC Secretariat in 2005. Because of the geographic nature of the global warming and climate change problems, a regional cooperation is also needed along with international initiatives. Climate impacts will affect the entire population of South Asia in the coming decades, in one form or other. Efforts to address sustainable development goals in every member state of the South Asian Association for Regional Cooperation (SAARC) could therefore be increasingly challenged by climate variability and change if regional initiatives are not taken soon. Fortunately, these countries have recently initiated collective actions to pursue climate-resilient development. Other vulnerable countries located beyond South Asia should also seriously consider how to overcome the impacts of global climate changes. In this regard, cordial cooperation is needed between countries responsible for producing greenhouse gases and countries widely considered to be victims of global warming and associated climate change.

References

Ahsan, R.M. and Hossain, M.K. (2004) Women and child trafficking in Bangladesh: a social disaster in the backdrop of natural calamities. In *Disaster and Silent Gender: Contemporary Studies in Geography* (eds R.M. Ahsan and H. Khatun). Dhaka: The Bangladesh Geographical Society (BGS), pp. 147–170.

Ahsan, R.M. and Khatun, H. (2004) *Disaster and the Silent Gender: Contemporary Studies in Geography*. Dhaka: The BGS.

Alam, M. and Rahman, A. (2008) Development and climate change policy-making process in Bangladesh. In *Climate Change in Asia: Perspectives on the Future Climate Regime* (eds Y. Kameyama, Y. et al.). Tokyo: United Nations University Press, pp. 51–65.

Archer, D. (2007) *Global Warming: Understanding the Forecast*. Malden, MA: Blackwell Publishing.

Athukorala, P. and Resosudarmo, B.P. (2005) The Indian Ocean tsunami: economic impact, disaster management, and lessons. *Asian Economic Papers* 4 (1): 1–39.

Bangladesh Bureau of Statistics (2003) *Statistical Yearbook of Bangladesh*. Dhaka: Bangladesh Bureau of Statistics.

Barnett, J. (2001) Adapting to climate change in Pacific island countries: the problem of uncertainty. *World Development* 29 (6): 977–993.

Barnett, J. and Adger, N. (2007) Climate change, human security and violent conflict. *Political Geography* 26: 639–655.

Barnett, J. and Webber, M. (2009) Accommodating migration to promote adaptation to climate change. A policy brief prepared for the Secretariat of the Swedish Commission on Climate Change and Development, Stockholm.

Bates, D.C. (2002) Environmental refugees? Classifying human migrations caused by environmental change. *Population and Environment* **23** (5): 465–477.

Biermann, F. and Boas, I. (2007) Preparing for a warmer world: towards a global governance system to protect climate refugees. Global Governance Working Paper No. 33. www.glogov.org.

Biermann, F. and Boas, I. (2008) Protecting climate refugees. The case for a global protocol. *Environment* **50** (6): 8–16.

Black, R. (1998) *Refugees, Environment and Development*. New York: Addison Wesley.

Black, R. (2001) Environmental refugees: myth or reality? UNHCR "New Issues in Refugee Research." Working Paper No. 34. UNHCR, Geneva.

BASC (Board on Atmospheric Science and Climate) (2011) America's Climate Change Choices. Washington, D.C.: Academic Press.

Burby, R.J. (2006) Hurricane Katrina and the paradoxes of government disaster policy: bringing about wise government decisions for hazardous areas. *Annals of the American Academy of Political and Social Sciences* **604**: 171–191.

Chowdhury A.M.R. *et al.* (1993) The Bangladesh cyclone of 1991: why so many people died. *Disasters* **17** (4): 291–303.

Conisbee, M. and Simms, A. (2003) *Environmental Refugees. The Case for Recognition*. London: New Economics Foundation.

Cupples, J. (2007) Gender and Hurricane Mitch: reconstructing subjectivities after disaster. *Disaster* **31** (2): 155–175.

Daily Star (2009) Allocation Fund for Population at Risk. December 13.

Doran, P.T. and Zimmerman, M.K. (2009) Examining the scientific consensus on climate change. *EOS* **90** (3): 22–23.

Dutt, G. (2010) Reaching a climate agreement: beyond the Copenhagen Accord. *Economic and Political Weekly* **45** (17): 32–37.

Dyer, G. (2008) *Climate Wars*. Toronto: Random House Canada.

Eakin, H. (2005) Institutional change, climate risk, and rural vulnerability: cases from Central Mexico. *World Development* **33** (11): 1923–1938.

Enarson, E. (2000) *Gender Issues in Natural Disasters: Talking Points and Research Needs*. Geneva: ILO.

Enarson, E. and Morrow, B.H. (1998) Why gender? Why women? An introduction to women and disaster. In *The Gendered Terrain of Disaster: Through Women's Eyes* (eds E. Enarson and B.H. Morrow). Westport, CT: Praeger Publishers, pp. 1–8.

Environmental Protection Agency (2009) Back to the Basics. http:///www.epa.gov/climate change/downloads/Climate_Basics.pdf (accessed April 21, 2010).

Felten-Biermann, C. (2006) Gender and natural disaster: sexualized violence and the tsunami. *Development* **49** (3): 82–85.

Fothergill, A. (1998) The neglect of gender in disaster work: an overview of the literature. In *The Gendered Terrain of Disaster: Through Women's Eyes* (eds E. Enarson and B.H. Morrow). Westport, CT: Praeger Publishers, pp. 11–25.

Gaan, N. (2005) Environmental scarcity, migration and future sea level rise in Bangladesh: security implications on India. *Asian Profile* **33** (6): 617–632.

Gautham, S. (2006) Teach the girls to swim tsunami, survival and the gender dimension. http://www.countercurrents.org/gen-gautham100806.htm (accessed August 5, 2008).

Government of Bangladesh (2008) *Cyclone Sidr in Bangladesh: Damage, Loss and Needs Assessment for Disaster Recovery and Reconstruction.* Dhaka: Government of Bangladesh.

Hafiz, A. (2005) The disaster continues. *Off Our Backs* 35 (9/10): 18.

Houghton, J. (2004) *Global Warming: The Complete Briefing*, 3rd edn. Cambridge: Cambridge University Press.

Hulme, M. (2008) Climate refugees: cause for a new agreement (commentary). *Environment* 50 (6): 50–52.

Hyndman, D. and Hyndman, D. (2011) *Natural Hazards and Disasters.* Belmont, CA: Brooks/Cole.

Ikeda, K. (1995) Gender differences in human loss and vulnerability in natural disasters: a case study from Bangladesh. *Indian Journal of Gender Studies* 2 (2): 171–193.

IPCC (Intergovernmental Panel on Climate Change) (1990) *Scientific Assessment of Climate Change – Report of Working Group 1.* Cambridge: University of Cambridge Press.

IPCC (2001) *Climate Change 2001: Impacts, Adaptation and Vulnerability. A Report of Working Group II of the IPCC.* Geneva: IPCC.

IPCC (2007) *Climate Change 2007: The Physical Science Basis.* New York: IPCC.

Islam, T. (2008) Strengthening community resilience as a measure of climate change adaptation in Bangladesh. Department of Geography and Environment, Jahangirnagar University. Unpublished paper.

Islam, M.S. (2009) The impact of climate change on Bangladesh: responding to a threat to food security for sustainable development. *Asian Profile* 37 (5): 509–519.

Jacobson, J.L. (1988) Environmental Refugees: A Yardstick of Habitability. Worldwatch Paper 86. Worldwatch Institute: Washington, D.C.

Khan, F. and Mustafa, D. (2007) Navigating the contours of the Pakistani hazardscapes: disaster experience versus policy. In *Working with the Winds of Change: Towards Strategies for Responding to the Risk Associated with Climate Change and Other Hazards* (eds M. Moench and A. Dixit). Kathmandu: ProVention Consortium, pp. 193–234.

Kibreab, G. (1997) Environmental causes and impacts of refugee movements: a critique of the current debate. *Disasters* 21: 20–38.

Kothari, U. (2002) Migration and Chronic Poverty. Chronic Poverty Research Center, Working Paper No. 16. Institute for Development Policy and Management, University of Manchester: Manchester.

Krug, G. et al. (2002) *World Report on Violence and Health.* Geneva: WHO.

Lal, M. and Aggarwal, D. (2000) Climate change and its impact on India. *Asia Pacific Journal on Environment and Development* 7 (1): 1–11.

Leighton, M. et al. (2009) The challenges of climigration. *Development & Cooperation* 9: 323–325.

Mahmood, A. (2007) *Earthquake Vulnerability Assessment – Pakistan: 2005–06.* Islamabad: The Population Council.

Marshall, L. (2005) Were women raped in New Orleans? Addressing the human rights of women in times of crisis. *Off Our Backs* 35 (9/10): 1414–1415.

McDonald, F. (2009) Climate experts urge engineering solutions to "directly cool the planet. *Irish Times*, March 12.

McDonald, R. (2005) How women were affected by the tsunami: a perspective from Oxfam. *PLoS Medicine* 2 (6): 474–475.

McGregor, J. (1993) Refugees and the environment. In *Geography and Refugee: Patterns and Processes of Change* (eds R. Black and V. Robinson). London: Belhaven.

McNamara, K.E. (2007) Conceptualizing discourses on environmental refugees at the United Nations. *Population and Environment* **29** (1): 12–24.

Myers, N. (1995) *Environmental Exodus: An Emergent Crisis in the Global Arena*. Washington, D.C.: Climate Institute.

Myers, N. (1997) Environmental refugees. *Population and Environment* **19**: 167–182.

Myers, N. (2002) Environmental refugees: a growing phenomenon of the 21st century. *Philosophical Transactions of the Royal Society B* **357**: 609–613.

Myers, N. (2005) *Environmental Refugees: An Emergent Security Issue*. Prague: Economic Forum.

Nag, S. (2008) '75 Million Bangladeshis may inundate India. *Hindustan Times*, March 26.

Neumayer, E. and Plumper, T. (2007) The gendered nature of disasters: the impact of catastrophic events on the gender gap in life expectancy, 1981–2002. *Annals of the Association of American Geographers* **97** (3): 551–566.

New Nation (2009a) Joint Bangladesh–UK parliamentary inquiry on climate change. *New Nation* August 9, 2009.

New Nation (2009b) Dhaka seeks 75pc of climate funds. *New Nation* December 9, 2009.

Ollenburger, J.C. and Tobin, G.A. (1998) Women and postdisaster stress. In *The Gendered Terrain of Disaster: Through Women's Eyes* (eds E. Enarson and B.H. Morrow). Westport, CT: Praeger Publishers, pp. 95–107.

Oreskes, N. (2004) Beyond the ivory tower: the scientific consensus on climate change. *Science* **306**: 1686–1686.

Oxfam (2005) The Tsunami's Impact on Women. Oxfam Briefing Note.

Paul, B.K. (1997) Flood research in Bangladesh in retrospect and prospect: a review. *Geoforum* **28** (2): 121–131.

Paul, B.K. (1999) Women's awareness of and attitudes towards the flood action plan (FAP) of Bangladesh: a comparative study. *Environmental Management* **23** (1): 103–114.

Paul, B.K. (2003) Relief assistance to 1998 flood victims: a comparison of the performance of the government and NGOs. *The Geographical Journal* **169** (1): 75–89.

Paul, B.K. (2005) Evidence against disaster-induced migration: the case of the 2004 tornado in north-central Bangladesh. *Disasters* **29** (4): 370–385.

Paul, B.K. (2010) Human injuries caused by Bangladesh's Cyclone Sidr: an empirical study. *Natural Hazards* **54**: 483–495.

Paul, B.K. and Hasnath, S.A. (2000) Out of Bangladesh: trafficking in women and girls. *Geographical Review* **90** (2): 268–276.

Perch-Nielsen, S. (2004) *Understanding the Effect of Climate Change on Human Migration: The Contribution of Mathematical and Conceptual Models*. Zurich: Department of Environmental Studies, Swiss Federal Institute of Technology.

Pikul, C. (2005) As tsunami receeds, women's risks appear. http://www.globalaging.org/elderrights/world/2004/tsunami.htm (accessed April 12, 2007).

Population Reference Bureau (2009) *2008 World Population Data Sheet*. Washington, D.C.: Population Reference Bureau.

Pulsipher, L.M. and Pulsipher A. (2011) *World Regional Geography: Global Patterns, Local Lives*. New York: W.H. Freeman and Company.

Raleigh, C. et al. (2009) *Assessing the Impact of Climate Change on Migration and Conflict*. Washington, D.C.: The World Bank.

Rashid, H. (2011) Media framing of public discourse on climate change and sea-level rise: social amplification of global warming versus climate justice for global warming impacts. In *Climate Change and Growth in Asia* (eds M. Hossain and E. Selvanathan). Cheltenham: Edward Elgar, pp. 232–260.

Rasid, S.F. and Michaud, S. (2000) Female adolescents and their sexuality: notions of honour, shame, purity and pollution during the floods. *Disasters* 24 (1): 54–70.

Rees, S. *et al.* (2005) Waves of violence – women in post-tsunami Sri Lanka. *The Australian Journal of Disaster and Trauma Studies* 2: 1–7.

Ross, L.J. (2005) A feminist perspective on Hurricane Katrina. *Off Our Backs* 35 (9/10): 11–13.

Rowntree, L. *et al.* (2011) *Globalization and Diversity: Geography of a Changing World*. Boston: Prentice Hall.

Sari, A.P. (2008) Introduction: climate change and sustainable development in Asia. In *Climate Change in Asia: Perspectives on the Future Climate Regime* (eds Y. Kameyama). Tokyo: United Nations University Press, pp. 3–17.

Sarwar, G.M. and Khan, M.H. (2007) Sea level rise: a threat to the coast of Bangladesh. *Internationales Asienforum* 38 (3–4): 375–397.

Shalizi, Z. and Lecocq, F. (2009) To mitigate or to adapt: is that the question? Observations on an appropriate response to the climate change challenge to development strategies. *The World Bank Research Observer* September 2, 2009.

Sommer, A. and Mosley, W.H. (1972) East Bengal Cyclone of November 1970: epidemiological approach to disaster assessment. *Lancet* 299: 1029–1036.

Stern, N. (2006) *Stern Review on the Economics of Climate Change*. London: HM Treasury.

Tichagwa, W. (1994) The effects of drought on the condition of women. *Focus Gender* 2 (1): 20–25.

UN (United Nations) (2006) Regional Workshop on Lessons Learned and Best Practices in the Response to the Indian Ocean Tsunami: Report and Summary of Main Conclusions. Medan, Indonesia.

UN High Commissioner for Refugees (UNHCR) (2009) *World Report*. Florence: UNHCR.

Warner, K. *et al.* (2010) Climate change, environmental degradation and migration. *Natural Hazards* 55: 689–715.

Wood, W.B. (2001) Ecomigration: linkages between environmental change and migration. In *Global Migrants, Global Refugees* (eds A.R. Zolberg and P.M. Benda). New York and Oxford: Berghahn, pp. 42–61.

World Bank (2000) *Bangladesh: Climate Change & Sustainable Development*, Report No. 21104 BD. Dhaka: World Bank.

World Bank (2009) *World Development Report 2010: Development and Climate Change*. Washington, D.C.: World Bank.

WHO (World Health Organization) (2005) *Gender and Health in Natural Disasters*. Geneva: WHO.

Author Index

Ackerman, E.A. 38, 61
Ackerman, R. 130, 152
Adams, B.J. 124, 151
Adger, W.N. 68, 70, 71, 87, 90, 91, 110, 295, 302
Aggarwal, D. 298, 299, 304
Aguirre, B. 177, 192, 200, 234
Ahern, M. 131, 133, 151
Ahsan, R.M. 134, 151, 272, 276, 302
Albala-Bertrand, J. 134, 135, 151, 259, 264
Alexander, D. 2, 27, 33, 39, 40, 51, 61, 93, 95, 110, 141, 148, 151, 158, 159, 192, 219, 222, 233, 249, 250, 257–259, 264
Alam, M. 299, 300, 302
Ali, M.H. 166, 192
Alwang, J. 96, 110
Ambrosio-Albala, M. 91, 110
Ansel, J. 94, 110
Archer, D. 280, 302
Arcury, T.A. 99, 110
Armenian, H.K. 133, 151
Armstrong, M.P. 57, 61
Arthukorala, P. 299, 302
Aspinall, F. 255, 264
Autier, P. 257, 264

Baggerly, J. 205, 233
Bandura, A. 181, 182, 192
Bangladesh Bureau of Statistics (BBS) 293, 302
Bankoff, G. 71, 110
Barnett, J. 298, 302, 303
Barsky, L.E. 216, 233
Barrows, H.H. 38, 61
Bateman, J.M. 184, 192
Bates, D.C. 296, 303
Batha, E. 238, 240, 241, 264
Bauer, R.A. 100, 110
Beaver, B. 201, 233
Bennish, M. 257, 267
Benson, C. 262, 264
Berke, P.R. 230, 233
Berkes, F. 74, 110
Bernett, J. 295, 302, 303
Bhuiyan, R.H. 99, 101, 115, 187, 195, 261, 266
Biddle, M.D. 184, 192
Biermann, F. 293, 297, 303
Blaikie, P. 40, 41, 46, 48, 61, 69, 71, 73, 77, 111
Blair, D. 184, 185, 194
Black, R. 293, 297, 299, 303
Blake, G. 127, 151

Environmental Hazards and Disasters: Contexts, Perspectives and Management, First Edition. Bimal Kanti Paul.
© 2011 John Wiley & Sons, Ltd. Published 2011 by John Wiley & Sons, Ltd.

Blanchard-Boehm, D.R. 182, 183, 192
Blanchard, W. 104, 111
BNIM Architects 224, 233
Boas, I. 293, 297, 303
Bohle, H.G. 73, 117
Bolin, R. 14, 33, 40, 41, 48, 61, 70, 73, 111, 183, 192, 237, 254, 264
Borden, K.A. 125–127, 151
Bourdieu, P. 75, 78, 111, 164, 192
Bowd, R. 59, 64
Boyd, E.C. 56, 61
Brammer, H. 149, 151
Brenner, S.A. 136, 151, 202, 233
Briere, J. 8, 33
Brock, V.T. 164, 165, 192, 221, 233
Brody, S.D. 69, 100, 111
Brookfield, H.C. 40, 61
Brooks, H.E. 125, 151
Brown, S. 136, 137, 151, 202, 234
Bruneau, M. 86, 88, 89, 111, 116
Buckland, J. 78, 111
Buekens, P. 146, 151
Burby, R.J. 299, 303
Burton, C.G. 84, 111
Burton, I. 2, 5, 8, 14, 32, 33, 38, 43, 61, 63, 94, 100, 106, 111, 160, 162, 192
Butler, A.S. 17, 33

Cacioppo, J.T. 183, 195
Cardona, O.D. 70, 111
Cartlidge, M.R. 225, 234
CBS News 134, 151
Chaiken, S. 183, 193
Chakraborty, J. 57, 61, 135, 155
Chambers, D.N. 139, 151
Chaplin, S. 182, 193
Chapin, F.S. 89, 111
Chapman, D. 3, 5, 14, 20, 33
Charles, N. 5, 33
Che, D. 221, 222, 224–226, 235
Choudhury, J.R. 166, 192
Chowdhury, A.M.R. 81, 111, 270, 303
Chowdhury, K.M.M.H. 28, 33
Chowdhury, R.K. 29, 35
Christoplos, I. 105, 111, 221, 234
Coates, L. 125, 133, 152
Coburn, A. 150, 154

Cochrane, J. 240, 267
Cohen, C. 8, 33, 135, 152
Collins, A. E. 45, 47, 52, 61
Company News 257, 264
Conisbee, M. 297, 303
Cook, M.J. 182, 183, 192
Coppola, D.P. 8, 14, 31, 33, 46, 53, 61, 70, 93, 94, 104, 111, 138, 142, 152, 159, 160, 162, 163, 167, 173, 176, 186, 187, 190–192, 193, 204, 205, 216, 217, 226, 234, 239, 241, 264
Cordasco, K.M. 75, 76, 111
Cordero, J.F. 146, 152
Correira, S. 87, 111
Corson, M.W. 44, 61
Cosgrave, J. 240, 250, 258, 267
Cotton, S. 130, 152
Crichton, D. 97, 111
Crooks, S. 85, 115
Cropper, M.L. 57, 61
Crozier, M.J. 44, 63
Csiki, S. 174, 175, 193
Cuny, F.C. 46, 51, 61, 74, 75, 111, 148, 152, 222, 234, 246, 249, 262, 264
Cupples, J. 237, 254, 257, 264, 269, 273, 276, 303
Curtis, A. 55–58, 61, 73, 112, 146, 152
Cutter, S.L. 3, 5, 6, 16, 30, 33–35, 37, 39, 42–44, 54–57, 62, 65, 67–70, 73, 78, 82–85, 91, 99, 112, 121, 125–127, 153, 182–184, 193, 197, 235

Daily Star 295, 297, 298, 303
d'Albe, F. 97, 112
Darlington, J.D. 187, 194
Dash, N. 55, 62
Davis, M. 101, 112
Dawn 254, 264
Degg, M. 8, 34
de Goyet, C.V. 206, 234
Delgado, M. 91, 110
del Ninno, C. 261, 264
Dercon, S. 74, 112
Dewan, A.A. 260, 264
Diacon, D. 178, 193
Dilmener, R.S. 177, 193
Dixon, R.W. 84, 85, 112

Dominici, F. 8, 34
Doran, P.T. 278, 303
Dorosh, P.A. 261, 264
Doswell, C.A. III. 125, 151
Dow, K. 182–184, 193
Drabek, T.S. 141, 152, 182, 189, 193
Dutt, G. 291, 292, 303
Dutt, S. 59, 64, 178, 180, 184, 185, 195
Duval, T.S. 102, 114, 188, 195
Dyer, G. 289, 303
Dynes, R.R. 253, 264

Eakin, H. 300, 303
Edgington, D.W. 222, 234
Edwards, B. 184, 192
Eikenberry, A.M. 198, 199, 234
Eisenman, D.P. 229, 234
Elliot, C. 261, 264
Elliott, D. 8, 33
El-Masri, S. 11, 34
El-Tawil, S. 234, 200
Emani, S. 56, 62
Emel, J. 38, 62
Emrich, C.T. 84, 112, 173, 193
Enarson, E. 269, 270, 273, 303
Enemarck, C. 90, 112
Erickson, N.J. 4, 34
Exum, H.A. 205, 233
Ewing, B.T. 150, 152

Fellmann, J.D. 19, 34
Felten-Biermann, C. 256, 264, 273, 274, 276, 303
Few, R. 74, 112, 147–149, 152
Flint, C. 54, 62
Flynn, J. 99, 112
Finco, M.V. 57, 62
Fischer, H.H. III. 206, 234
Fischhoff, B. 100, 112
Fisher, S. 141, 152
Fitzsimons, D.E. 84, 85, 112
Folke, C. 89, 91, 112
Foster, H. 85, 112
Fothergill, A. 77, 112, 133, 152, 198, 234, 269, 270, 303

Fowler, A. 248, 261, 265
Frank, K.L. 31, 34
Franzino, R. 198, 234
French, S.P. 121, 150, 152
Frerks, G.E. 48, 62, 199, 219, 234
Fritz, C. 7, 8, 34
Fritz, H.M. 128, 152

Gaan, N. 298, 303
Gad-el-Hak, M. 12, 34
Gallay, C. 201, 234
Gardner, P.D. 100, 113
Garrick, B.J. 100, 113
Gautham, S. 129, 152, 258, 265, 274, 303
Gaydos, J.C. 247–249, 257, 265
Gillespie, D. 189, 193
Glewwe, P. 74, 113
Glickman, T.S. 9, 34
Gillespie, T.W. 58, 62
Godsil, R. 146, 152
Goranson, H.T. 253, 265
Granovetter, M. 75, 113
GrazulisT.P. 29, 34
Greenberg, L. 182, 193
Guha-Sapir, D. 9, 34
Gruntfest, E. 169, 184, 193

Haas, J.E. 14, 36
Hafiz, A. 275, 304
Hagan, P. 90, 114
Hagenbaugh, B. 150, 152
Hagman, G. 51, 62
Haider, R. 259, 265
Hall, G. 74, 113
Hanson, L. 92, 114
Haque, C.E. 2, 34, 40, 62, 135, 152, 184, 185, 193, 194, 262, 265
Harrington, L. 89, 113, 177, 179, 181, 194, 223, 234
Hasnath, S.A. 298, 305
Hastings, J.V. 53, 54, 62
Heath, S.E. 183, 194, 201, 234
Heeger, B. 109, 113, 250, 265
Heijman, W. 91, 113
Helfert, M.R. 57, 62
Hepner, G.F. 57, 62

Hewitt, K. 5, 9, 14, 34, 35, 39, 40, 43, 48, 52, 62, 63, 99, 113, 237, 252, 265
Hill, A.A. 121, 153
Hodgson, E. 55, 57, 63
Hofer, T. 33, 34, 132, 153
Holling, C.S. 87, 89, 113
Holtgrave, D. 102, 113
Horney, J. 187, 194
Hossain, M.K. 276, 302
Houghton, J. 278, 304
Hughey, E.P. 173, 193
Hulme, M. 297, 304
Huyck, C.K. 124, 151
Huang, B. 79, 115
Hyndman, D. 279, 281–283, 285–289, 304

Ikeda, K. 59, 63, 81, 113, 133, 153, 184, 194, 269, 270, 272, 273, 304
India Today 258, 265
Islam, M.S. 298, 304
Islam, T. 300, 304

Jacobson, J.L. 297, 304
Jacoby, T. 255, 266
Jensen, J.R. 57, 63
Jia, Z. 206, 234
Johnson, C. 212, 214, 215, 234
Johnson, R.W. 75, 76, 111
Jonkman, S.N. 125, 131–133, 153

Kahn, M.E. 134, 153
Kaplan, S. 100, 113
Kapur, A. 41, 63
Kasperson, R.E. 43, 63
Kates, R.W. 2, 8, 33, 38, 63
Kaufman, M. 222, 234
Kelman, I. 125, 131–133, 153
Kennedy, C.H. 245, 265
Khan, F. 44, 63, 271, 304
Khan, M.H. 293, 306
Khan, M.M.I. 213, 235
Khatun, H. 134, 151, 272, 302
Kibreab, G. 297, 304
Kimhi, S. 90, 113
King, P.S. 137, 155

Kingdon, J.W. 221, 235
Klein, R.J.T. 86, 113
Klinenberg, E. 76, 113
Kniesner, T.J. 139, 153
Korten, D.C. 261, 265
Kothari, U. 296, 304
Krishnan, P. 74, 112
Krauss, E. 255, 265
Krug, E.G. 141, 153, 274, 304
Kruse, J.B. 102, 115
Kulig, J. 92, 114
Kuni, O. 131, 133, 145, 153
Kuo, A.G. 183, 194

Lal, M. 298, 299, 304
Leach, B. 39, 63
Lecocq, F. 298, 299, 306
Leighton, M. 296, 304
Lein, H. 70, 114
Leven, J. 212, 235, 251, 266
Levy, J.I. 34
Lewis, D.J. 48, 63, 248, 254, 261, 265
Lindell, M.K. 124, 138, 142, 146, 153, 157, 167, 171, 172, 182, 188, 190, 191, 194, 195, 206, 220, 221, 235
Liverman, D.M. 42, 63, 69, 73, 114, 183, 194
Loose, C. 148, 153
Lougeay, R. 57, 63
Louis, T.A. 34
Lowrance, W.W. 95, 114
Lowry, Jr. J.H. 56, 63
Lu, J.C. 179, 180, 194
Lu, X.X. 58, 65
Lu, Y. 56, 65
Lulla, K., P. 57, 62
Luna, E.M. 262, 265
Luther, L. 209, 210, 235
Luz, G.A. 247–249, 257, 265

MacDonald, R. 144, 153
McDonald, F. 270, 271, 274, 289, 304
Madden, L.V. 85, 114
Mahmood, A. 271, 304
Maskrey, A. 159, 194
Massimo, S. 127, 153
McAdoo, B.G. 15, 34

AUTHOR INDEX

McCaffrey, S. 101, 114
McClelland, G.H. 101, 114
McEntire, D.A. 18, 19, 34, 53, 63, 109, 114, 172, 179, 185, 194, 206, 227, 235, 245, 265
McGregor, J. 293, 297, 304
McMaster, R.B. 57, 63
McNamara, K.E. 293, 304
McVerry, G. 95, 117
Maguire, B. 90, 114
Malhotra, N. 183, 194
Marshall, B.K. 141, 154
Marshall, L. 275, 304
Maskrey, A. 253, 265
Matin, N. 262, 263, 265
Mauro, A. 11, 34
Menoni, A. 97, 114
Mesjasz, C. 7, 34
Messerli, B. 33, 34, 132, 153
Michaud, S. 277, 306
Middleton, N. 46, 48, 63, 218, 235, 262, 265
Mileti, D.S. 14, 35, 103, 104, 110, 114, 119, 138, 140, 147, 148, 153, 157, 164, 165, 167, 168, 182–184, 187, 194, 195, 197, 205, 215, 217, 218, 220–223, 229, 235
Miller, G. 141, 153, 207, 235
Mills, J.W. 55, 57–59, 62, 63, 73, 112, 146, 152
Mirza, M.M.Q. 132, 153
Mitchell, B. 38, 64
Mitchell, J.K. 38, 42, 43, 52, 54, 60, 64, 73, 96, 114, 249, 266
Mitchell, J.T. 2, 6, 16, 30, 35, 124, 127, 155, 197, 235
Moore, M.T. 259, 266
Morgan, O. 206, 207, 235
Morrow, B.H. 269, 303
Montz, B.E. 2, 7, 11, 16, 20–22, 28, 31–33, 35, 36, 39, 45, 55, 56, 58, 60, 64, 65, 67, 95, 96, 101, 103, 106, 109, 114, 132, 139, 155, 158, 162, 169, 177, 196, 229, 230, 236, 244, 251, 255, 259, 260, 267
Mosley, W.H. 133, 155, 270, 306
Moszynski, P. 250, 266

Mulilis, J-P. 102, 114, 188, 195
Murphy, K. 160, 161, 166, 195
Mustafa, D. 40, 44, 64, 68, 85, 114, 252, 253, 255, 260, 261, 266, 271, 304
Myers, N. 293, 297, 305

Nag, S. 298, 305
Nathan, F. 71, 114
Natural Hazards Observer 198, 235
Neff, R. 115
Neumayer, E. 133, 135, 153, 269, 270, 272, 305
New Nation 301, 305
Nishikiori, N. 125, 132, 133, 154
Noji, E.K. 20, 35, 131, 133, 136, 137, 144, 154, 202, 233
Norris, F.H. 15, 35, 140, 154, 187, 195
Noticias.info 150, 154

O'Keefe, P. 46, 48, 63, 218, 235, 262, 265
Olick, J.K. 90, 114
Oliver, S. 3, 11, 35
Oliver-Smith, A. 39, 64, 70, 114
Ollenburger, J.C. 272, 305
Olsen, M.R. 250, 260, 266
Olshansky, R.B. 221, 235
Oreskes, N. 278, 305
O'Riordon, T. 73, 114
Osborn, S. 223, 224, 235
Osti, R.T. 58, 64
Overton, I. 58, 65
Oxfam 199, 238, 244, 270, 271, 305
Ozdemir, O. 102, 115
Ozerdem, A. 50, 59, 64, 130, 154, 201, 235–255, 258, 266

Palka, E. 245, 249, 266
Palm, R. 41, 43, 55, 64, 100, 115
Pan, Q. 138, 154
Parker, D. 51, 64, 80, 81, 115, 140, 148, 154
Paton, D. 188, 195
Pattan, A. 168, 195
Pattison, W.D. 37, 64
Paul, A.K. 161, 195

Paul, B.K. 22, 35, 48, 50, 59, 64, 65, 79, 81, 93, 99, 101, 115, 123, 127, 129, 130, 132, 135, 136, 137, 143, 145, 149, 154, 155, 164, 165, 178, 180, 184, 185, 187, 192, 195, 200, 202, 203, 207, 212, 215, 221, 222, 225, 226, 233, 235, 237, 243, 245–248, 251, 257, 260–262, 266, 273, 276, 296, 298, 305
Payne, C.F. 109, 115
Peek, L. 183, 194, 198, 234
Peek-Asa, C. 137, 155
Peet, R. 38, 62
Pelling, M. 76, 115, 222, 223, 235, 254, 261, 266
Penton, D. 58, 65
Perch-Nielsen, S. 299, 305
Perry, R.W. 7, 35, 172, 178, 182, 190, 191, 194, 195
Pethick, J.S. 85, 115
Petty, R.E. 183, 195
Phillips, B.D. 30, 35, 180, 195, 209, 210, 215, 217, 224, 226–228, 232, 233, 235
Picou, J.S. 141, 154
Picture This 9, 35
Pilkey, O. 150, 154
Pikul, C. 273, 305
Pimm, S.L. 87, 115
Plate, E.J. 105, 115
Platt, R. 40, 65
Plumper, T. 133, 134 153, 269, 270, 272, 305
Poitra, M. 161, 195
Polsky, C. 70, 115
Population Reference Bureau (PRB) 300, 305
Porfiriev, B.N. 12, 35
Pradhan, E.K. 125, 133, 154
Prater, C.S. 124, 138, 142, 146, 153, 157, 157, 188, 195, 206, 220, 221, 235
P'Rayan, A. 148, 155
Pulsipher, A. 279, 282, 305
Pulsipher, L.M. 279, 282, 305
Pushchak, R. 94, 111
Pyszczynski, T. 182, 195

Quarantelli, E.L. 7, 12, 35, 182, 195, 213, 215, 217, 235, 236

Rahman, A. 299, 300, 302
Rahman, M.O. 39, 65, 78, 111, 257, 267
Rahman, M.K. 165, 195, 222, 236
Raleigh, C. 143, 155, 296, 299, 305
Ramirez, M. 137, 155
Rashid, H. 280, 281, 305
Rasid, H. 149, 155
Rasid, S.F. 277, 306
Rees, S. 274–276, 306
Resosudarmo, B.P. 299, 302
Robbins, J. 90, 114
Rockstrom, J. 91, 115
Rodrigue, C.M. 54, 65
Roger, R.C. 244, 267
Rosenberg, M. 131, 155
Ross, L.J. 249, 267, 275, 306
Rowntree, L. 279, 280, 290, 306

Saarinen, T.F. 38, 65
Sahin, S. 57, 61
Salkowe, R.S. 135, 155
Sanyal, J. 58, 65
Sarewitz, D. 69, 115
Sari, A.P. 289, 290, 306
Sarwar, G.M. 293, 306
Schmidlin, T.W. 131, 137, 155, 216, 236
Schmidtlein, M.C. 10, 35, 78, 115
Schouten, M. 91, 115
Schreurs, M.A. 239, 267
Scott, M.S. 57, 65
Sen, A.K. 17, 35, 70, 115, 134, 135, 155
Shalizi, Z. 298, 299, 306
Shamai, M. 90, 113
Shamsuddoha, M. 29, 35
Sheehan, L. 9, 35
Sherry, M. 160, 161, 166, 195
Showalter, P.S. 56, 65
Shultz, M. 136, 155
Siddique, A.K. 131, 155
Siegel, J.M. 141, 155
Simmons, K.M. 125, 128, 131, 137, 155

Simms, A. 297, 303
Skidmore, M. 134, 150, 155
Slater, F. 5, 33
Slovic, P. 98–101, 115, 116
Smith, K. 3, 9, 10, 14, 19, 20, 35, 107, 116, 138–140, 147, 149, 155, 160, 195, 251, 253, 254, 259, 267
Solecki, W.D. 43, 62
Sommer, A. 133, 155, 270, 306
Sottili, C. 148, 153
Sollis, P. 48, 65, 252, 267
Sorensen, J.H. 55, 57, 65, 176, 178, 180, 182–184, 195
Spiegel, P.B. 12, 17, 18, 35
Stanford, L. 14, 33, 40, 41, 61, 70, 73, 111, 183, 192, 237, 254, 264
Starr, C. 100, 103, 116
Stephen, L. 85, 116
Stern, N. 294, 306
Stimers, M. 211, 286
Streeten, P. 48, 65262, 267
Streeter, C.L. 189, 193
Sultana, F. 44, 65
Sultana, P. 59, 65
Susman, P. 14, 36, 40, 48, 65, 74, 116, 237, 252, 267
Sutter, D. 125, 128, 131, 137, 155, 161, 195
Swift, J. 73, 116

Tabor, N. 240, 241, 267
Taher, M. 262, 263, 265
Tamamoto, M. 240, 267
Tapsell, S. 144, 155
Taylor, A.J. 73, 116, 242, 267
Telford, J. 240, 258, 267
Thacker, M.T.F. 127, 155
Thomas, D.S.K. 54, 65, 124, 127, 155
Thompson, P.M. 59, 65
Thywissen, K. 2, 7, 36, 68, 86, 88, 89, 97, 98, 116
Tichagwa, W. 276, 306
Tierney, K. 14, 36, 86, 88, 116, 157, 170, 188, 189, 190, 196, 218, 220, 232, 236, 253, 264
Timmerman, P. 69, 116

Tipple, G. 11, 34
Tobin, G.A. 2, 11, 16, 21, 22, 28, 31–33, 36, 39, 55, 58, 60, 65, 86, 95, 96, 101, 104, 106, 109, 116, 132, 139, 155, 158, 162, 169, 177, 196, 229, 230, 236, 244, 251, 255, 259, 260, 267, 272, 305
Tokioka, T. 58, 64
Townsend, J.G. 260, 267
Toya, H. 134, 150, 155
Turner, B.A. 8, 36
Turner, R.H. 187, 196
Twigg, J. 95, 116

Uitto, J.I. 188, 196
Underwood, B.J. 102, 116
United Nations Development Program (UNDP) 15, 36, 46–51, 65, 94, 97, 116, 239

Vakil, A.C. 261, 267
Valley, P. 239, 267
Valpreda, E. 56, 66
Van Dissen, R. 95, 117
Varma, D. 109, 117
Varma, R. 109, 117
Viscusi, W.K. 139, 156
Vogt, B. 178, 180, 195

Waldo, B. 246, 267
Walker, B.H. 86, 117
Walker, P. 246–249, 267
Walsh, B. 69, 117
Ward, R. 139, 140, 147, 149, 155, 251, 253, 259, 267
Warner, K. 282, 283, 295, 306
Washington Post 129, 156
Watts, M. 39, 40, 66, 73, 117
Weber, E.U. 98, 99, 100, 102, 113, 116, 117
Webber, M. 295, 303
Weber, M. 169, 193
Wehrfritz, G. 240, 267
Wenger, D.E. 13, 36
Werker, E. 8, 33, 135, 152
Wharton, F. 94, 110

Wheelis, M. 85, 114
White, G.F. 14, 21, 36, 38, 66, 106, 117, 160, 196
White, S.C. 261, 268
Whitney, D.J. 188, 194
Whyte, A.V. 38, 66, 96, 117
Wisner, B. 2, 3, 11, 12, 14, 15, 36, 40, 48, 66, 70–73, 76, 78, 117, 182, 196, 218, 220, 222, 223, 236, 237, 252–254, 262, 268
Witte, K. 100, 117, 183, 196
Wood, W.B. 297, 306

World Commission on Environment and Development (WCED) 86, 117, 223, 236

Yarnal, B. 115
Yonder, A. 252, 253, 256, 268

Zahran, S. 69, 77, 117
Zaman, M.Q. 42, 66
Zhou, H. 86, 87, 89, 91, 92 117
Zimmermann, M.K. 278, 303
Zurick, D. 124, 156

Subject Index

Abnormal flooding 4, 149
Accepted risk 104
Acute stress disorder (ASD) 205
Adaptive capacity 70, 74
Adaptive strategies 88
Aggregate debris 210
Ambulances 203
American Red Cross 212, 244
Anthrax 18
Anthropogenic 19
Armed conflicts 17, 52
Armenian earthquake 7
Asian Development Bank (ADB) 120, 241
Atlantic Hurricane Basin 29
Atmospheric hazards 16
Avalanches 16
Avian "bird" flu 18

Bacterial hazards 20
Bam earthquake 256
Bangladesh National Building Code (BNBC) 166
Behavioral approach 38, 39, 42
Behavioral paradigm 38
Behavioral response 77
Behavioral studies 39, 40
Bhopal disaster 19, 80, 109
Biological hazards 18

Biohazards 18
Biophysical vulnerability 82
Bioterrorism 19
Bird flu 18
Blizzards 16, 20, 28, 31, 32, 131, 132, 216
Blunt trauma 136
Board on Atmospheric Science and Climate (BASC) 278, 303
Building codes 14, 164, 165
Built environment 69

Calamities 36, 267, 302
CAMEO/ALOHA 185
Catholic Relief Service (CARE) 95, 111, 239
Caribbean Community and Common Market (CARICOM) 247
Caribbean Disaster Relief Unit (CDRU) 247
Cash donations 263
Catastrophe 12
Causal factors 71
Cell phones 175
Centers for Disease Control and Prevention 18
Center for Epidemiology and Animal Health (CEAH) 130, 151

Environmental Hazards and Disasters: Contexts, Perspectives and Management, First Edition. Bimal Kanti Paul.
© 2011 John Wiley & Sons, Ltd. Published 2011 by John Wiley & Sons, Ltd.

SUBJECT INDEX

Center for Research on the Epidemiology of Disasters (CRED) 9, 124, 125, 135
Chemical hazards 57
Chernobyl accident 19, 29, 231
Chlorofluorocarbons 280
Chronic hazards 19, 20
Climate change 277, 295
Climate justice 297
Climate migrants 296
Climate refugees 293, 295, 298
Civil disorders 17
Civil strife 20
Coastal embankments 22, 162
Coastal erosion 16
Coastal floods 17
Coastal vulnerability 84
Coastal vulnerability index (CVI) 85
Cold waves 16
Communication chain 43
Complex emergencies 17, 18
Complex humanitarian emergencies 17
Compound disasters 16
Command and control 54
Computer hazards 19
Confirmation behavior 181
Coping capacity 86, 88
CNN effect 250
CNN factor 250
Countdown interval 22
Creeping disasters 31
Critical incident stress (CIS) 205
Cry wolf effect 184
Cultural recovery 230
Cyber-terrorism 53
Cyclone Gorky 22, 127
Cyclone Nargis 2, 126, 128
Cyclone Sidr 81, 121, 122, 123, 127, 130, 135–137, 180, 198, 202, 214, 238, 273
Cyclone warning systems 22

Damage 22
Damage assessment 151
Damage and Loss Assessment (DaLa) methodology 120, 121
Damage, loss, and needs assessment (DLNA) 120
Dam failure 20

Deaths 8, 10, 12–15, 119, 124, 128, 130, 135, 137, 138
Debris Flows 27
Debris management 191
Debris removal 10, 207, 212
Decade for Natural Disaster Reduction (IDNDR) 67
Deforestation 2, 80, 279, 292, 297, 299
Democracy 134, 135, 147
Department of Homeland Security (DHS) 53, 85, 147, 192
Department of Human Services 87, 112
Desertification 20
Development 46, 48
Direct impact 52, 138
Disabled person(s) 40, 133, 176, 180, 183, 184
Disaster assistance 10, 34, 73, 221, 226, 227, 244, 246, 247, 249, 253, 254, 261
Disaster Cartography 57, 58
Disaster cycle 77, 157
Disaster damage 119–124, 138, 151
Disaster declaration 10, 226, 227
Disaster duration 5
Disaster Effects 119
Disaster Impacts 119, 138, 142
Disaster management 46
Disaster management cycle 167
Disaster mitigation 8, 55
Disaster plan 176, 190
Disaster prevention 8, 64, 115, 116, 134
Disaster recovery 35, 48, 147, 152, 157, 193, 195, 211, 217–236, 263
Disaster recovery cycle 198, 199, 218, 219
Disaster reduction 36, 44, 47, 63, 67, 86, 116, 158
Disaster resilience of loss-response of location (DRLRL) 91–93
Disaster resilience of place (DROP) 91
Disaster response 18, 22, 198
Disaster victims 10, 14, 200, 202
Disaster and Emergency Assistance Act 10
Disaster risk reduction (DRR) 10, 14, 200
Diurnal factor 31
Diurnal patterns 22
Doppler Radar 58
Dread 101, 102

Drill 186
Drought 5, 8, 9, 16, 17, 20, 21, 23, 24, 27, 29, 31, 32
Duration 22, 33, 80, 96
Dyke 87

Earthquakes 2, 5, 6, 8, 15–17, 20, 21, 23, 24, 27, 29, 31, 32, 46, 56–58, 107, 120, 123–128, 131, 133, 141, 143, 148, 160, 167, 171–173, 188, 201–203, 207, 212, 219, 257
Ecological perspective 37, 38
Ecological resilience 87
Economic disaster injury loans (EIDL) 232
Economic impacts 141, 142
Economic vulnerability 79, 82
Economists 68, 150
Effective response 10, 170
Elderly people 73
Elevated models 58
Elusive hazards 19
Emergency Events Database (EM-DAT) 9, 124
Emergency alert system (EAS) 174
Emergency assistance 9, 10, 58, 88, 143, 179, 197–199, 210, 211, 216, 220, 223, 226
Emergency management 57, 59, 114, 151, 152, 155, 159, 170, 171, 185, 186, 189, 192, 193, 195, 198, 204, 205, 211, 216, 217, 229, 234
Emergency management agency 57, 152, 193, 211
Emergency medical service (EMS) 189
Emergency operation plan (EOP) 190, 191
Emergency response 8, 28, 109, 135, 170, 186, 190, 191, 195, 198, 199, 207
Emergency shelter 213
Engineering resilience 87
Enhanced Fujita Scale 24, 26
Entitlements 74
Environmental control 107
Environmental degradation 3, 70, 80
Environmental determinism 38
Environmental disasters 11, 41, 62, 84, 119

Environmental hazards 35, 37, 39, 41, 61, 65, 67, 68, 84, 99, 112–114, 116, 119, 152–155, 157, 182, 190, 193, 195, 197, 214, 226, 227, 229, 235
Environmental migrants 296
Environmental Protection Agency (EPA) 208, 211, 234, 244, 263, 283, 284, 303
Environmental refugees 253, 263–265, 293
Environmental vulnerability 80, 82
Epidemics 18, 20, 153, 206, 234
Evacuation 7, 25, 35, 51, 56, 61, 74, 77, 81, 89, 90, 109–111, 113, 125, 127, 136, 154, 162, 170, 176–187, 189, 192–195, 201, 232, 233, 235
Evacuation orders 51, 58, 81, 177, 179–185, 193, 195
Evacuation route 187
Evacuation program 74
Exercise 185, 186, 190
Exposure 19, 26, 33, 40, 43, 56, 57, 59, 68, 70, 74, 75, 83–85, 95–98, 101, 102, 106–108, 132, 146, 164, 165, 167, 212, 234

Famine 8, 17, 19, 73, 126, 140, 218
Fatalities 6, 9, 10, 14, 22, 33, 98, 104, 125–135, 137, 157, 161, 201, 231
Federal Emergency Management Agency (FEMA) 57, 139, 152, 159, 160, 165, 166, 168, 185, 192, 193, 211, 217, 221, 222, 227, 234, 255
Flash Flood Alley 30
Flash floods 16, 17, 27, 30, 31, 131, 139, 219
Floods 2, 5, 6–8, 16, 17, 20–23, 27–29, 30, 31, 33, 34, 38, 40, 47, 57–59, 66, 80, 107, 120–127, 131, 133, 134, 141, 144, 145, 147–150, 207, 212, 215, 219
Flood stage 23
Fire 9, 14, 16, 17, 50, 55, 57, 58, 100, 197, 200, 206
First-order impact 138
First responders 150, 198, 202–205
First-round impact 140
Fog 16
Forest fires 14, 16, 50, 239

Foot-and-Mouth disease 18
Free Aceh Movement 18
Frequency 22, 27, 28, 33
Frost 16
Fujita Scale 24
Functional exercise 186
Full-scale exercise 186

Gambler fallacy 100
Gender perspectives 269
Gender response 269
Geographical Dimensions of Terrorism project (GDOT) 54
Geographic Information System (GIS) 45, 54–57, 59, 60, 169, 177
General hazards risk communication model (GHRCM) 182, 183
General risk communication model (GRCM) 182, 183
Geologic hazards 16
Global climate change 6, 45, 223, 269, 278, 281, 284, 292, 294
Global Initiative on Community-based Adaptation to Climate Change (GIBACC) programs 301
Global Positioning System (GPS) 58–60
Global warming 45, 70, 107, 269, 277–281, 283–285, 287–290, 299, 302
Globalization 53, 54
Going green 223, 224
Google Earth 58
Government Accounting Office (GAO) 10, 34, 255, 265
Government of Bangladesh 119, 121, 123, 130, 135, 152, 162, 193, 273, 303
Government of Yemen 119–122, 124, 152
Great Bengal famine 17
Great Midwestern floods 259
Great Plains 215
Greenhouse gas 280
Gross Domestic Product (GDP) 79, 121
GNP 134, 148

Hail 16
Hailstorms 16
Hantaviruses 18
Hazard-affected bodies (HABs) 87

Hazard duration 21
Hazard frequency 21
Hazards mitigation 14, 86
Hazards-of-place model 82, 83
Hazards typologies 1
Hazardscape 44
Hazards of place 43
HAZUS 57, 120, 124, 185
HAZUS-MH 57, 120
Health Protection Agency (HPA) 144, 153
Heat waves 16, 28
Hemorrhagic fever 18
Heuristic cues 183
Horizontal evacuation 179, 185
Horizontal integration 229, 230
Household hazardous waste (HHW) 210
Household vulnerability 76
Human adjustments 39
Human ecology perspective 5
Human-generated disasters 20
Human-induced hazards 20
Human rights 47, 75, 254, 256, 263, 264, 270, 274
Human response 15
HIV/AIDS 18, 275
Hurricane Andrew 55, 148, 198, 208
Hurricane Agnes 7
Hurricane Bonnie 184
Hurricane Charley 150, 222
Hurricane Floyd 178
Hurricane Gustav 201
Hurricane Georges 47
Hurricane Hugo 148
Hurricane Ike 201, 216
Hurricane Katrina 3, 56, 58, 77, 84, 86, 89, 109, 125, 135, 141, 142, 177, 179, 180, 198, 201, 202, 208, 244, 274, 293
Hurricane Katrina Community Advisory Group (HKCAG) 181, 194
Hurricane Lili 179, 180
Hurricane Mitch 46, 47, 143, 144, 257
Hurricane Rita 178, 179
Hurricane Vulnerability Index (HVI) 84
Hurricanes 6, 16, 17, 27–29, 31
Hybrid model 102

SUBJECT INDEX

Ice storms 16, 29
IEM 142, 153
Impact Ratio 14
Indian Ocean Tsunami 8, 18, 124, 128, 129, 143, 201, 206, 207, 239, 240, 256, 272, 274
Indirect impact 52, 138
Indirect losses 15
Individual behavior 39
Individual vulnerability 76
Industrial accidents 9, 28
Industrial disasters 9
Industrial hazards 19
Industrial pollution 20
Influenza 18
Injuries 8, 12–15, 22, 103, 104, 119, 135–137
Institutional vulnerability 78, 80
Insurance 40, 55, 107, 108, 147, 148, 160, 163, 164, 166, 187, 215, 225–227, 259
Intangible impact 139
Internally displaced persons 18
Intentional hazards 17
Inter-Agency Standing Committee 17, 18, 34
International Campaign on Climate Refugees' Right (ICCR), 298
International Decade for Natural Disaster Reduction (IDNDR), 44, 45, 47, 158, 159, 169
International Federation of Red Cross and Red Crescent Societies (IFRC and RCS) 21, 34, 86, 113, 121, 125, 126, 128, 153, 240, 255–258, 265
International Monetary Fund (IMF) 241
International nongovernmental organizations (INGOs) 199
International Organization for Migration 142–144, 153
International Panel on Climate Change (IPCC) 87, 113, 277, 280, 289, 293, 294, 299, 304
Internet 53

Kashmir earthquake 201, 254
Kyoto Protocol 289, 290

Lacerations 136
Land reform 91
Landslides 2, 16, 17, 26, 27, 30, 31
Leadership in Energy and Environmental Design (LEED) 224, 225
Lead time 173
Lightning 16, 99
Likert scale 103
Local government 10
Logging 2
Loma Prieta earthquake 7
Lousiana Coalition Against Domestic Violence (LCADV) 275

Malaria 47, 144, 284
Magnitude 7, 10, 12, 13, 15, 22–24, 27, 31, 33
Media attention 18, 22
Medical care 53, 197, 202, 204
Mental health 47, 146
Mercalli Intensity scale 23
Millennium Development Goals (MDGs) 47, 51, 75
Mitigation measures 14, 21, 22, 159, 164
Mobile homes 25, 83
Mobile medical teams (MMTs) 202
Monetary losses 10
Morbidity 18
Mortality 18
Mozambique flood 250
Mudflows 3, 16, 242
Mudslides 3, 6, 30
Multidisciplinary Center for Earthquake Engineering Research (MCEER) 88
Multi-Donor Trust Fund (MDTF) 300
Multiple hazards 56
Municipal solid waste (MSW) 210
Myths 109, 206

National Adaptation Program of action (NAPA) 300
National Climatic Data Center 127
National Flood Insurance Program (NFIP) 40, 163
National Hurricane Center (NHC) 171
National Oceanic and Atmospheric Administration (NOAA) 25, 174, 175

National Weather radio (NWR) 175
National Weather Service (NWS) 23, 172, 174
Natural disasters 1, 8, 10, 11, 20, 28, 55, 147, 148, 202
Natural forces 70
Non-intentional hazards 19
Normal flooding 5
Natural hazards 1, 3, 4, 6, 16, 18, 20, 38, 44, 94
Neumayer 133, 134, 153, 269, 270, 272, 303
Non-government Organizations (NGOs) 14, 48, 61, 78, 150, 237, 238, 240, 245, 254, 258, 260–263, 273, 274, 301, 302
Non-structural mitigation measures 167
Normal flood 149
Normalcy bias 181
North Atlantic Treaty Organization (NATO) 248
North Carolina Division of Emergency Management (NCDEM), 159, 195
Northridge earthquake 148
Nuclear accident 29
Nuclear hazards 19

Oil spills 19, 57
Optimistic bias 100

Palmer Drought Severity Index 23
Pandemics 20
Participatory research methodologies 59
Pathogens 18
Perceived risk 99, 101, 102
Permanent housing 212, 214
Personal protective equipment (PPE) 191
Person-Relative-to-Event (PrE) theory 102
Pest infestations 20
Pets 141, 183, 184, 201
Physical characteristics of 21, 22, 32
Physical dimensions of natural hazards 1, 21
Physical mechanism 22
Physical parameters 21, 31, 33
Physical vulnerability 75, 79, 80
Place vulnerability 82, 84

Plague 18
Political ecological perspective 40–42
Political economic perspective 43
Pollution 19
Post-traumatic stress disorder (PTSD) 205, 206
Poverty 41, 46–50, 73–75, 77, 79, 90, 142, 143, 169, 205, 252, 262, 264, 290, 292
Preparedness 3, 8, 14, 22, 60, 77, 97
Pressure and Release (PAR) model 71, 72
Primary data 59
Primary hazards 16, 17
Principal component analysis (PCA) 83
Private sector 108, 123, 124, 185, 226
Protozoan hazards 20
Psychometric paradigm 100
Public assistance 10, 55, 119, 211, 226, 227
Public cyclone shelters 22, 48, 178
Public facilities 10, 169, 212, 222, 226
Public sectors 124
Purdah 272

Range of adjustments 39
Rapid Rural Appraisal 59
Recovery 15, 77, 90
Recurrence interval 27
R4 framework 89
Refugees 18
Remote sensing 55
Rescue operation 200
Resilience triage 86, 87, 88
Resiliency 49, 67
Response 13, 60
Resources For the Future 9
Retrofitting 167
Return period 27, 96
Reverse 911, 175
Richter scale 7, 23, 222
Rickettsiae 18
Risk acceptance 162
Risk assessment 55, 94, 103, 105, 106, 172
Risk avoidance 162
Risk communication 94, 108

Risk factors 69
Risk ladder 101
Risk management 105, 108
Risk mapping 56
Risk perception 94, 99, 100, 102
Risk perspective 43
Riskscape 44
River floods 27

Saffir-Simpson Hurricane Scale 23, 25
Salvation Army 200, 212
Satellite imagery 55, 58
Save the Children 199, 239
Sea level rise (SLR) 296, 299
Search and rescue (SAR) operation 200
Seasonality 22, 28
Secondary causes 15
Secondary data 59
Secondary hazards 16, 17
Seismic event 23
Self-efficacy 181
Sensitivity 70
September 11, 2001 attacks 54
Severe acute respiratory syndrome 18
Shadow evacuation 180
Sherman landslides 26
Shock 8, 68, 86
Simple Triage and Rapid Transport (START) 204
Sirens 174, 175
SLOSH 185
Small Business Administration (SBA) 232
Smallpox 18
South Asian Association for Regional Cooperation (SAARC) 302
Snowfall 5
Snowstorms 16, 28, 29
Social Amplification of Risk Framework (SARE) 43
Social capital 75, 76, 90
Social cognitive theory (SCT) 181, 182
Social development 49
Social-ecological system (SES) 89, 91
Social hazards 17
Social impact 140
Social networks 75, 164

Social reliance 90
Social vulnerability 77, 78, 82
Social Vulnerability Index (SoVI) 83, 84
Spatial Analysis 55
Spatial data 59
Spatial distribution 22, 29, 30
Spatial dispersion 29, 30
Spatial extent 22, 29
Spatial Hazard Events and Losses Database for the United States (SHELDUS) 127
Spatial location 29
Spatial Video Acquisition System (SVAS) 58
Spontaneous evacuation 180
Stress 7, 11, 76, 87, 131, 139, 140, 141, 146, 148, 159, 186, 205, 216, 220, 245, 256, 282
Structural approach to natural hazards 40, 42
Structural mitigation measures 167
Structural vulnerability 79
Storm surges 17, 23, 28
System vulnerability 80, 81

Table-top exercise 186
Tangible impact 139
Technical risk 95, 96, 100, 103
Technological fixes 39, 40
Technological hazards 19, 20, 44
Technological risk 95
Temporal data 59
Temporal dimension 27
Temporal distribution 22
Temporal spacing 28
Temporary housing 212–214
Temporary shelter 197, 213
Territoriality 54
Terror Management Theory (TMT) 182
Terrorism 17, 19, 45, 52–54
Tertiary impact 139
Thunderstorms 16
Tornado Alley 30
Tornado safe room 168
Tornadoes 16, 17, 20, 21, 27–31, 127, 150

Toxins 18
Transportation hazards 19
Tropical cyclone 7, 16, 20, 22, 198
Triage 203
Tsunamis 8, 16, 17, 20, 27
Tuberculosis 18
Typhoon Rita 253
Typhoons 8

Union Carbide Plant 19, 109
United Nations (UN) 3, 47, 65, 120, 128, 129, 155, 259, 267, 274, 275, 306
United Nations Center for Human Settlements (UNCHS) 11, 36
UN Children's Fund (UNICEF) 199, 239, 241
United Nations Development program – Department of Humanitarian Affairs (UNDP-DHA) 51, 66
United Nations Disaster Relief Organization (UNDRO) 51, 66, 96, 97, 116
UN Economic Commission for Latin America and Caribbean (UN-ECLAC) 120, 156
UN Framework Convention on Climate Change (UNFCCC) 300, 302
UN High Commissioner for Refugees (UNHCR) 298, 306
United Nations International Strategy for Disaster Reduction (UNISDR) 3, 36, 104, 116
United States Department of Homeland Security – Office for Domestic Preparedness (USDHS-ODP) 85, 116
US Assistance for International Development (USAID) 244, 267
US Department of Agriculture (USDA) 259
US Department of Energy 223, 225, 236
US Green Building Council 225, 236

United States Geological Survey (USGS) 23, 172
United States Office of Foreign Disaster Assistance 10

Value of a statistical life (VSL) 139
Varley A. 51, 66
Vertical evacuation
Vertical integration 229, 230
Violence 20
Volcanic activities
Volcanic eruptions 2, 16, 17, 20, 28, 29
Vulnerable conditions 11
Vulnerability 3, 12, 22, 26, 41–43, 46, 49, 51, 60, 67–71, 73–77, 83, 85, 93, 95–97, 107
Vulnerability and Capacity Index (VCI) 85
Vulnerability approach 43
Vulnerability science 67, 68
Vulnerability of place 43
Vulnerability hotspots 59

War 20
Warfare 17
Warning system 74
Watch 173
Waterscape 44
West Nile virus 18
Wildfires 2, 5, 6, 17, 20
Windstorms 16, 28, 215
Window-of-opportunity 51
Windows of opportunity 120–122, 159, 220, 221
World Bank 119, 120, 121, 227, 238, 293, 295–297, 306
World Food Program (WFP) 241
World Health Organization (WHO) 20, 36, 207, 220, 236, 277, 306
World Trade Center 52, 54
World War I 18
World Wide Web 58, 59